Lecture Notes in Computer Scier

Edited by G. Goos, J. Hartmanis and J. van L

Springer
Berlin
Heidelberg
New York
Barcelona
Hong Kong
London
Milan
Paris
Singapore
Tokyo

Tatyana Yakhno   (Ed.)

Advances in Information Systems

First International Conference, ADVIS 2000
Izmir, Turkey, October 25-27, 2000
Proceedings

 Springer

Series Editors

Gerhard Goos, Karlsruhe University, Germany
Juris Hartmanis, Cornell University, NY, USA
Jan van Leeuwen, Utrecht University, The Netherlands

Volume Editor

Tatyana Yakhno
Dokuz Eylul University
Computer Engineering Department
Bornova, 35100, Izmir, Turkey
E-mail: yakhno@iis.nsk.su

Cataloging-in-Publication Data applied for

Die Deutsche Bibliothek - CIP-Einheitsaufnahme

Advances in information systems : first international conference ;
proceedings / ADVIS 2000, Izmir, Turkey, October 25 - 27, 2000.
Tatyana Yakhno (ed.). - Berlin ; Heidelberg ; New York ; Barcelona ;
Hong Kong ; London ; Milan ; Paris ; Singapore ; Tokyo : Springer, 2000
(Lecture notes in computer science ; Vol. 1909)
  ISBN 3-540-41184-4

CR Subject Classification (1998): H.2, H.3, H.4, I.2, C.2, H.5

ISSN 0302-9743
ISBN 3-540-41184-4 Springer-Verlag Berlin Heidelberg New York

Springer-Verlag Berlin Heidelberg New York
a member of BertelsmannSpringer Science+Business Media GmbH
© Springer-Verlag Berlin Heidelberg 2000
Printed in Germany

Typesetting: Camera-ready by author, data conversion by PTP-Berlin, Stefan Sossna
Printed on acid-free paper   SPIN: 10722701   06/3142   5 4 3 2 1 0

Organization

Honorary Chair: Irem Ozkarahan, Turkey

Program Committee Co-chairs: Abdullah Uz Tansel, USA
Tatyana Yakhno, Turkey

Program Committee:

Adil Alpkocak, Turkey   Victor Malyshkin, Russia
Erden Basar, North Cyprus   Mario A. Nascimento, Canada
Frederic Benhamou, France   Erich Neuhold, Germany
Cem Bozsahin, Turkey   Selmin Nurcan, France
Ufuk Caglayan, Turkey   Kemal Oflazer, Turkey
Fazli Can, USA   Gultekin Ozsoyoglu, USA
Paolo Ciaccia, Italy   Meral Ozsoyoglu, USA
Yalcin Cebi, Turkey   Tamer Ozsu, Canada
Oguz Dikenelli, Turkey   Marcin Paprzycki, USA
Asuman Dogac, Turkey   Yusuf Ozturk, USA
Opher Etzion, Israel   Malcolm Rigg, Luxemburg
Yakov Fet, Russia   Suryanarayana Sripada, USA
Lloyd Fosdick, USA   Martti Tienari, Finland
Forouzan Golshani, USA   Turhan Tunali, Turkey
Malcolm Heywood, Turkey   Ozgur Ulusoy, Turkey
Alp Kut, Turkey   Nur Zincir-Heywood, Turkey

Additional Referees

Ilaria Bartolini   Mustafa Inceoglu   Stefano Rizzi
Peter van Beek   Vitaly Morozov   Carine Souveyet
Erbug Celebi   Anna Nepomniaschaya   Murat Tasan
Cenk Erdur   Cagdas Ozgenc   Arif Tumer
Georges Grosz   Marco Patella   Osman Unalir
Necip Hamali   Wilma Penzo

Conference Secretary

Adil Alpkocak (Turkey)

Local Organizing Committee

Erbug Celebi   Bora Kumova   Evren Tekin
Gokhan Dalkilic   Tanzer Onurgil   Serife Yilmaz
Emine Ekin   Cagdas Ozgenc
Durdane Karadiginli   Tugba Taskaya

Preface

This volume contains the proceedings of the First International Conference on Advances in Information Systems (ADVIS) held in Izmir (Turkey), 25–27 October, 2000.

This conference was dedicated to the memory of Professor Esen Ozkarahan. He was great researcher who made an essential contribution to the development of information systems. This conference was organized by the Computer Engineering Department of Dokuz Eylul University of Izmir. This department was established in 1994 by the founding chairman Professor Ozkarahan and there he worked for the last five years of his live.

The main goal of the conference was to bring together researchers from all around the world working in different areas of information systems to share new ideas and to represent their latest results. We received 80 submissions from 30 countries. The Program Committee selected 44 papers for presentation at the conference.

The invited and accepted contributions cover a large variety of topics: general aspects of information systems, data bases, data warehousing, computer networks, Internet technologies, content-based image retrieval, information retrieval, constraint programming and artificial intelligence.

The success of the conference was dependent upon the hard work of a large number of people. We gratefully acknowledge the members of the Program Committee who helped to coordinate the process of refereeing all submitted papers. We also thank all the other specialists who reviewed the papers.

We would like to thank our sponsors for their financial and organizational support. We also appreciate the cooperation we have received from Springer-Verlag.

We thank Natalia Cheremnykh and Olga Drobyshewich for their professional work in producing the final version of these proceedings.

August 2000   Tatyana Yakhno

Preface

In Memoriam

Professor Ozkarahan is one of the pioneers of database machine research and a founder of database systems research in Turkey. Professor Ozkarahan is well known for his pioneering work on the RAP (relational associative processor) database machine. He designed RAP for the hardware implementations of relational databases. RAP includes all the relational algebra operations and the explicit form of the semi-join operation that later became the cornerstone of query optimization in relational databases.

Professor Ozkarahan conducted research in various aspects of RAP database machines: hardware, VLSI implementation, RAP operating system, query languages, performance evaluation, network of RAP machines, and its use in information retrieval. In the 1980s he worked on building a prototype of a RAP database machine with support from Intel Corporation towards its eventual commercialization. However, due to a sudden major management change in Intel at the time, the commercialization was not realized.

Professor Ozkarahan published numerous articles in major database journals and conference proceedings. He was also the author of two books published by Prentice-Hall:

- Database Management: Concepts, Design, and Practice
- Database Machines and Database Management

Professor Ozkarahan completed his B.Sc. in electrical engineering at Middle East Technical University (METU) and received his Master's Degree in operations research at New York University. After completing his Ph.D. thesis at the University of Toronto in 1976, he returned to METU in Turkey, where he established the first database research group in the country. He graduated a record number of M.S. and Ph.D. students until 1981, when he returned to the United States and joined the faculty at Arizona State University, where he worked on building a prototype of RAP. In 1988, he started to work at Pennsylvania State University as a full professor. In 1993, he returned to Turkey and established the Computer Engineering Department of Dokuz Eylul University where he was a chair until 1999.

In his personal life, Professor Ozkarahan was a very attentive and caring family man, both as father and husband. Professor Ozkarahan was very concerned with the lives of his daughters and kept close track of their needs, thus instilling in them the drive to be successful hardworking individuals, just as he was.

Table of Contents

Data Warehouses

Databases

Computer Networks

Constraint Programming

Content-Based Image Retrieval

Internet Technologies

Information Systems

Information Retrieval

Artificial Intelligence

Model of Summary Tables Selection for Data Warehouses

Bogdan Czejdo[1*], Malcolm Taylor[1], and Catherine Putonti[2]

[1] MCC
3500 West Balcones Center Drive, Austin, TX 78759
czejdo@loyno.edu, mtaylor@mcc.com
[2] Loyola University
New Orleans, LA 70118
{cputonti}@loyno.edu

Abstract. In this paper, we consider the problem of selecting summary tables for a Data Warehouse, in which aggregate queries are frequently posed. We introduce the concept of a complete aggregation graph for defining the possible summary tables. We present a model for determining an optimal set of summary tables. The selection algorithm based on the model takes into consideration query response time constraints and cost/benefit evaluation.

1 Introduction

Today's businesses require the ability to not only access and combine data from a variety of sources, but also summarize, drill down, and slice the information across subject areas and business dimensions. Consequently, the Data Warehouse has become essential to growing businesses ready to compete in today's market [2]. One of the main problems in building the Data Warehouse is to determine the optimal set of summary tables. The warehouse designer might decide to keep a single table containing all the data, or might instead use one or more summary tables, which are derived from the original one by means of various forms of aggregation. In general, tables with less aggregation can be used to answer more queries. On the other hand, tables with more aggregation occupy less space and, when they can be used to answer a query, they will do so more efficiently.

In this paper, we consider the problem of selecting summary tables for a Data Warehouse, in which aggregate queries are frequently posed. We introduce the concept of a complete aggregation graph for defining the possible summary tables. We present a model for determining an optimal set of summary tables. The selection algorithm based on the model takes into consideration query response time constraints and cost/benefit evaluation.

The problem of using materialized views to answer aggregate queries has been addressed in the literature. Gupta et al [4] present a set of syntactic transformations for queries and views involving grouping and aggregation. Srivastava et al [8]

---

* Visiting MCC from Loyola University, Department of Mathematics and Computer Science, New Orleans, LA 70118

T. Yakhno (Ed.): ADVIS 2000, LNCS 1909, pp. 1-13, 2000.

propose a more general semantic approach, which identifies the conditions under which a view can be used to answer a query. However, these prior works do not consider the relative costs of different views which could be used to answer the same query, nor do they address the question of selecting the optimal set of views for a Data Warehouse. Our principal contribution is the development of a cost-based approach to evaluating the benefit of each view, and thus selecting the optimal set of views for the Data Warehouse.

The paper is organized as follows. In Section 2, we describe a Data Warehouse on the example of TurboMachine company. We assume that the fact table in the Data Warehouse has the finest granularity (no summary tables). In Section 3, we define various summary tables that could be used in the Data Warehouse, and the association of each query with a specific table. Section 4 introduces the components of the model including constraints on query execution time and cost and benefit functions. Section 5 presents an algorithm for selecting the optimal set of summary tables based on the model. Conclusions are given in Section 6.

2 Data Warehouse with Single Fact Table

In the process of creation of a Data Warehouse typically the heterogeneous database schemas are integrated into a star schema containing dimension tables and a fact table. The discussion about semantic heterogeneity of schemas [5] for different databases and process of schema integration [2] is not a topic of this paper. Here, we concentrate on determining granularity of fact table(s) [2]. The chosen level of granularity should support all queries for the Data Warehouse. Once the granularity of a base fact table is determined, the question of including additional summary tables needs to be answered.

Let us consider a Data Warehouse for TurboMachine company that has several branches with separate databases. Each branch lends machines to other companies (other companies referred here as locations) and collects (usually once a week) the leasing fee based on the meter decribing the extent of use of each machine.

The Data Warehouse schema for TurboMachine company is a relatively typical star schema as shown in Figure 1. The model includes dimension tables: Machine, Location, and Time, and a fact table T11. The role of each table is as follows. **Machine** contains and keeps records of the information for a specific machine. **Location** contains the information about each company which has/had leased a machine. Here, we assume that **Time** is only a conceptual table since all time attributes can be identified in the fact table T11.

T11 has a log of all collections made at a location for a date and for a specific machine. T11 has five key attributes LNumber, Year, Month, Day, MID referred to also as index attributes. The additional attributes are called aggregate attributes. In our example for simplicity we have only one aggregate attribute Money.

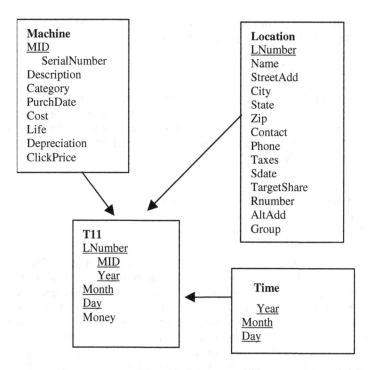

Fig. 1. The initial star schema for Data Warehouse TurboMachine company

3 Data Warehouse with Summary Tables

Very often, the performance of Data Warehouse with only one fact table is not acceptable. Summary tables need to be introduced in addition to the main fact table T11 in order to improve performance. Therefore, it is important to develop and follow some rules in the design of a Data Warehouse with frequent aggregate queries. Let us first look at an example of summary tables T23, T24, T35, T36, and T43 as shown in Figure 2.

The additional summary tables are obtained by grouping T11 by various index attributes of T11. Table T23 is obtained by grouping T11 by index attributes LNumber, Year, Month, Day and aggregating Money to obtain SumOfMoney.

```
CREATE TABLE T23(LNumber, Year, Month, Day, SumOfMoney) AS
SELECT LNumber, Year, Month, Day, SUM(Money)
FROM T11
GROUP BY LNumber, Year, Month, Day;
```

Similarly we can create table T24 by grouping T11 by index attributes LNumber, MID, Year and Month, and aggregating Money to obtain SumOfMoney.

```
CREATE TABLE T24 (LNumber, Year, Month, MID, SumOfMoney) AS
SELECT LNumber, MID, Year, Month, SUM(Money),
FROM T11
GROUP BY LNumber, MID, Year, Month
```

The next level (third level) tables can be created from table T11 from the first level or from some already computed tables from the second level. For example to create table T35 we can use table T11, T23, or T24. If we choose table T24 then we need to do grouping T24 by index attributes Lnumber, Year and Month, and aggregating Money to obtain SumOfMoney.

```
CREATE TABLE T35 (LNumber, Year, Month, SumOfMoney) AS
SELECT LNumber, Year, Month, SUM(SumOfMoney),
FROM T24
GROUP BY LNumber, Year, Month
```

Similarly we can create tables on all lower levels. This model offers four levels of granularity. At a level 1 (highest level), the granularity is the finest; there are many records due to the fact that each is unique by a specific date. When these records are summarized, the level of granularity is coarser. For example T24, a monthly summary, is at a lower level then T11 since its granularity is coarser. Furthermore, T36 and T43 are even lower. In general, coarser levels of granularity do not require as many records to be stored. Transformation of a table from higher level to lower level is accomplished by grouping by all but one index attribute. Since during this operation the singled out index attribute is removed, we will refer to this operation as R with the appropriate argument what is also shown in Figure 2.

Generally, the summary tables define a directed graph which we will refer to as an aggregation graph. If the aggregation graph contains all possible summary tables then we call it the complete aggregation graph. The complete aggregation graph for TurboMachine Data Warehouse is shown in Figure 3. It has a root node corresponding to a fact table with the smallest granularity (T11) — we will refer to this node as level 1. The other nodes correspond to other summary tables and each arc has a label The graph has n+1 levels where n is the number of index attributes in the fact table T11. The root has also n children referred to as T21, T22, etc. Each node on the second level has n-1 children. More generally, each node on a k level has n-k+1 children. The lowest level has always only one node. In our case, since T11 has 5 index attributes, our complete aggregation graph has 6 levels and root has 5 children as shown in Figure 3. Transformation of a table from higher level to lower level is accomplished by grouping by all but one index attribute. As discussed before, Since during this operation the singled out index attribute is removed, we will refer to this operation as R with the appropriate argument as also shown in Figure 3.

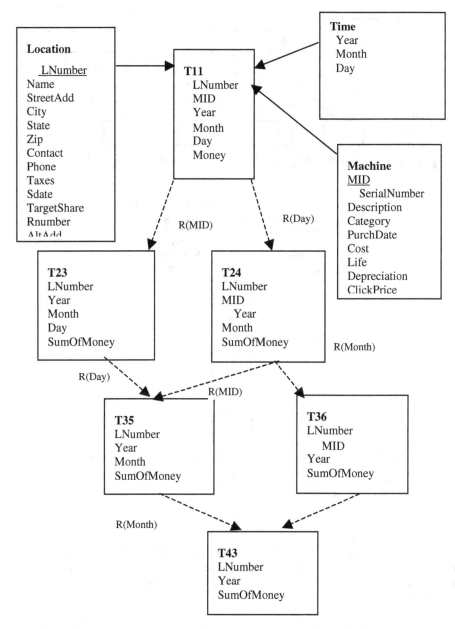

Fig. 2. The extended star schema with summary tables for TurboMachine company

Each query can be associated with the specific node of a complete aggregation graph. For example, let us consider a sample query Q35: "Find names for locations and the sum of money they made in December 1999".

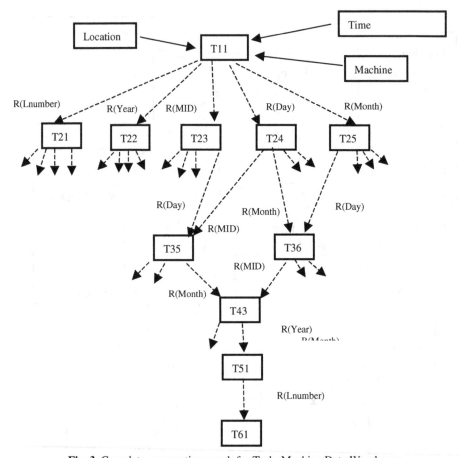

Fig. 3. Complete aggregation graph for TurboMachine Data Warehouse

    SELECT Location.Name, SumOfMoney
    FROM T35, Location
    WHERE Location.Lnumber=T35.LNumber
    AND Year = 1999
    AND Month = 'December';

This query can be associated with the summary table T35 and therefore was called Q35. Similarly, sample query Q36: "Find machine categories for all machines that made more than $5000 in 1999 while being in location with location number 234", shown below

    SELECT Machine.Category
    FROM T36, Machine
    WHERE T36.MID = Machine.MID
    AND Year = 1999
    AND Lnumber = 234
    AND SumOfMoney > 5000;

is associated with table T36.

Also sample query Q51: "Find names of locations that made in their lifetime more than $1000000", shown below

```
SELECT Location.Name
   FROM T51, Location
   WHERE Location.Lnumber=T51.LNumber
   AND SumOfMoney > 1000000;
```

is associated with table T51. If all the tables T35, T36, and T51 are available then, obviously, the execution of the above queries would be much quicker. If these tables are not available then the query processing would require grouping. Generally, we assume that either the table exists or that the appropriate view is defined. For example let us assume that table T51 is not present and that the view is defined based on available tables in Figure 2, as shown below

```
DEFINE VIEW T51(Lnumber, SumOfMoney) AS
   SELECT Lnumber, SUM(SumOfMoney)
   FROM T43
   GROUP BY Lnumber;
```

Complete aggregation graph and mapping of queries into this graph is used as a basis to create a cost model for materialization of views.

4 Summary Table Model

The complete aggregation graph defines the possible summary tables T_{kl} for the warehouse. For each table T_{kl}, we compute the table's cardinality. We assume that the cardinality of the original table T11 is known, and the cardinality of any other table can be derived in straightforward fashion from that of its parent in the complete aggregation graph. Let S_{kl} denote the cardinality of table T_{kl}. Then, for example, $S_{23} = S_{11}/$(average number of machines per location).

We can associate with each aggregate query a table that is the lowest table in the complete aggregation graph that is capable of answering the query. If a given query, can be processed using table T_{kl}, and can not be processed using any children in the aggregation graph, it will be referred to as query Q_{kl}. For simplicity of presentation we assume that there is at most one query that corresponds to each table. The approach easily generalizes to a case with many queries per table. We assume that each query Q_{kl} will be submitted with frequency f_{kl}.

Since the all nodes from the complete aggregation graph might not be available, in many practical situations we need to use a table from the higher level to answer the query. Generally, the time to process query Q_{ij} using table T_{kl} is denoted by $t_{ij}(k,l)$. This time is largely determined by the cardinality of T_{kl}, but is also influenced by whether or not the query involves aggregation on the table. In general, the tables that are lower in the aggregation graph are more efficient to use because (i) they are smaller, and (ii) they may already have pre-computed aggregates. In our model,

$t_{ij}(k,l) = \alpha(S_{kl})$ if k=i and l=j

$t_{ij}(k,l) = \beta\ (S_{kl})$ if k!=i or l != j

If table Tij exists for query Qij then the processing time is dependent on the cardinality of the table Tij (typically linearly). If table Tij does not exist for query Qij then the processing time is dependent on the cardinality of the lowest level table Tkl, however that time is much longer (the value of $\beta\ (S_{kl})$ is much higher than $\alpha(S_{kl})$ for the same table Tkl).

For simplicity, we assume that the time is dependent on the size of the operand table, and on whether or not aggregation is required on that table, but is not significantly affected by the precise form of the query.

For each query Q_{ij}, there is a constraint that the system must be able to generate an answer to the query within time C_{ij}. Hence, the warehouse must contain at least one table that is capable of satisfying the constraint, for any given query.

The benefit function B_{kl} for table T_{kl} is defined with respect to an existing aggregation graph as follows:

$$B_{kl} = \Sigma\ (t_{mr} - t_{kl}).f_{ij},$$

where t_{mr} is the time required to answer query Q_{ij} using the table T_{mr}, and t_{kl} is the time to answer the query using T_{kl}. T_{mr} is the table used to answer query Q_{ij} based on the existing aggregation graph.

We further assume that the warehouse designer would generally prefer to have a smaller number of tables rather than a larger number. We could consider the overhead (additional cost) of adding and maintaining the extra table with respect to an existing aggregation graph as O_{kl}.

The described above components of our model of summary tables are good basis for the algorithm(s) to select the optimal set of summary tables in Data Warehouses.

5 Identifying an Optimal Set of Summary Tables

5.1 Phase 1 — Preliminary Elimination of Summary Tables from Complete Aggregation Graph

Since we are working with a potentially very large number of tables in the complete aggregation graph, a mechanism is required to rapidly eliminate those tables which are most clearly not going to be needed. In order to identify the least relevant tables, we begin by associating each known query with the table that can be used to directly answer that query (i.e., the table that has all the necessary aggregate values pre-computed). Then, a table can be eliminated if 1) there are no direct queries associated with it, and 2) for each child C of this table in the complete aggregation graph C has another parent P in the complete aggregation graph, such that P has a direct query associated with it. For our example let us assume that we have identified queries Q35, Q36 and Q51.Then the aggregation graph after preliminary elimination is shown in Fig. 4.

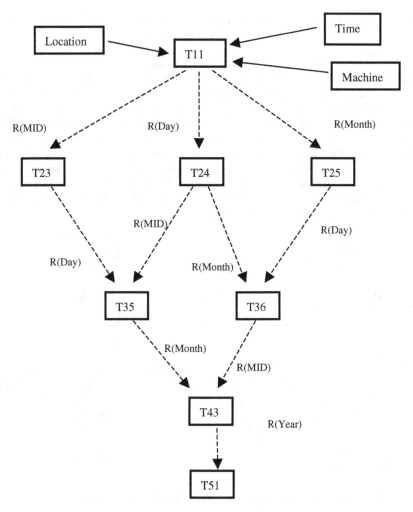

Fig. 4. The aggregation graph after preliminary elimination phase.

5.2 Phase 2 — Elimination of Summary Tables from Aggregation Graph Based on Time Constraints

Having eliminated the least useful tables from consideration, we apply a more detailed cost analysis on the remaining tables. As defined above, we have cost functions which model the time required to answer query Q_{ij} using table T_{kl}. In addition, for each query Q_{ij}, we associate a time constraint C_{ij}. This means that the warehouse must provide the capability of answering query Q_{ij} in time less than or equal to C_{ij}. By evaluating the cost function for each query using each table, we derive a set of subgraphs $\{CG_{ij}\}$ of the aggregation graph (called constraint subgraph) as follows:

For each table T_{mn} in the aggregation graph, T_{mn} is in CG_{ij} if there exists T_{kl} such that
1) $t_{ij}(k,l) < C_{ij}$, and
2) $T_{kl} = T_{mn}$ or T_{mn} is an ancestor of T_{kl} in the aggregation graph.

From the set of graphs $\{CG_{ij}\}$, further analysis is required in order to determine the best set of tables. There are different approaches to that task. One possibility is to compute the intersection of all of the graphs $\{CG_{ij}\}$. If this intersection is non-empty then it means that any of the tables in the intersection is capable of answering all the queries without violating the specified constraints. From that intersection we choose the table on the lowest level. Let us consider the example of aggregation graph from Figure 4. Figures 5, 6, and 7 show the aggregation graph (from Figure 4) with three different constraint subgraphs for three queries Q51, Q36, and Q35 respectively. In each of these figures the oval specifies the constraint subgraph. The intersection of these subgraphs would result in a single tableT24.

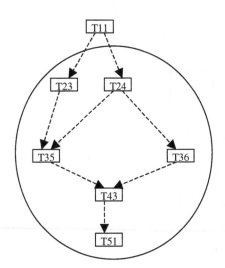

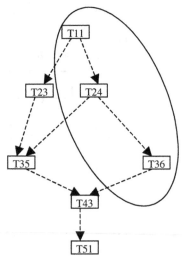

Fig. 5. Constraint subgraph for query Q51   **Fig. 6.** Constraint subgraph for query Q36

In this case, we can choose the smallest table from the intersection as the only one required for the warehouse. This solution is not necessarily optimal, though it is guaranteed to satisfy the constraints. It seeks to minimize the number of tables, and chooses the cheapest option which achieves that minimum. However, Figure 8 illustrates a case where it may not be the optimum solution.

In this case, {T24} is the theoretical minimum, but {T35, T36} will probably be the practical minimum because two aggregate tables will often have a combined size that is significantly less than the size of one non-aggregate table. If there are many queries, and strict constraints on response times, it is quite likely that the intersection will be empty. Figures 9 and 10 illustrate that possibility.

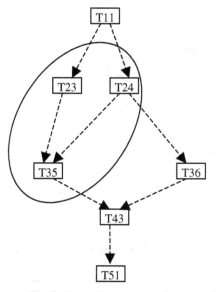

Fig. 7. Constraint subgraph for query Q35

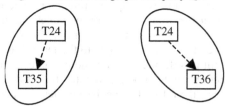

Fig. 8. Constraint subgraph for hypothetical queries Q35' and Q36'

Because of these reasons we will use an alternative method which we call top-down elimination method. Since the tables at the top of the aggregation graph are the largest and most expensive ones to store, we seek to eliminate the largest tables first. We start from the top of the aggregation graph, and consider each table in turn. A table can be eliminated if no graph CG_{ij} will be made empty by eliminating that table.

For our example of the constraint graphs (Fig. 5, 6 and 7) the top-down elimination method in the first step will eliminate T11 because all queries can be computed on time without using T11. In the next step T23 and T24 will be eliminated because of the same reason. In the next step we decide to keep T35 and T36 since these are the only remaining tables that can be used to answer queries Q35 and Q36.

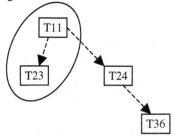

Fig. 9. Constraint subgraph for a hypothetical queries Q23'

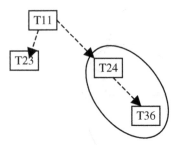

Fig. 10. Constraint subgraph for a hypothetical query Q36'

In the next step tables T43 and T51 will be eliminated because the remaining query Q51 can be computed on time without using them. As a result our aggregate graph would contain only two tables T35 and T36.

5.3   Phase 3 — Adding Summary Tables Based on Benefit Function

The objective of phase 2 was to arrive at a minimum set of tables that would satisfy the time constraints for all known queries. In phase 3, we consider expanding the number of tables in order to improve response times for some of the queries. To do this, we compute a benefit function for each of the tables eliminated during phase 2. This enables us to rank the tables in terms of their benefit. We do this by computing $MAX(B_{kl})$ and adding the most beneficial table to the Data Warehouse if desired. The process can continue until it is decided to stop adding tables to the warehouse. In this process the benefit function needs to be recomputed for all tables (except those with value zero already) since their benefit can be smaller for the extended aggregate graph.

Ideally, the process of selecting additional table for the existing aggregation graph should include the overhead O_{kl} as well as B_{kl} resulting in need to compute $MAX(B_{kl} - O_{kl})$. However, in practice it is very hard to obtain an accurate measure of O_{kl}, and it will usually be quite satisfactory to make decisions based on benefit alone.

For our example of aggregation graph obtained in phase 2, table T51 would have the highest value of the benefit function. If we decide to add this table, the benefit function of tables T11, T23, T24 is equal to zero since they are not going to be used to compute queries anyway.

6   Conclusions

In this paper, we considered the problem of determining an optimal set of summary tables for a Data Warehouse, in which aggregate queries are frequently posed. We introduced the concept of a complete aggregation graph for defining the possible summary tables. We presented a model for selecting summary tables. The algorithm based on the model takes into consideration query response time constraints and cost and benefit functions.

Although this work was done in the context of a Data Warehouse system, the ideas are also applicable to multidatabase systems [1, 3, 6] when applications require fast answers for requests involving aggregations. In order to increase the performance of a multidatabase system in that respect, the summary tables (materialized views) can be created and, if necessary, dynamically modified.

The use of materialized views introduces the need for maintaining consistency between the views and the underlying databases. We have not addressed this issue, but anticipate that an existing technique such as 2VNL [7] will be used for that purpose.

References

[1] R. Bayardo et al. "InfoSleuth: Agent-based Semantic Integration of Information in Open and Dynamic Environments", *Proceedings of the SIGMOD*, ACM Press, 6 1997.

[2] Bischoff J. and T. Alexander, *Data Warehouse: Practical Advice from the Experts*. New Jersey: Prentice-Hall, Inc., 1997.

[3] A. Elmagarmid, M. Rusinkiewicz, and A. Sheth, eds. *Management of Heterogeneous and Autonomous Database Systems*. San Francisco, CA: Morgan Kaufmann Publishers, Inc., 1999.

[4] A. Gupta, V. Harinarayan, and D. Quass "Aggregate-Query Processing in Data Warehousing Environments", *Proceedings of the VLDB*, 1995.

[5] Hammer, J. and D. McLeod. "An approach to resolving semantic heterogeneity in a federation of autonomous, heterogeneous database systems." *International Journal of Intelligent and Cooperative Information Systems* 2(1):51-83, March 1993.

[6] M. Nodine et al. "Active Information Gathering in InfoSleuth" *International Journal of Cooperative Information Systems*, 5(1/2), 2000.

[7] D. Quass, J. Widom "On-Line Warehouse View Maintenance", *Proceedings of the SIGMOD*, 1997.

[8] D. Srivastava, S. Dar, H. Jagadish, A. Levy "Answering Queries with Aggregation Using Views", *Proceedings of the VLDB*, 1996.

Modeling the Behavior of OLAP Applications Using an UML Compliant Approach

Juan Trujillo, Jaime Gómez, and Manuel Palomar

Departamento de Lenguajes y Sistemas Informáticos
Universidad de Alicante. SPAIN
{jtrujillo,jgomez,mpalomar}@dlsi.ua.es

Abstract. We argue that current approaches in the area of conceptual modeling of OLAP applications are lacking in some basic multidimensional (MD) semantics as well as they do not consider the behavior of these systems properly. In this context, this paper describes an OO conceptual modeling approach (based on a subset of UML) to address the above-mentioned issues. Thus, the structure of the system is specified by a class diagram, and from then, we define cube classes for specific user requirements. Then, the behavior of the system is considered by means of state and interaction diagrams (based on certain OLAP operations) on these cube classes.

Keywords: Conceptual Modeling, OLAP, UML, Multidimensionality

1 Introduction

There has recently been an increased interest in how to properly design OLAP [3] scenarios. These applications impose different requirements than On-line Transactional Processing (OLTP) systems, and therefore, different data models and implementation methods are required for each type of system.

Many design methods as well as conceptual models [4], [5], [13] are mainly driven by the structure of the operational data sources. Furthermore, they are not able to express all the underlying semantics of multidimensional (MD) data models, such as the many-to-many relationships between facts and some dimensions or the strictness and completeness of classification hierarchies [10].

In this context, there have lately been some approaches [6], [11] that try to apply OO techniques in OLAP areas. These approaches address the design of MD models taking advantage of the benefits provided by the OO paradigm, such as inheritance, code reuse, schema evolution, polymorphism and so on. Unfortunately, these approaches are centered in the solution space (implementation level) and do not provide a modeling language to specify the peculiarities of OLAP applications in the problem space (conceptual model level).

The MD schema of an OLAP application is directly used by the end user to formulate queries. Thus, the MD schema is crucial as it determines the type of queries the user can formulate and the set of OLAP operations to be applied on each query [12]. Therefore, a proper conceptual approach should consider the user query behavior.

T. Yakhno (Ed.): ADVIS 2000, LNCS 1909, pp. 14–23, 2000.

In this context, our research efforts have been focused on an OO conceptual modeling approach [15] to consider the underlying semantics of OLAP applications. This approach (based on a subset of UML, [2]) introduces a class diagram to capture the relevant properties associated to the structural aspects of these systems. In this paper, we extend this approach to model the behavior of OLAP applications by considering the semantics of cube classes, used to specify OLAP requirements in the class diagram, with a minimal use of constrains and extensions on UML. The behavior of a cube class is specified by means of state diagrams to model the user queries by the application of certain OLAP operations that allow users to navigate along the same cube class. Interaction diagrams are used to specify how cube classes can interact each other. In this sense, an interaction diagram shows the intentions of a final user when a new projection or coarser conditions on existing cube classes are specified.

This paper is organized as follows. Section 2 starts by summarizing how to build the class diagram based on an example to show its power of expression and defines cube classes for three requirements that will be handled throughout the paper. Section 3 focuses on modeling the user query behavior by means of state and interaction diagrams. In section 4, we compare our conceptual approach with the related work presented so far. Finally, in section 5, we present the conclusions and sketch some works that are currently being carried out.

2   Class Diagram

This proposal can be viewed as a star/snowflake approach [7] in which the basic components are classes. Two kinds of static relationships are used:

- associations: to represent relationships between instances of classes, e.g. a store is placed in a sales area; a sales area may contain a number of stores
- subtypes: to represent the relationship between and object and its type, e.g. snacks are a kind of food

Figure 1 shows the specification of a class diagram for a *sales system*. In this diagram the fact sales of products is analyzed along four dimensions (product, customer, store and time).

Facts are considered as classes in a shared aggregation relation (a special case of association in which the association relationship between the classes is considered as a "whole-part" and the parts may be parts in any whole) of n dimension classes. Derived measures can also be represented ("/" derived measure_name). Their derivation rules are placed between brackets around the class. All measures are additive by default. Semi-additivity and non-additivity is considered by constraints on measures and also placed around the class. These constraints define in a non-strict syntax the allowed operations along dimensions.

With respect to dimensions, each level of a classification hierarchy is considered as a class. These classes must form a Directed Acyclic Graph (DAG)

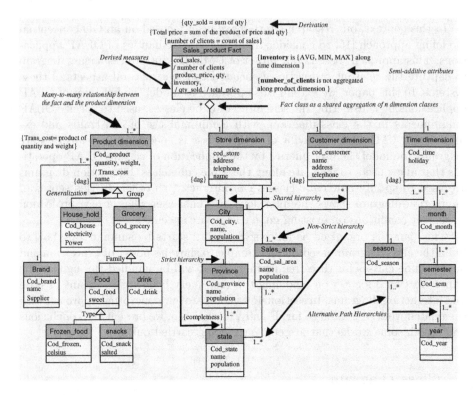

Fig. 1. The class diagram of the sales system

(constraint dag) starting from each dimension class; a class B (level b) of a hierarchy is considered as an association of a class A (level a). DAG's allow us to consider alternative path and multiple classification hierarchies.

The peculiarities of classification hierarchies such as the strictness (an object of a lower level of a hierarchy belongs to only one of a higher level) and completeness (all members belong to one higher-class object and that object consists of those members only) are also considered. In concrete, these features are specified by means of the cardinality of the roles of the associations and the constraint {completeness} respectively, as seen in the store dimension (see Figure 1). Dimensions can be categorized by means of Generalization/Specialization hierarchies, as observed in the product dimension.

2.1   On Querying the Conceptual Model

The class diagram provides the context to represent user's requirements in the conceptual model by means of object collections named cube classes. The basic components of these classes, as observed on top of Figure 2, are as follows:

– Head area: name of the cube class

- Fact area: representation of the measures to be analyzed
- Slice area: conditions to be satisfied
- Dice area: dimensions to address the analysis and their grouping conditions

To have a complete view of the approach presented here, let us suppose the following three base requirements (considered as three different cube classes, see Figure 3) that must be satisfied by the sales system:

1. the quantity_sold of products where the year_of_sales is "1998" must be grouped according to the store_state
2. the quantity_sold of products where the group_of_products is "Grocery" and the store_state is "Valencia" must be grouped according to the product_family brand, store_province and city
3. the quantity_sold of products where the year_of_sales is "1998" must be grouped according to customers and products

3   Modeling User Query Behavior

A set of OLAP operations can be applied to cube classes to consider the user query behavior. These operations are classified depending on whether they allow users to navigate along the same cube class (i.e. the user is still interested in the same parameters, and therefore, in the same cube class) or switch into others (the user changes the current intentions). To specify this behavior, our conceptual modeling approach uses state and interaction diagrams respectively. It should be taken into account that both diagrams are only defined for cube classes, as we are only interested in considering the behavior of these classes.

3.1   State Diagrams

State diagrams are used to describe the behavior of a system by establishing valid object life cycles for every cube class. By valid life, we mean a right sequence of states that characterizes the correct behavior of the objects that belong to a specific class.

Transitions represent valid changes of states that can be constrained by introducing conditions. They follow the syntax "event [guard]/action". This means that when a certain event reaches a state, if the condition [guard] is true, the action must lead the object into the corresponding target state.

A state diagram must be used to specify the behavior for each specified cube class. These diagrams will show the different states that cube objects reach based on certain OLAP operations. The allowed operations are as follows:

- roll-up: switching from a detailed to an aggregated level within the same classification hierarchy
- drill-down: switching from an aggregated to a more detailed level within the same classification hierarchy

– combine: showing more detail about the current objects, which can be applied in two different ways:
- showing an attribute of the current level for aggregation (e.g. population of cities)
- switching from a classification hierarchy into a different one within the same dimension (e.g. showing the different brands of the products aggregated by family)

– divide: is the reverse operation of combine and hides details of the current objects. This operation can be applied in two different ways:
- hiding an attribute of the current level for aggregation (e.g. hiding population of cities)
- hiding a classification hierarchy within the same dimension (e.g. hiding the brand property of products)

– slice: sub-selecting a subset of the current object collection as long as the new condition is smaller than the current one (e.g. sub-selecting a certain family or group of products for a particular analysis)[1]

Figure 2 shows the corresponding state diagram for the cube class 2. This diagram is organized in two areas clearly different. On top of Figure 2, all the details of the current cube class are depicted at all time, distinguishing the following areas: Fact (qty_sold of products is the measure to be analyzed), Slice(products whose group is grocery and stores in the state of Com. Valenciana) and Dice (object collection will be grouped by the product family, and brand, store province and city).

Let us now focus on describing the state diagram itself. It may first be distinguished that a dashed line, which indicates that OLAP operations can be applied to both dimensions concurrently, separates the states corresponding to the different dimensions included in the Dice area (product and store). For each dimension a valid object life cicle is defined. The state that represents the finest level of aggregation is labeled by "Dicing by code" and is considered as the greatest lower bound for every path of every classification hierarchy.

Then, one state is defined for every class (level) that belongs to a classification hierarchy and are labeled by "Dicing by" plus the name of the class as seen in the store dimension in Figure 2. For every classification hierarchy, there is one state labeled by "Dicing by ALL" to consider the least upper bound for every classification hierarchy path and is used to aggregate objects to a single value, i.e. to show data in the coarsest level of aggregation. The initial state is represented by a black circle and indicates the finest grouping condition for every dimension ("" Dicing by brand" and "Dicing by city"). Finally, the final state is represented by a bull's-eye and indicates that the destroy transition will finish the process of the analysis.

With reference to OLAP operations, roll-up/drill-down are used to move into states to aggregate/de-aggregate data along classification hierarchies respectively. Moreover, combine/divide can be used to switch into different levels

---

[1] A statistical table might be used for showing a multidimensional view of the cube class as shown in [14]

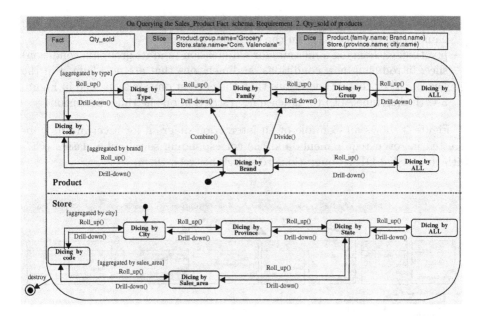

Fig. 2. The state diagram for the cube class of the requirement 2

of different classification hierarchies along the same dimension (e.g. product dimension). These operations must check some conditions with respect to the state come from and go to for a correct execution (not shown in the diagram for the sake of simplicity). Finally, the slice and combine/divide (applied on the first case) operations do not change the current state of the cube class objects, and therefore, are not shown in the diagram to avoid redundancy.

3.2   Interaction Diagrams

In UML, interaction diagrams are models that describe how groups of objects collaborate in some behavior. More concretely, an interaction diagram captures the behavior of a single use case. Our approach has adopted sequence diagrams [2] for their clarity and low complexity. In these diagrams, an object is represented as a box at the top of a vertical line. This vertical line is called the object's lifeline. The lifeline represents the object's life during interaction. Then, arrows between dashed lines describe which operations move from one cube class into another.

Thus, these diagrams show the different intentions of final users, which are mainly considered when users wish a new projection or coarser conditions on cube classes. Thus, the set of OLAP operations that lead users either to new cube classes or to existing ones are as follows:

- rotate: projecting on new dimensions (with its grouping conditions) and eliminating dimensions from the current Dice area[2]
- pivoting: exchanging a measure for a dimension (and its grouping condition)
- slice: introducing new conditions on dimensions that are not defined in the Dice area[3] or coarser conditions than the current ones (e.g. as seen in Figure 3 with slice(time,ALL) to indicate no restriction on the time dimension)

Figure 3 shows an example of an interaction diagram for a certain user interested in seven requirements and the corresponding seven cube classes (CC) with the needed OLAP operations to switch between them.

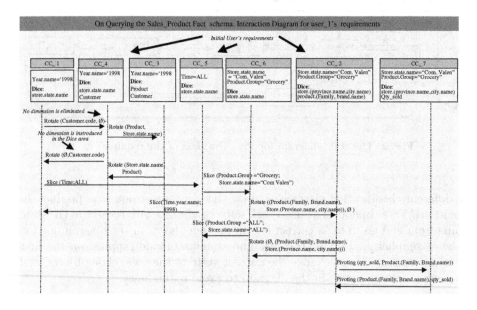

Fig. 3. The interaction diagram that models user's current intentions

It may first be distinguished that there are three cube classes that have already been defined (initial user's requirements) as they correspond to the three base requirements above-presented. Every cube class is represented as a box labeled with the corresponding requirement. This diagram shows that the final user wishes to start the process in the requirement 1. From there, the user wishes to apply the Rotate(customer.code,ϕ) OLAP operation to introduce a new grouping condition, which leads the user to a new cube class (CC_4).

It may be observed that in all operations, the reverse operation has been defined as it is quite reasonable for users to go back to the previous point in the

---

[2] It may be noticed that any of both attributes can be omitted.

[3] If the new condition was on dimension included in the Dice area of the cube class, the slice operation would be subselecting objects, and therefore, this would hold users within the same cube class.

analysis, and therefore, the Rotate(ϕ;Customer.code) operation has been defined from CC_4 to CC_1. From CC_4, the user would also like to apply the following operation: Rotate(Product;Store.state.name), which leads the user to CC_3, so the user has switched into a cube class that has already been defined to specify the requirement 3. The last operation applied to CC_1 is Slice(Time,ALL) to introduce a coarser condition on objects, which leads users to a the CC_5.

4 Comparison with Other Related Work

There have lately been published some works in the area of conceptual modeling of OLAP applications. The methodology presented in [5] and its underlying Dimensional Fact (DF) model [4] are based on the construction of the multidimensional (MD) schema from the entity-relationship (ER) schemes of the operational systems and they provide mechanisms to represent query patterns graphically. However, they do not take into consideration for example the correlation between successive queries and the evolution of queries along classification hierarchies, issues that are the aim of this work. The approach presented in this paper provides conceptual modeling constructs to specify query patterns (cube classes), which provide further details about the elements to be analyzed.

The proposal presented in [16] extends the ER model to consider multidimensional properties. This approach has some similarities with ours in representing the exact cardinality between facts and dimensions as well as the strictness and completeness along classification hierarchies. Nevertheless, as it is based on the ER model, neither functionality nor behavior can be considered. Therefore, user's requirements, OLAP operations and the evolution of queries cannot be considered. In our approach, cube classes allow us to enrich the power of expression of the model and make the design process much more interactive with the final user.

Another relevant work in this context is the Multidimensional ER (M/ER) model presented in [13], in which graphical elements are defined to consider facts and roll-up relationships. This work has been extended in [12] by means of graphical structures to specify the user query behavior by means of the correlation of successive queries through OLAP operations. Nevertheless, as the base M/ER model cannot consider data functionality or behavior, there is not a total integration between the M/ER model and the graphical structures considered to capture the evolution of queries. In our opinion, this situation makes difficult to clarify the method semantics, and therefore, an effective tool for checking consistency between both models becomes necessary.

The work presented in [6] defines easy and powerful algorithms to translate the Star/Snowflake schema into an OO schema based on the further knowledge provided by the user's queries to facilitate the maintenance of views. However, the paper refers to pointers as basic connections between entities (classes), and therefore, it is clearly centered in the implementation issues.

In [11] there is a proposal of complex OLAP reports supported by an OO model described in UML. This approach is the best one in considering complex

OLAP reports. Nevertheless, the description of relationships between entities, and therefore, basic semantics of the MD data models (e.g. the additivity of fact attributes, cardinality between facts and dimensions,...) are not considered.

The proposal of Lehner in [8] considers MD objects in the proposed Nested Multidimensional Model and defines a group of OLAP operations to be applied to these objects to permit a subsequent data analysis. However, basic MD data semantics (as the additivity of measures along dimensions) are not considered.

5   Conclusions

In this paper, we have presented an OO conceptual modeling approach for the design of OLAP applications. This approach (based on a subset of UML) takes advantage of the well-known OO conceptual modeling techniques to specify a class diagram that allows us to consider the underlying semantics of a MD data model, such as the exact cardinality between facts and dimensions, and the strictness and completeness along classification hierarchies. User's requirements are considered in the model by means of object collections (cube classes). State diagrams and interaction diagrams are used to model the evolution of cube classes by OLAP operations. In this way, the relevant properties (static and dynamic) of OLAP applications are considered in the same conceptual model.

The works that are currently being carried out include the implementation of a CASE tool prototype according to the approach presented in this paper. Furthermore, this prototype will provide a user interface for querying cube classes that allows users to execute correct OLAP operations defined in both state and interaction diagrams. Finally, other works that are currently being carried out are the application of OO techniques for a proper schema evolution and the automatic definition of workloads.

References

[1] Proceedings of the ACM 2nd International Workshop on Data warehousing and OLAP, Kansas City, Missouri, USA, November 1999.

[2] Grady Booch, James Rumbaugh, and Ivar Jacobson. The Unified Modeling Language User Guide. Addison-Wesley, 1998.

[3] E.F. Codd, S.B. Codd, and C.T. Salley. Providing OLAP (On-Line Analytical Processing) to User Analyst: An IT Mandate. Available from Arbor Software's web site. http://www.arborsoft.com/OLAP.html, 1996.

[4] M. Golfarelli, D. Maio, and S. Rizzi. Conceptual Design of Data Warehouses from E/R Schemes. In Proceedings of the 31st Hawaii Conference on System Sciences, pages 334–343, Kona, Hawaii, 1998.

[5] M. Golfarelli and S. Rizzi. A methodological Framework for Data Warehouse Design. In Proceedings of the ACM 1st International Workshop on Data warehousing and OLAP, pages 3–9, Washington D.C., USA, November 1998.

[6] V. Golpalkrishnan, Q. Li, and K. Karlapalem. Star/Snow-Flake Schema Driven Object-Relational Data Warehouse Design and Query Processing Strategies. In Mohania and Tjoa [9], pages 11–22.

 [7] R. Kimball. *The data warehousing toolkit*. John Wiley, 1996.

 [8] W. Lenher. Modelling Large Scale OLAP Scenarios. In *Advances in Database Technology - EDBT'98*, volume 1377, pages 153–167. Lecture Notes in Computer Science, 1998.

 [9] Mukesh Mohania and A. Min Tjoa, editors. *In Proceedings of the First International Conference On Data Warehousing and Knowledge Discovery*, volume 1676 of *Lecture Notes in Computer Science*. Springer-Verlag, 1999.

[10] TB. Pedersern and CS. Jensen. Multidimensional Data Modeling of Complex Data. In *Proceedings of the 15th IEEE International Conference on Data Engineering*, Sydney, Australia, 1999.

[11] T. Ruf, J. Goerlich, and T. Reinfels. Dealing with Complex Reports in OLAP Applications. In Mohania and Tjoa [9], pages 41–54.

[12] C. Sapia. On Modeling and Predicting Query Behavior in OLAP Systems. In *Proceedings of the International Workshop on Design and Management of Data Warehouses*, pages 1–10, Heidelberg, Germany, June 1999.

[13] C. Sapia, M. Blaschka, G. Höfling, and B. Dinter. Extending the E/R Model for the Multidimensional Paradigm. In Yahiko Kambayashi, Dik Lun Lee, Ee-Peng Lim, Mukesh K. Mohania, and Yoshifumi Masunaga, editors, *In Proceedings of the First International Workshop on Data Warehouse and Data Mining*, volume 1552 of *Lecture Notes in Computer Science*, pages 105–116. Springer-Verlag, 1998.

[14] J. Trujillo, M. Palomar, and J. Gómez. Detecting patterns and OLAP operations in the GOLD model. In *Proceedings of the ACM 2nd International Workshop on Data warehousing and OLAP* [1], pages 48–53.

[15] J. Trujillo, M. Palomar, and J. Gómez. Applying Object-Oriented Conceptual Modeling Techniques To The Design of Multidimensional Databases and OLAP Applications. In Hongjun Lu and Aoying Zhou, editors, *In Proceedings of the First International Conference On Web-Age Information Management*, volume 1846 of *Lecture Notes in Computer Science*, pages 83–94. Springer-Verlag, 2000.

[16] N. Tryfona, F. Busborg, and J.G. Christiansen. starER: A Conceptual Model for Data Warehouse Design. In *Proceedings of the ACM 2nd International Workshop on Data warehousing and OLAP* [1], pages 3–8.

Conceptual Multidimensional Data Model Based on MetaCube

Thanh Binh Nguyen, A. Min Tjoa, and Roland Wagner

Institute of Software Technology, Vienna University of Technology
Favoritenstrasse 9-11/188, A-1040 Vienna, Austria
{binh,tjoa}@ifs.tuwien.ac.at
Institute of Applied Knowledge Processing, University of Linz
Altenberger Strasse 69, A-4040 Linz, Austria
wagner@ifs.uni-linz.ac.at

Abstract. In this paper, we propose a conceptual multidimensional data model that facilitates a precise rigorous conceptualization for OLAP. First, our approach has strong relation with mathematics by applying a new defined concept, i.e. *H-set*. Afterwards, the mathematic soundness provides a foundation to handle natural hierarchical relationships among data elements within dimensions with many levels of complexity in their structures. Hereafter, the multidimensional data model organizes data in the form of metacubes, the concept of which is a generalization of other cube models. In addition, a metacube is associated with a set of groups each of which contains a subset of the metacube domain, which is a H-set of data cells. Furthermore, metacube operators (e.g. *jumping*, *rollingUp* and *drillingDown*) are defined in a very elegant manner.

1 Introduction

Data warehouses and OLAP are essential elements of decision support [5], they enable business decision makers to creatively approach, analyze and understand business problems [16]. While data warehouses are built to store very large amounts of integrated data used to assist the decision-making process [9], OLAP, which is first formulated in 1993 by [6], is a technology that processes data from a data warehouse into multidimensional structures to provide rapid response to complex analytical queries. Online Analytical Processing (OLAP) data is frequently organized in the form of multidimensional data cubes each of which is used to examine a set of data values, called *measures*, associated with multiple *dimensions* and their multiple levels. Moreover, dimensions always have structures and are linguistic categories that describe different ways of looking at the information [4]. These dimensions contain one or more natural hierarchies, together with other attributes that do not have a hierarchy's relationship to any of the attributes in the dimensions [10]. Having and handling the predefined hierarchy or hierarchies within dimensions provide the foundation of two typical operations like *rolling up* and *drilling down*. Because unbalanced and multiple hierarchical structures (Fig. 1,2) are the common structures of dimensions, the two current OLAP technologies, namely ROLAP and MOLAP, have limitations in the handling of dimensions with these structures [15].

T. Yakhno (Ed.): ADVIS 2000, LNCS 1909, pp. 24-33, 2000.

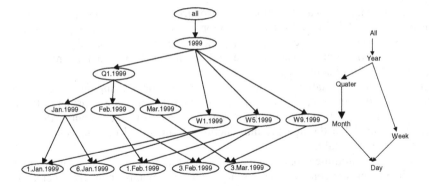

Fig. 1. An instance of the dimension *Time* with unbalanced and multiple hierarchical structure

Fig. 2. A schema of the dimension *Time* with multihierarchical structure.

The goal of this paper is the introduction of a conceptual multidimensional data model that facilitates a precise rigorous conceptualization for OLAP. Our approach has strong relation with mathematics by applying a set of new defined mathematic concepts, i.e. *H-relation, H-element, H-set, H-graph* and *H-partition*. The mathematic soundness provides a foundation to handle natural hierarchical relationships among data elements within dimensions with many levels of complexity in their structures. Afterwards, the multidimensional data model organizes data in the form of metacubes. Instead of containing a set of data cells, each metacube is associated with a set of groups each of which contains a subset of the data cell set. Furthermore, metacube operators (e.g. *jumping, rollingUp* and *drillingDown*) are defined in a very elegant manner.

The remainder of this paper is organized as follows. In section 2, we discuss about related works. Then in section 3, we introduce a conceptual data model that begins with the introduction of H-set concepts and ends with the definition of metacube concepts. The paper concludes with section 4, which presents our current and future works.

2 Related Works

Since Codd's [6] formulated the term Online Analytical Processing (OLAP) in 1993, many commercial products, like Arborsoft (now Hyperion) Essbase, Cognos Powerplay or MicroStrategy's DSS Agent have been introduced on the market [2]. But unfortunately, sound concepts were not available at the time of the commercial products being developed. The scientific community struggles hard to deliver a common basis for multidimensional data models ([1], [4], [8], [11], [12], [13], [21]). The data models presented so far differ in expressive power, complexity and formalism. In the followings, some research works in the field of data warehousing systems and OLAP tools are summarized.

In [12] a multidimensional data model is introduced based on relational elements. Dimensions are modeled as "dimension relations", practically annotating attributes with dimension names. The cubes are modeled as functions from the Cartesian product of the dimensions to the measure and are mapped to "grouping relations" through an applicability definition.

In [8] n-dimensional tables are defined and a relational mapping is provided through the notation of completion. Multidimensional database are considered to be composed from set of tables forming denormalized star schemata. Attribute hierarchies are modeled through the introduction of functional dependencies in the attributes of dimension tables.

[4] modeled a multidimensional database through the notations of dimensions and f-tables. Dimensions are constructed from hierarchies of dimension levels, whereas f-tables are repositories for the factual data. Data are characterized from a set of roll-up functions, mapping the instance of a dimension level to instances of other dimension level.

In statistical databases, [17] presented a comparison of work done in statistical and multidimensional databases. The comparison was made with respect to application areas, conceptual modeling, data structure representation, operations, physical organization aspects and privacy issues.

In [3], a framework for Object-Oriented OLAP is introduced. Two major physical implementations exist today: ROLAP and MOLAP and their advantages and disadvantages due to physical implementation were introduced. The paper also presented another physical implementation called O3LAP model.

[20] took the concepts and basic ideas of the classical multidimensional model based on the Object-Oriented paradigm. The basic elements of their Object Oriented Multidimensional Model are dimension classes and fact classes.

[15] gives a first overview of our approach. In this paper, we address a suitable mutidimensional data model for OLAP. The main contributions are: (a) a new set of *H-set* concepts that provide a foundation to define multidimensional components, i.e. dimensions, measures and metacubes; (b) the introduction of a formal multidimensional data model, which is used for handling dimensions with any complexity in their structures; (c) the very elegant manners of definitions of three metacube operators, namely *jumping*, *rollingUp* and *drillingDown*.

3   Conceptual Data Model

In our approach, a multidimensional data model is constructed based on a set of dimensions $\mathcal{D} = \{D_1,..,D_x\}, x \in \mathbf{N}$, a set of measures $\mathcal{M} = \{M_1,..,M_y\}, y \in \mathbf{N}$ and a set of metacubes $\mathcal{C} = \{C_1,..,C_z\}, z \in \mathbf{N}$, each of which is associated with a set of groups $Groups(C_i) = \{G_1,..,G_p\}, p,i \in \mathbf{N}, 1 \le i \le z$. The following sections formally introduce the descriptions of dimensions with their structures, measures, metacubes and their associated groups.

3.1   The Concepts of *H-Set*

Definition 3.1.1. [*H-relation*] A binary relation $\prec^*$ on a set S is called *a H-relation* if $\prec^*$ is:
- *irreflexive:* if $(x,x) \notin R$, for all $x \in S$.
- *transitive:* if $(x,z) \in R$ and $(z,y) \in R$ then $(x,y) \in R$, for all $x, y, z \in S$.

Definition 3.1.2. [*H-set and H-element*] A *H-set* is a pair $(S, \prec^*)$ and satisfies the following conditions:
- Each element of S is called *H-element*, therefore the set S is now denoted by $S = \{he_i, .., he_j\}$.
- Having only one *root* element: $\exists! root \in S : (\exists\, he_p \in S : he_p \prec^* root)$,
- Having leaf elements: $\{(he_t \in S) \mid (\exists\, he_u \in S : he_t \prec^* he_u)\}$,
- An *H-graph* $G_H = (V_H, E_H)$ representing the H-set $(S, \prec^*)$ is a directed graph that is defined as follows:
- $V_H = S$. A vertex represents each element of the set S. There exists only one *root* vertex, which is the vertex for the *root* element.
- $E_H \subseteq S \times S$. If $he_i \prec^* he_j$ for distinct elements he_i and he_j, then the vertex for he_i is positioned higher than the vertex for he_j; and if there is no he_k different from both he_i and he_j such that $he_i \prec^* he_k$ and $he_k \prec^* he_j$, then an edge is drawn from the vertex he_i downward to the vertex he_j.

Definition 3.1.3. [*H-path*] Let G_H be a *H-graph* representing a *H-set* $(S, \prec^*)$. A *H-path* hp in G_H is a nonempty sequence $hp = (he_0, .., he_l)$ of vertexes such that $(he_i, he_{i+1}) \in E_H$ for each $i \in [0, l-1]$. The H-path hp is from he_0 to he_l and has length l.

Definition 3.1.4. [*H-partition*] A collection of distinct non-empty subsets of a set S, denoted by $\{S_1, .., S_l\}, l \in \mathbf{N}$, is a *H-partition* of S if the three following conditions are satisfied:
- $S_i \cap S_j = \phi, \ \forall i, j \in \mathbf{N}, 1 \le i < j \le l$,

- $\displaystyle\bigcup_{i=1}^{l} S_i = S$,

- $\forall S_i, 1 \le i \le l : (\exists\, he_t, he_u \in S_i) \mid ((he_t \prec^* he_u) \text{ or } (he_u \prec^* he_t))$.

3.2   The Concepts of Dimensions

First, we introduce hierarchical relationships among dimension members by means of one hierarchical domain per dimension. A hierarchical domain is a *H-set* of dimension elements, organized in hierarchy of levels, corresponding to different levels of granularity. It also allows us to consider a dimension schema as a *H-set* of levels. In this concept, a dimension hierarchy is a *H-path* along the dimension schema, beginning at the root level and ending at a leaf level. Moreover, the definitions of two dimension operators, namely $O_E^{ancestor}$ and $O_E^{descendant}$, provide abilities to navigate along a dimension structure. In a consequence, dimensions with any complexity in their structures can be captured with this data model.

Definition 3.2.1. [*Dimension Hierarchical Domain*] A dimension hierarchical domain, denoted by $dom(D) =< DElements(D), \prec_E^* >$, is a *H-set*, where:

- $DElements(D)= \{all\} \cup \{dm_1,.., dm_n\}$ is a set of dimension elements of the dimension D, e.g. *1999, Q1.1999, Jan.1999,* and *1.Jan.1999,* etc are dimension members within the dimension *Time* (Fig. 1),
- The *all* is the *root*,
- The binary relation $\prec_E^*$ on the set *DElements*(D) is a *H-relation*.

Definition 3.2.2. [*Dimension Schema*] A dimension schema is a *H-set* of levels, denoted by $DSchema(D)=\langle Levels(D), \prec_L^* \rangle$, where:

- $Levels(D) = \{All\} \cup \{l_1,.., l_h\}, h \in \mathbf{N}$ is a finite set of levels of a dimension D, where:
- $\forall l_i \in Levels(D), \ l_i =< Lname_i, dom(l_i))$:
    - $Lname_i$ is the name of a level, e.g. *Year, Month, Week, and Day* are level names of the dimension *Time*,
    - $dom(l_i)$ is one of $\{dom(All), dom(l_1),.., dom(l_h)\}, h \in \mathbf{N}$, the collection of which is a *H-partition* of *DElements*(D),
    - The *All* is the *root level*, where: $dom(All) = \{all\}$,
- Leaf levels: $\{ \forall l_i \in Levels(D) | (\forall dm_j \in dom(l_i)$ is a *leaf element*)$\}$.

- The binary relation $\prec_L^*$ on the set *Levels*(D) is a *H-relation* and satisfies the following condition:

$\forall l_i, l_j \in Levels(D)$, $l_i \prec_L^* l_j$ is given if there exits a map $f_{ances} : dom(l_j) \rightarrow dom(l_i)$:

$(\forall dm_p \in dom(l_j))$, $(\exists! dm_q \in dom(l_i) | dm_q = f_{ances}(dm_p))$, such: $dm_p \prec_E^* dm_q$.

Definition 3.2.3. [*Dimension Hierarchy*] Let G_H^L be a *H-graph* representing the *H-set* $DSchema(D)=\langle Levels(D), \prec_L^* \rangle$, which is the schema of a dimension D. A hierarchy is a *H-path* $hp = (All,.., l_{leaf})$ that begins at the *All* (*root*) level and ends at a leaf level.

Let $H(D) = \{h_1,.., h_m\}, m \in \mathbf{N}$ be a set of hierarchies of a dimension D. If *m*=1 then the dimension has single hierarchical structure, else the dimension has multihierarchical structure.

Definition 3.2.4. [*Dimension Operators*] Two dimension operators (*DO*), namely $O_E^{ancestor}$ and $O_E^{descendant}$, are defined as follows:

$\forall l_c, l_a, l_d \in Levels(D)$, $\forall dm_i \in dom(l_c)$:

$$O_D^{ancestor}(dm_i, l_a) = \begin{cases} dm_j \in dom(l_a): dm_j \prec_E^* dm_i & \text{If } (l_a \prec_L^* l_c) \\ undefined & \text{Else} \end{cases}$$

$$O_D^{descendant}(dm_i, l_d) = \begin{cases} \{dm_t \in dom(l_d) \mid dm_i \prec_E^* dm_t\} & \text{If } (l_c \prec_L^* l_d) \\ undefined & \text{Else} \end{cases}$$

3.3   The Concepts of Measures

In this section we introduce measures, which are the objects of analysis in the context of multidimensional data model. First, we introduce the notion of measure schema, which is a tuple $MSchema(M) = \langle Fname, O \rangle$. In case a measure that O is "NONE", then the measure stands for a fact, otherwise it stands for an aggregation.

Definition 3.3.1. [*Measure Schema*] A schema of a measure M is a tuple $MSchema(M) = \langle Fname, O \rangle$, where:

- *Fname* is a name of a corresponding fact,
- $O \in \Omega \cup \{NONE, COMPOSITE\}$ is an operation type applied to a specific fact [2].
  Furthermore:
  - $\Omega = \{SUM, COUNT, MAX, MIN\}$ is a set of aggregation functions.
  - COMPOSITE is an operation (e.g. average), where measures cannot be utilized in order to automatically derive higher aggregations.
  - NONE measures are not aggregated. In this case, the measure is the fact.

Definition 3.3.2. [*Measure Domain*] Let $\mathcal{N}$ be a numerical domain where a measure value is defined (e.g. **N, Z, R** or a union of these domains). The domain of a measure is a subset of $\mathcal{N}$. We denote by $dom(M) \subset \mathcal{N}$.

3.4   The Concepts of MetaCubes

First, a metacube schema is defined by a triple of a metacube name, an *x* tuple of dimension schemas, and a *y* tuple of measure schemas. Afterwards, each data cell is an intersection among a set of dimension members and measure data values, each of which belongs to one dimension or one measure. Furthermore, data cells of within a metacube domain are grouped into a set of associated granular groups, each of which expresses a mapping from the domains of *x*-tuple of dimension levels (independent variables) to *y*-numerical domains of *y*-tuple of numeric measures (dependent variables). Hereafter, a metacube is constructed based on a set of dimensions, and consists a metacube schema, and is associated with a set of groups.

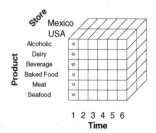

Fig. 3. *Sales* metacube is constructed from three dimensions: *Store, Product* and *Time* and one fact: *TotalSale*.

Let a metacube C be constituted from x dimensions $D_1,..,D_x, x \in \mathbf{N}$, and y measures $M_1,..,M_y, y \in \mathbf{N}$.

Definition 3.4.1. [*MetaCube Schema*] A metacube schema is tuple $CSchema(C) = \langle Cname, DSchemas, MSchemas \rangle$:

- *Cname* is the name of a metacube,
- $DSchemas =< DSchema(D_1),..,DSchema(D_x) >$ is a x-tuple of schemas of x dimensions $D_1,..,D_x, x \in \mathbf{N}$ with $DSchemas(i) = DSchema(D_i), 1 \le i \le x$,
- $MSchemas =< MSchema(M_1),..,MSchema(M_y) >$ is a y-tuple of schemas of y measures $M_1,..,M_y, y \in \mathbf{N}$ with $MSchemas(j) = MSchema(M_j), 1 \le j \le y$.

Definition 3.4.2. [*MetaCube Hierarchy Domain*] A metacube hierarchy domain, denoted by $dom(C) = \langle Cells(C), \prec_C^* \rangle$ is a H-set, where:

- Given a function $f : \overset{x}{\underset{i=1}{\times}} dom(D_i) \times \overset{y}{\underset{j=1}{\times}} dom(M_j) \to \{true, false\}$, *Cells(C)* is determined as: $Cells(C) = \{ c \in \overset{x}{\underset{i=1}{\times}} dom(D_i) \times \overset{y}{\underset{j=1}{\times}} dom(M_j) | f(c) = true \}$

- The binary relation $\prec_C^*$ on the set *Cells(C)* is a *H-relation*.

Definition 3.4.3. [*Group*] A group is triple $G = \langle Gname, GSchema(G), dom(G) \rangle$ where:

- *Gname* is the name of the group,
- $GSchema(G) = \langle GLevels(G), GMSchemas(G) \rangle$:

  - $GLevels(G) =< l_{D_1},...,l_{D_x} > \overset{x}{\underset{i=1}{\times}} Levels(D_i)$ is a x-tuple of levels of the x dimensions $D_1,..,D_x, x \in \mathbf{N}$.
  - $GMSchemas(G) =< MSchema(M_1),..,MSchema(M_y) >$ is a y-tuple of measure schemas of the y measures $M_1,..,M_y, y \in \mathbf{N}$.

- $dom(G) = \{c \in \overset{x}{\underset{i=1}{\times}} dom(l_{D_i}) \times \overset{y}{\underset{j=1}{\times}} dom(M_j) \in Cells(C)\}$

Let h_i be a number of levels of each dimension D_i $(1 \leq i \leq x)$. The total set of groups over

a metacube C is defined as $Groups(C) = \{G_1,..,G_p\}, p = \prod_{i=1}^{x} h_i$ [18].

Definition 3.4.4. [*MetaCube Operators*] Three basic navigational metacube operators (*CO*), namely *jumping*, *rollingUp* and *drillingDown*, which are applied to navigate along a metacube C, corresponding to a dimension D_i, are defined as follows:
$\forall G_c \in Groups(C)$, $l_c \in Levels(D_i)$ and $l_c \in GLevels(G_c)$, $\forall l_j, l_r, l_d \in Levels(D_i)$.

jumping:
$$jumping(G_c, l_j, D_i) = G_j = < GLevels(G_j), GMSchemas(G_j) >$$
Where:
$GMSchemas(G_j) = GMSchemas(G_c)$,
$GLevels(G_j)(i) = l_j$, $GLevels(G_j)(k) = GLevels(G_c)(k), \forall k \neq i$.

rollingUp:
$\forall dm \in dom(l_c)$, $G_r = jumping(G_c, l_r, D_i)$.
$$rollingUp(G_c, dm, l_r, D_i) = G_r^{sub} = < GSchema(G_r^{sub}), dom(G_r^{sub}) >$$
Where:
$GSchema(G_r^{sub}) = GSchema(G_r)$,
$dom(G_r^{sub}) = \{c_r \in dom(G_r) \mid \exists c \in dom(G_c): c.dms(i) = dm,$
$\qquad c_r.dms(i) = O_d^{ancestor}(dm, l_r, D_i), c_r.dms(j) = c.dms(j), \forall j \neq i\}$

drillingDown:
$\forall dm \in dom(l_c)$, $G_d = jumping(G_c, l_d, D_i)$.
$$drillingDown(G_c, dm, l_d, D_i) = G_d^{sub} = < GSchema(G_d^{sub}), dom(G_d^{sub}) >$$
Where:
$GSchema(G_d^{sub}) = GSchema(G_d)$,
$dom(G_d^{sub}) = \{c_d \in dom(G_d) \mid \exists c \in dom(G_c): c.dms(i) = dm,$
$\qquad c_d.dms(i) \in O_d^{descendant}(dm, l_d, D_i), c_d.dms(j) = c.dms(j), \forall j \neq i\}$

Definition 3.4.5. [*Metacube*] A metacube is a tuple C=$< CSchema, \mathcal{D}, Groups, CO >$, where:
- *CSchema* is a metacube schema,
- $\mathcal{D} = D_1,.., D_x, x \in \mathbf{N}$ is the set of dimensions,
- *Groups* is a total set of groups of the metacube.
- *CO* is a set of metacube operators.

4 Conclusion and Future Works

In this paper, we have introduced the conceptual multidimensional data model, which facilitates even sophisticated constructs based on multidimensional data elements such as dimension elements, measure data values and then cells. The data model has strong relationship with mathematics by using a set of new mathematic concepts, namely H-set to define its multidimensional components, i.e. dimensions, measures, and metacubes. Based on these concepts, the data model is able to represent and capture natural hierarchical relationships among dimension members. Therefore, dimensions with complexity of their structures, such as: unbalanced and multihierarchical structures [15], can be modeled in an elegant and consistent way. Moreover, the data model represents the relationships between dimension elements and measure data values by mean of data cells. In consequence, the metacubes, which are basic components in multidimensional data analysis, and their operators are formally introduced.

In context of future works, we are investigating two approaches for implementation: pure object-oriented orientation and object-relational approach. With the first model, dimensions and metacube are mapped into an object-oriented database in term of classes. In the other alternative, dimensions, measure schema, and metacube schema are grouped into a term of metadata, which will be mapped into object-oriented database in term of classes. Some useful methods built in those classes are used to give the required Ids within those dimensions. The given Ids will be joined to the fact table, which is implemented in relational database.

Acknowledgment

This work is partly supported by the EU-4th Framework Project GOAL (Geographic Information Online Analysis)-EU Project #977071 and by the ASEAN European Union Academic Network (ASEA-Uninet), Project EZA 894/98.

References

1. Agrawal, R., Gupta, A., Sarawagi, A.: Modeling Multidimensional Databases. IBM Research Report, IBM Almaden Research Center, September 1995.
2. Albrecht, J., Guenzel, H., Lehner, W.: Set-Derivability of Multidimensiona Aggregates. First International Conference on Data Warehousing and Knowledge Discovery. DaWaK'99, Florence, Italy, August 30 - September 1.
3. Buzydlowski, J. W., Song, II-Y., Hassell, L.: A Framework for Object-Oriented On-Line Analytic Processing. DOLAP 1998
4. Cabibbo, L., Torlone, R.: A Logical Approach to Multidimensional Databases. EDBT 1998
5. Chaudhuri, S., Dayal, U.: An Overview of Data Warehousing and OLAP Technology. SIGMOD Record Volume 26, Number 1, September 1997.
6. Codd, E. F., Codd, S.B., Salley, C. T.: Providing OLAP (On-Line Analytical Processing) to user-analysts: An IT mandate. Technical report, 1993.
7. Connolly, T., Begg, C.: Database system: a practical approach to design, implementation, and management. Addison-Wesley Longman, Inc., 1999.

8. Gyssens, M., Lakshmanan, L.V.S.: A foundation for multi-dimensional databases, Proc. VLDB'97.
9. Hurtado, C., Mendelzon, A., Vaisman, A.: Maintaining Data Cubes under Dimension Updates. Proc IEEE/ICDE '99.
10. Kimball, R.: The Data Warehouse Lifecycle Toolkit. John Wiley & Sons, Inc., 1998.
11. Lehner, W.: Modeling Large Scale OLAP Scenarios. 6th International Conference on Extending Database Technology (EDBT'98), Valencia, Spain, 23-27, March 1998.
12. Li, C., Wang, X.S.: A Data Model for Supporting On-Line Analytical Processing. CIKM 1996.
13. Mangisengi, O., Tjoa, A M., Wagner, R.R.: Multidimensional Modelling Approaches for OLAP. Proceedings of the Ninth International Database Conference "Heterogeneous and Internet Databases" 1999, ISBN 962-937-046-8. Ed. J. Fong, Hong Kong, 1999
14. McGuff, F., Kador, J.: Developing Analytical Database Applications. Prentice Hall PTR, 1999.
15. Nguyen, T.B., Tjoa, A M., Wagner, R.R.: *Conceptual Object Oriented Multidimensional Data Model for OLAP*. Technical Report, IFS, Vienna 1999.
16. Samtani, S., Mohania, M.K., Kumar, V., Kambayashi, Y.: Recent Advances and Research Problems in Data Warehousing. ER Workshops 1998.
17. Shoshani, A.: OLAP and Statistical Databases: Similarities and Differences. Tutorials of PODS 1997.
18. Shukla A., Deshpande, P., Naughton, J. F., Ramasamy, K.: Storage Estimation for Multidimensional Aggregates in the Presence of Hierarchies. VLDB 1996: 522-531
19. Thomsen, E.: OLAP solutions: Building Multidimensional Information Systems. John Wiley& Sons, Inc., 1997.
20. Trujillo, J., Palomar, M.: An Object Oriented Approach to Multidimensional Database. Conceptual Modeling (OOMD), DOLAP 1998.
21. Vassiliadis, P.: Modeling Multidimensional Databases, Cubes and Cube operations. In Proc. 10th Scientific and Statistical Database Management Conference (SSDBM '98), Capri, Italy, June 1998.
22. Wang, M., Iyer, B.: Efficient roll-up and drill-down analysis in relational database. In 1997 SIGMOD Workshop on Research Issues on Data Mining and Knowledge Discovery, 1997.

Using Portfolio Theory for Automatically Processing Information about Data Quality in Data Warehouse Environments

Robert M. Bruckner and Josef Schiefer

Institute of Software Technology (E188)
Vienna University of Technology
Favoritenstr. 9-11 /188, A-1040 Vienna, Austria
{bruckner, js}@ifs.tuwien.ac.at

Abstract. Data warehouses are characterized in general by heterogeneous data sources providing information with different levels of quality. In such environments many data quality approaches address the importance of defining the term "data quality" by a set of dimensions and providing according metrics. The benefit is the additional quality information during the analytical processing of the data. In this paper we present a data quality model for data warehouse environments, which is an adaptation of Markowitz's portfolio theory. This allows the introduction of a new kind of analytical processing using "uncertainty" about data quality as a steering factor in the analysis. We further enhance the model by integrating prognosis data within a conventional data warehouse to provide risk management for new predictions.

1 Introduction

In the past data quality has often been viewed as a static concept, which represents the write-once read-many characteristics of many data warehouses: The quality of a piece of information (fact) is evaluated only once (e.g. during the cleansing process) and stored in the data warehouse. From the data consumers' point of view, data quality is an additional property of a fact, which can be physically stored by an attribute or data quality dimensions. Depending on the context, the refinement (e.g. more details) of a fact can result in either of two possibilities 1) the creation of a new fact (with adapted data quality settings), or 2) updating the old one.

The trend toward multiple uses of data, exemplified by the popularity of data warehouses, has highlighted the need to concentrate on dynamic approaches, where the quality depends on the context in which data is used. Furthermore, data is viewed as a key organizational resource and should be managed accordingly [1].

In this paper we propose the adaptation of Markowitz's portfolio theory [3] to data warehouse environments. Our approach can be viewed as an additional layer above an existing representation of data quality in a data warehouse, where information about the "uncertainty" (of the quality evaluation) is used to select "optimal" portfolios, described by weight vectors for the base facts. The model considers the "fitness for use" [11], because data quality is an integral part during the analytical processing of

T. Yakhno (Ed.): ADVIS 2000, LNCS 1909, pp. 34-43, 2000.
© Springer-Verlag Berlin Heidelberg 2000

the data. Furthermore the examined data quality can become the steering factor in each analysis.

The remainder of this paper is organized as follows. Section 2 contains a description of the goals of this approach. Section 3 gives a short overview of Markowitz's portfolio theory, which is the basis for the adaptation, described in section 4. In section 5 we apply the approach to "conventional" data warehouses. The integration of prognosis data is discussed in section 6 and followed by an investigation of the general limitations of this approach (section 7).

The paper concludes with section 8 (related works) and section 9, where we give a summary and conclusion of this work.

2 Goal of This Approach

The automatic consideration of data quality by using portfolio selection is suitable for "traditional" requirements of data warehouse users, such as ad-hoc queries and predefined reports, as well as analytical processing where the data quality of the involved facts is an important issue.

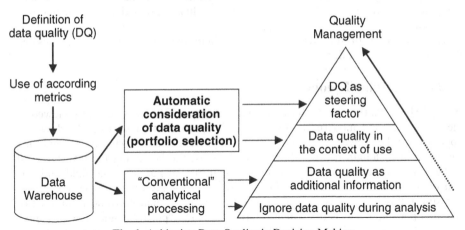

Fig. 1. Achieving Data Quality in Decision Making

Besides *conventional* analytical processing, the approach described in detail in sections 4 to 6, enables the *automatic consideration* of data quality, as shown in Fig. 1. In the first case the examined data quality is just an additional property of a fact. It is the user's responsibility to take advantage of the provided data quality information. Consequently only experienced data warehouse users (experts) are able to "exploit" this kind of knowledge. Similar to the far-reaching effects and consequences of poor data quality [7], the ignorance of that information can have diverse impacts, such as poorer decision making and difficulties to set and execute enterprise strategies.

The *automatic consideration* of data quality by using portfolio selections introduces a new layer between the data warehouse and its users by considering the context in which the data is used. Decision making often requires the combination of many aspects (represented by facts in a data warehouse) within an analysis. Therefore it is necessary to

- define data quality according to the "fitness for use" principle [11].
- consider and store "uncertainty" of the evaluated data quality.

This enables new perspectives on analytical processing of data for decision making:

- automatically calculate (optimal) weights that describe the impact of a specific fact on a decision according to its data quality and the involved uncertainty.
- consider the user preferences due to risk aversion, e.g. a venturesome user, who has high requirements on the data quality of the involved facts, nevertheless accepts high uncertainty of the assessed data quality of the facts.
- use the derived weights to combine the facts to a decision (with a resulting data quality and uncertainty assessment).
- introduce *"quality-driven" decisions*, where the user requirements due to the resulting data quality and uncertainty assessment of the decision are pre-specified.
- users need no further background knowledge to apply this approach for analytical processing. Even inexperienced users are able to take advantage of this approach.

Unfortunately it is not always possible to gain all the information from the data sources that are necessary to make good decisions. If there is a strong dependency between predictions about the future and those decisions (made to affect the future), it will be advantageous to integrate additional regular and prognosis data into a data warehouse, where all users have access to these valuable predictions represented in a consistent manner.

The integration of such data requires no alterations in the analytical processing through portfolio selection.

3   The Portfolio Theory

In 1952 Harry Markowitz published a landmark paper [3] that is generally viewed as the origin of the "modern portfolio theory" approach to investing. Markowitz's approach can be viewed as a *single-period[1]* approach, where money is invested only for a particular length of time (holding period). At the beginning the investor must make a decision regarding what particular securities to purchase and hold until the end of the period. Since a *portfolio* is a collection of securities (in the context of investing), this decision is equivalent to selecting an optimal portfolio from a set of possible portfolios and is thus often referred to as the "portfolio selection problem". Each security is characterized by the *expected return* and the *uncertainty* of this estimation (that is done by the investor).

Markowitz notes that the typical investor, in seeking both to maximize expected return and minimize uncertainty (= *risk*), has two conflicting objectives that must be balanced against each other when making the purchase decision at the beginning of the holding period. The investor should view the rate of return associated with any one of the various considered securities to be non-deterministic. In statistics this is known as a *random variable W_i*, described by its moments: its expected value or *mean*

---

[1] Markowitz shows in [5] that investing is generally a multi-period activity, where at the end of each period, part of the investor's wealth is consumed and part is reinvested. Nevertheless, this one-period approach can be shown to be optimal under a variety of reasonable circumstances (see also [9]).

r_i and the *standard deviation* σ_i (formula 1). Therefore a portfolio W_p is a collection of the weighted random variables.

$$W_i = (r_i, \sigma_i) .\tag{1}$$

$$W_p = a_1 W_1 + ... + a_n W_n \text{ with } \sum_{i=1}^{n} a_i = 1 .\tag{2}$$

The calculated expected return r_p of a portfolio is the weighted average of the expected returns of its component securities. The calculated standard deviation σ_p (variance σ_p^2) of a portfolio is given by formula 4, where σ_{ij} denotes the covariance of the returns between security i and security j and ρ_{ij} indicates the correlation factor.

$$r_p = a_1 r_1 + ... + a_n r_n .\tag{3}$$

$$\sigma_p^2 = \sum_{i=1}^{n}\sum_{j=1}^{n} a_i a_j \sigma_{ij} \text{ with } \sigma_{ij} = \rho_{ij}\sigma_i\sigma_j .\tag{4}$$

The *covariance* is a measure of the relationship between two random variables. A positive value for covariance indicates that the securities' returns tend to go together. For example, a better-than-expected return for one is likely to occur along with a better-than-expected return for the other.

The method that should be used in selecting the most desirable portfolio utilizes *indifference curves* [9]. These curves represent the investor's preferences for risk and return and thus can be drawn on a two-dimensional figure. In general two assumptions [3] hold for investors:

- *Nonsatiation*: It is assumed that investors, when given a choice between two otherwise identical portfolios, will select the one with the higher level of expected return.
- *Risk Aversion*: This means that the investors, when given a choice, will not want to take fair gambles, where a fair gamble is defined to be one that has an expected payoff of zero with an equal chance of winning or losing.

The *feasible set* simply represents the set of all portfolios that could be formed from a group of n securities. In general, this set will have an umbrella-type shape similar to the one shown by Fig.2. The *efficient set* can be located by applying the efficient set theorem (given by the assumptions above) to the feasible set. From this (infinite) set of efficient portfolios[2] the investor will find the *optimal* one by:

- Plotting indifference curves
  Plot the indifference curves on the same figure as the efficient set and then choose the portfolio that is farthest northwest along the indifference curve (Fig.2).
- Critical line method [4] for the identification of *corner portfolios*[3].
  Utilizing the information about the corner portfolios it becomes a simple matter for a computer to find the composition, and in turn the expected return and standard deviation, of the optimal portfolio by the approximation of the tangency point between the (calculated) efficient set and one of the investor's indifference curves.

---

[2] All other portfolios are "inefficient" and can be safely ignored.
[3] A corner portfolio is an efficient portfolio where any combination of two adjacent corner portfolios will result in a portfolio that lies *on the efficient set* between these two portfolios.

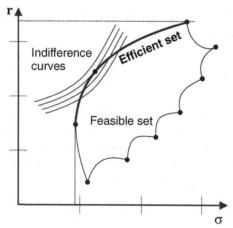

Fig. 2. The Selection of the Optimal Portfolio

4 Adaptation of the Portfolio Theory for Data Warehouses

There are many approaches in the literature that can be applied to studying data quality (classified by a framework in [13]), but there is no commonly accepted model providing metrics to precisely assess data quality. One reason for this is the different context in which the data is produced (operational systems) and used (data warehouses). As a consequence every data quality assessment will be characterized by a kind of "*uncertainty*", which should be evaluated and stored as well.

The equivalences that can be identified between finance markets and data warehouse environments are presented in Table 1.

Table 1. Identified Equivalences: Portfolio Theory for Investments and Data Warehouses

Data Warehouse Environment	Portfolio Theory for Investments
data warehouse user	investor
fact	security, random variable
data quality	expected return on security
uncertainty of the data quality assessment	standard deviation
covariance	covariance
analytical processing objectives: minimize uncertainty and maximize data quality of the involved facts	selection of an optimal portfolio

The adaptation of the portfolio theory (according to [5]) requires the *validation of the assumptions* and requirements for a data warehouse environment.

1. single-period decisions

   Analytical processing in data warehouses usually takes place when the update window is closed (all available facts are not changing within this period). New information provided at the next update window can be used to rerun the analysis and compare the result with the old one.

2. risk aversion and nonsatiation

As already mentioned decision making in every context (e.g. investments, analytical processing) is dependent on the quality of its input parameters. Therefore the two objectives (risk aversion and nonsatiation) apply to data warehouse environments as well.

3. methods for the approximation of all input parameters

Many of these parameters can be derived automatically in data warehouse environments (e.g. data quality assessments), but in general the correlation factors (needed for the covariance matrix) between facts have to be approximated manually.

Based on these assumptions and identified equivalences the area of application of portfolio selection can be extended to decision support in data warehouse environments. Analytical processing of data in "conventional" data warehouses using this new approach is explained in section 5. The integration of prognosis data and automatic consideration through portfolio selection is discussed in section 6.

5 Portfolio Selection in "Conventional" Data Warehouses

"*Conventional*" data warehouses contain data from operational and legacy systems but do not include prognosis data as proposed in the next section. During the data cleansing process the data quality and its uncertainty is assessed (in general the quality of data depends on the applied context). The adapted portfolio selection method can be used two-fold in data warehouse environments:

1. Additional quality information for whole analysis results

The result of the portfolio selection - the calculated weight vector (which describes the influence that each involved fact should have to achieve the optimal portfolio) is used to derive the data quality and uncertainty assessment for the whole analysis result. Therefore the components (a_i, $i=1,...,n$) of the weight vector are inserted into formulas 3 and 4 mentioned in section 0. The derived quality statement for the whole analysis gives the user an idea of the quality of the involved facts.

2. *Risk-driven* decisions

The indifference curves describing the user requirements on the quality properties of the analysis result are replaced by only one straight line that is equivalent to the user's requirement (e.g. resulting data quality should be high = 0.9 with minimized uncertainty in the quality assessment). We call this case a *risk-driven* decision because the data quality and the uncertainty assessments of the involved facts are the steering factor of such an analysis. If it is possible due to the quality of the involved facts, the user's quality requirement will always be met precisely by the approximation of the point of intersection between the (calculated) efficient set and the straight line (which represents the user's quality requirement). Furthermore the user's uncertainty of choosing the right weights for the influence of facts in a complex analysis will disappear, because Markowitz showed that the calculated weight vector is optimal (under the assumption of correct input parameters).

Example for using portfolio selection in the context of a risk-driven decision:

The analysis, which is the basis for the risk-driven decision considers three facts. They are given by their data quality and uncertainty assessments (see Table 2) and their correlation factors (ρ_{AB}=0.53, ρ_{AC}=0.706 and ρ_{BC}=0.209). All together there are 9 input parameters which are utilized to identify the *corner portfolios C(i)* by using the *critical line method* [4]. The result of this step is illustrated in Table 3.

Table 2. Characterization of the Three Facts Involved in the Analysis

Fact	Data quality	Uncertainty
A	76.2%	12.08%
B	84.6%	29.22%
C	82.8%	17.00%

Table 3. Computed Corner Portfolios

Corner portfolio	Assessments[4]		Weight vector of the corner portfolio		
	Data quality	Uncertainty	A	B	C
C(1)	84.60%	29.22%	0.00	1.00	0.00
C(2)	83.20%	15.90%	0.00	0.22	0.78
C(3)	77.26%	12.22%	0.84	0.00	0.16
C(4)	76.27%	12.08%	0.99	0.00	0.01

At this point in the analysis the user's data quality requirement is taken into consideration. In general a data quality requirement r° leads to a linear equation (formula 5), where r^C and r^D denote the data quality assessments of two suitable adjacent corner portfolios. The factor Y in this equation describes the influence ratio of the corner portfolios to meet the user's data quality requirement.

$$r^\circ = Y * r^C + (1-Y) * r^D. \tag{5}$$

In our example the user's data quality requirement is 0.8 which lies between the two adjacent corner portfolios: *C(2)* and *C(3)*. Solving the linear equation (formula 5) then yields $Y = 0.46$. The combination of this result with the weight vectors of the two adjacent corner portfolios, derives a new weight vector that describes the necessary influence of the considered facts to meet the user's data quality requirement exactly:

$$Y * C(2) + (1-Y) * C(3) = 0.46 * \begin{pmatrix} 0.00 \\ 0.22 \\ 0.78 \end{pmatrix} + 0.54 * \begin{pmatrix} 0.84 \\ 0.00 \\ 0.16 \end{pmatrix} = \begin{pmatrix} \mathbf{0.45} \\ \mathbf{0.10} \\ \mathbf{0.45} \end{pmatrix}$$

The combination of the three considered facts with the weights A=0.45, B=0.10 and C=0.45 will achieve an assessed data quality of 0.8 and *minimized uncertainty*.

We call this a *risk-driven decision* because the uncertainty about data quality is the steering factor in the analysis.

---

[4] The data quality and uncertainty assessments are computed using formulas 3 and 4 (mentioned in section 3) with the according weight vectors of the corner portfolios.

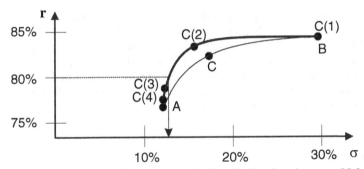

Fig. 3. Analysis under Consideration of a Data Quality Requirement of 0.8 and Minimized Uncertainty.

6 Integration of Prognosis Data

As already mentioned in section 2 it will be advantageous to integrate additional regular and prognosis data into a data warehouse, if there is a strong dependency between predictions about the future and decisions made to affect the future.

The characterization of regular data through data quality and uncertainty assessments (section 4) is necessary to apply the presented approach and will be used to integrate prognosis data as well:

The *data quality* of a prediction is equivalent to the trustworthiness from the data warehouse users' viewpoint. This assessment is dependent on criterions similar to that of regular data [10], for instance completeness, reliability, consistency and believability of the provided predictions.

The *uncertainty* of data quality assessments can be provided by external sources, where each reasonable prognosis contains information about the (statistical) risk of deviations due to the assumptions (e.g. survey results on a specified target group). If such additional information is not available, the uncertainty has to be evaluated in the context of the data warehouse environment manually.

Table 4. Regular vs. Prognosis Data (in general)

	Regular Data	Prognosis Data
Data quality assessment	high	low - high
Uncertainty assessment	low	medium - high

- **Exclusive usage** of prognosis data in analytical processing: The adapted portfolio selection approach can be applied to regular and prognosis data as well without any alterations, because both are assessed by their data quality and uncertainty.
- The integration of **regular and prognosis** data in the same analysis to generate new predictions requires further investigation to achieve reasonable results: In general prognosis data is distinguishable from regular data by their mean quality and uncertainty assessments, as described in Table 4. If we apply the presented portfolio selection approach to an analysis involving both kinds of information at the same time without any alteration, we will achieve an

unintended result: Regular data will be privileged, because in general the mean data quality assessment will be higher than the one of the predicted data. A solution for this situation is the introduction of a *displacement* that is added temporarily to the data quality assessments of prognosis data (only during the portfolio selection). This causes an increase in the data quality assessment, but the uncertainty assessment remains unchanged. Small displacements will favor regular data (suitable for short-term predictions).

7 General Limitations

The presented method for identifying an optimal portfolio for analytical processing has the following limitations:

- Complexity
  The number of input parameters that are needed to perform portfolio selection according to Markowitz increases in quadratic order with the number of the involved facts: $O(n^2)$. In general a portfolio selection with n facts involved requires

$$\frac{n(n+1)}{2} + 2n = n\left(\frac{n-1}{2} + 2\right) \text{ input parameters.} \tag{6}$$

- Metrics for measuring data quality
  After more than a decade of intensive research in the area of data quality [13] there is still no consensus about a "fully featured" data quality model providing precise metrics. However data quality assessments are an important input parameter for the portfolio selection.
- Determination of the correlation between facts
  Data warehouses are characterized by a huge amount of facts that is continuously increasing. Whereas there are several methods [11] to examine data quality aspects, there is no one to automatically determine the correlation between facts.
- Application of the results
  In general weight vectors are only applicable to continuous value domains (e.g. numbers) but not to discrete domains, like strings or booleans.

8 Related Work

This work is related to [2], where a first draft of the approach is described and is further extended in this paper.

In the area of data warehouse research the management of data quality is an emerging topic, where approaches from other research fields (as we did in this paper) are investigated [12], [6] to provide better information for data consumers.

In the context of investments there are some other models that require fewer input parameters (e.g. Index Model [8]: n securities $\rightarrow$ $3n+2$ parameters), but they are not directly applicable to data warehouse environments. Furthermore they require additional information, which can be provided by finance markets but not by data warehouses (at this point in time, because of the lack of appropriate methods).

9 Conclusion

Based on the portfolio theory we identified equivalences in decision making in the context of finance markets and data warehouses and adapted Markowitz's approach.

The introduction of this new kind of analytical processing provides two benefits: 1) a derived quality statement for the whole analysis, that gives an idea of the properties of the involved facts and 2) risk-driven decisions, where uncertainty about data quality is a steering factor in the analysis. The approach can be applied to regular and prognosis data as well.

The presented approach for identifying an optimal portfolio that minimizes the assessed uncertainty and maximizes the data quality of the involved facts during analytical processing requires the further estimation of all covariances between the facts, which is in general difficult to automate.

Our future work will concentrate on discovering efficient solutions to reduce the high number of required input parameters to apply this approach.

References

1. Ballou, Donald P., Tayi Giri K.: *Enhancing Data Quality in Data Warehouse Environments*. Communications of the ACM, January 1999, Vol. 42(1), 73–78
2. Bruckner, Robert M.: *Datenqualitäts- und Evaluierungsaspekte von Data Warehouses (in german)*. Master Thesis, Vienna University of Technology, Institute of Software Technology (E188), 1999
3. Markowitz, Harry M.: *Portfolio Selection*. Journal of Finance, Vol. 7(1), March 1952, 77–91
4. Markowitz, Harry M.: *The Optimization of a Quadratic Function Subject to Linear Constraints*. Naval Research Logistics Quarterly 3, nos. 1–2 (March–June 1956), 111–133
5. Markowitz, Harry M.: *Portfolio Selection*. Yale University Press, New Haven, Connecticut, 1959
6. Redman, Thomas C.: *Data Quality for the Information Age*. Artech House Publishers, Boston, 1996
7. Redman, Thomas C.: *The Impact of Poor-Data Quality on the Typical Enterprise*. Communications of the ACM, February 1998, Vol. 41(2), 79–82
8. Sharpe, William F.: *Capital Asset Prices: A Theory of Market Equilibrium under Conditions of Risk*. The Journal of Finance, Vol. 19(3), September 1964, 425–442
9. Sharpe, William F., Alexander, Gordon J., Bailey, Jeffery V.: *Investments*. 6th Edition; Prentice-Hall Inc., Englewood Cliffs, New Jersey, 1998
10. Strong, Diane M., Lee, Yang, Wang, Richard Y.: *Data Quality in Context*. Communications of the ACM, May 1997, Vol. 40(5), 103–110
11. Tayi Giri K., Ballou, Donald P.: *Examining Data Quality*. Communications of the ACM, February 1998, Vol. 41(2), 54–57
12. Wang, Richard Y.: *A Product Perspective on Total Data Quality Management*. Communications of the ACM, February 1998, Vol. 41(2), 58–65
13. Wang, Richard Y., Storey Veda C., Firth Christopher P.: *A Framework for Analysis of Data Quality Research*. IEEE Transactions on Knowledge and Data Engineering, August 1995, Vol. 7(4), 623–639

Representative Sample Data for Data Warehouse Environments

Thanh N. Huynh, Binh T. Nguyen, J. Schiefer, and A.M. Tjoa

Institute of Software Technology (E188)
Vienna University of Technology
Favoritenstr. 9 - 11 /188, A-1040 Vienna, Austria.
{thanh, binh, js, tjoa}@ifs.tuwien.ac.at

Abstract. The lack of sample data for data warehouse or OLAP systems usually makes it difficult for enterprises to evaluate, demonstrate or benchmark these systems. However, the generation of representative sample data for data warehouses is a challenging and complex task. Difficulties often arise in producing familiar, complete and consistent sample data on any scale. Producing sample data manually often causes problems that can be avoided by an automatic generation tool that produces consistent and statistically plausible data. In this paper, we determine requirements for sample data generation, and introduce the BEDAWA tool, a sample data generation tool, designed and implemented in a 3-tier CORBA architecture. A short discussion on the sample data generating results proves the usability of the tool.

1 Introduction

In the field of data warehouses and OLAP technologies [1], [3], [5], sample data is needed for different purposes, such as benchmarking, testing, and demonstrating. Unfortunately there exists a lack of tools available to generate sample data that is familiar, consistent, scalable, and reflective of multiple degrees of freedom. Referring to sample data generation problem, [16] analysed the samples for data warehouses and OLAP products of well-known companies, and concluded that the generation of sample data for their products is not yet sufficiently solved.

In this paper, we propose BEDAWA — a *tool to generate large scalable sample data for data warehouse systems*. In this approach, the statistical model is used to define relationships between dimensions and facts in the context of a star schema [11], [13].

The BEDAWA tool has a separate sample data generation engine (SDGE), which can be used as a service by any sample data consumer. It has a 3-tier architecture to improve the access and performance of the sample data generation service. For connecting transparently sample data consumers with the SDGE, the tool uses CORBA [8] as middleware. CORBA allows distributing the sample data production to several servers. It is completely implemented in Java and because of its distributed architecture it can be easily used and accessed by a web browser.

T. Yakhno (Ed.): ADVIS 2000, LNCS 1909, pp. 44-56, 2000.

The sample data are generated in the XML format, which allows sample data consumers a very flexible application and data handling. The XML sample data can be customized and transformed into any required format.

The remainder of this paper is organized as follows. In section 2, we discuss related works. Afterwards, we propose the requirements of sample data generation process in section 3. Section 4 introduces the statistical model of our prototype. In section 5, we present the BEDAWA sample data generation process. The 3-tier CORBA architecture of the BEDAWA tool is presented in section 6. Evaluating the results of the tool is shown in section 7. Finally we provide the conclusion and future works.

2 Related Works

Benchmarks are used for measuring and comparing the operations of systems or applications working on these systems. This especially applies to databases where many efforts have been made to develop tools for benchmarking purposes.

The Wisconsin benchmark [6] has been proved as suitable benchmark for measuring the queries against relational database systems. AS^3AP is another benchmark that tests multi-user models by different types of database workloads [19] for relational databases. There are many other benchmarks that have been developed for testing and comparing different systems in various application domains such as "The Set Query Benchmark", "The Engineering Database Benchmark", and "Engineering Workstation-Server" [2].

For data warehouses and OLAP, few approaches are known in the literature. The TPC-R and TPC-H benchmarks that were introduced in 1998 intended to replace the TPC-D [18] benchmark [20]. The TPC-R and TPC-H benchmarks are specified for "reporting" and "ad hoc query" on decision support systems, respectively. These specifications allow more precise and domain specific performance measurement. In [15], the APB-1 was issued to measure the overall performance of an OLAP server. In this benchmark, the committee defines the AMQ (Analytical Queries per Minute) term to present the number of analytical queries processed per minute and suggests 10 kinds of "ad-hoc queries with parameters" to best simulate a realistic operating environment.

3 Requirements of Sample Data Generation Process

Depending on the purpose, sample data consumers should be able to specify individual sample data for their applications. Furthermore they should have an open interface to the generation tool to easily control the generation process. For the generation of sample data for data warehouse and OLAP systems we identified following requirements:

Separation of Sample Data Design and Production. The sample data generation is an iterative process that starts with the modelling of the required sample data, and ends with the automatic production of this data. A sample data generation service should separate the design and the production of the data, to allow a clear distinction between the modelling (build-time) and the production (run-time).

Full Flexibility in the Sample Data Design. The sample data is characterised by its data structures and data values. The sample data design should be flexible in defining and changing the structure of sample data. The data structures of sample data for various data warehouse or OLAP systems can be very different. To generate sample data that is usable in variant systems, the generative process has to allow defining and changing the structure of the desired sample data.

Statistical Correctness. The statistical information about the real-world data can be applied to generate large amounts of sample data, which can be used for analysis applications. The statistical correctness allows us to not only generate very large consistency data but also to generate a likelihood of real world data from the statistical point of view.

Generation of "Real World" Sample Data. "Real world" sample data can be differently defined, primarily depending on the application context. The following aspects should be considered:

- For data viewing and browsing, the sample data should be easy to read and look familiar to the viewer.
- For data relations and structures, the distribution of generated sample data values should satisfy the desire of the sample data designer.

Consistency. For sample data generation, in some cases, inconsistent data can be used to test the side-effect actions of a system. However, in most other cases, the data consistency is required for testing, demonstrating and benchmarking a system. Therefore, we intend to generate a sample data that is consistent.

Scalability. The generation service should be able to generate data on a large scale. Especially for sample data of data warehouse environments there is the need for enormously large amounts of data (Gigabytes and Terabytes). Therefore a sample data generation service should be able to handle such amounts of data and to optimise the performance for the sample data production (i.e. by using several servers).

Support of Data Customisations. Sample data consumers often see sample data from different views. It is often necessary to customise the sample data according to these views by performing data transformations on the originally generated sample data. Sample data in the XML format is especially appropriate for data customisations.

Support of Any Data Format. It should be possible to generate generic sample data and to convert it into any target format. Standard formats, like XML, DDL, or CSV should be directly supported. But sample data consumers should also have the possibility to generate individual sample data with propriety formats.

Control of the Sample Data Production. Actors should have full control of the sample data generation process. Actors can be either human users or applications that are able to control the sample data generation by interactions like executing, cancelling or adjusting the sample data generation. Especially during the sample data production, actors often need mechanisms to 1) *allow further customisations* to adapt the sample data to their applications, and 2) *ensure* that the produced sample data is corresponding to their requirements.

Security/Authorization. The sample data production can be a very resource and time intensive service. Therefore the service should allow only an authorized usage by the sample data consumers to avoid overloads, prohibited use of the service or denial-of-service attacks.

Remote/Local Transparency. A sample data generation service should be able to run as standalone application or distributed on several servers. When very large amounts of sample data should be generated, it is necessary to execute the sample data production on one or several servers. For providing the logical glue that will connect the distributed services, a middleware like CORBA or RMI is necessary.

Platform Independency. A sample data generation service should be able to run on any platform and any operating system. Sample data consumers should be able to use the service via WWW browsers without assuming any special configuration, or plug-ins. Therefore, platform-independent languages like Java are advantageous for the implementation of the service.

4   Statistical Model

4.1 Need of Statistical Model

The definition of a database benchmark can be a very time consuming task, especially, to enter data into a database. This is particularly true for data warehouse and OLAP systems, where the required data volume can become huge.

Each data warehouse and OLAP system has its own data and data structures. The data is usually stored in dimensional structures with respect to the different granularity levels and the interdependencies between different dimensions. To generate sample data for these systems, a rigorous calculation model which is based on a sound statistical approach is needed. The model allow users not only to present the relationships between the data elements, but also to take into consideration the limitation given by the degrees of freedom of generated sample data. In the process of generating such large and complex data without the help of an appropriate model, unpredictable consistency errors can appear.

4.2 The Statistical Model

In dimensional modelling (star or snowflake schema), a fact value y is dependent on a tuple of dimension members from the fact dimension components. We can present this relationship by the function:

$$y = f(\alpha_i, \beta_j, \gamma_k, \ldots) \tag{1}$$

where:

y : a value of the fact,

$\alpha_i, \beta_j, \gamma_k, \ldots$: variables that stand for the effects of dimension members i^{th}, j^{th}, k^{th} of dimensions D_α, D_β, $D_\gamma, \ldots$, respectively, to the fact y.

In fact, this relation is very complex, and extremely difficult to determine. A common way to determine this function $f(\alpha_i, \beta_j, \gamma_k, \ldots)$ is using statistical methods. At first, we collect as much as possible data of the relation, e.g., the fact values of the corresponding tuple of dimension members. Based on the collected data, we try to find out functions that can be used to present relationships between the collected data values.

For simulating realistic OLAP situations, depending on the structure of future sample data, a suitable calculating model can be selected. Various calculating models can be applied on different structures of the sample data with different meaning. We apply a linear statistical model to calculate facts in the context of the star schema [11]. This linear statistical model is applied to present the relations between the fact and the dimensions. Each dimension member will have an effect on the fact, called "fact effect value". A fact value of a fact is calculated based on the fact effect values of dimensions members. Considering the relationship between the fact and the dimensions, which have an effect on the fact, the actor who interacts with the BEDAWA tool will define these relationships by using the statistical linear model of n-way classification with fixed effects. For simplicity and readability reason we omit the interaction. In the following, we will use the n-way classified model without interaction.

The linear model is also extremely suitable for presenting the relationship between the dimensions and the fact in star schema. The linear relationship can be found often in the real world.

The linear model used in the BEDAWA tool is described in the following [4]:

$$y_{ijk\ldots} = \mu + \alpha_i + \beta_j + \gamma_k + \ldots + \varepsilon_{ijk\ldots} \tag{2}$$

where:

$\alpha, \beta, \gamma, \ldots$:   fact effect value sets of dimensions D_α, D_β, D_γ that have effect on the fact y. Assuming that these dimensions have numbers of members in their domains are n, m, p, ... respectively,

$i, j, k, \ldots$:   index numbers that are in the range [1..n], [1..m], [1..p], ..., respectively,

μ :   average mean of the fact,

α_i, β_j, γ_k,... : fact effect values of the ith, j^{th}, k^{th} dimension members, respectively,

$\varepsilon_{ijk...}$: error random value, using normal distribution function, $N(0,\sigma^2)$.

$y_{ijk...}$: a value of the fact.

And the conditions:

$$\sum_{i=1}^{n} \alpha_i = 0 \quad , \quad \sum_{j=1}^{m} \beta_j = 0 , \quad \sum_{k=1}^{p} \gamma_k = 0 \quad , \cdots \tag{3}$$

In this model, we assume that there is no interaction between the fact effect values from each couple of dimensions. This means that the fact effect values of a dimension do not depend on the fact effect values of any other dimensions.

In context of the hierarchy structure of a dimension, the dimension data can be presented in a tree with different levels of its hierarchy. A branch of this tree presents a dimension member. Consequently, the fact effect value of the dimension applied to a fact can be calculated dependent on the effect of all nodes from this branch. For each node of the dimension tree, the node's fact effect value can be defined for the fact. The fact effect value in this context is the effect of the branch from the root to the leaf of the dimension tree. We calculate the fact effect value of a branch by summing up all fact effect values of all nodes from that branch. For instance, the fact effect value α_1 of the dimension α is calculated by the formula (4):

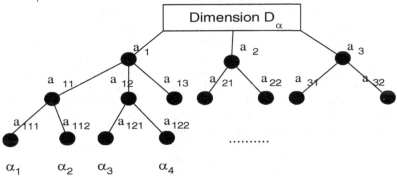

Fig. 1. The fact effect value tree of the dimension D_α

$$\alpha_i = a_j + a_{j_k} + a_{j_{k_l}} + ... \tag{4}$$

where a_j, a_{j_k}, $a_{j_{k_l}}$ ... are the fact effect values of the nodes of the branch α_1.

To satisfy the condition (3) we can apply to fact effect value tree a condition that at any node of the dimension fact effect value tree, sum of all fact effect values of its sub-nodes is equal to zero:

$$\sum_{j} a_j = 0, \quad \sum_{k} a_{j_k} = 0, \quad \sum_{l} a_{j_{k_l}} = 0,... \tag{5}$$

5   The Sample Data Generation Process of BEDAWA

The sample data generation is in general an iterative process, which starts with the specification of the desired sample data, and ends with the automatic generation of this data. The iterations of the generation process are dependent on the quality of the generated data. The user decides based on the generated results, whether modifications of the sample data specification are necessary and repeats the sample data generation until (s)he receives the expected result. Typically, the user begins with a smaller amount of data and after assuring that all requirements are reasonably met, (s)he can start with the production with large amounts of data for data warehouses.

Generally the sample data generation process consists of three functional areas: 1) The *build-time functions*, concerned with the definition and the modelling of the sample data requirements, 2) The *run-time* functions, concerned with the actual execution of the sample data generation and 3) the *run-time interactions* functions, concerned with users and other applications for controlling the generation process. Figure 2 shows the three functional areas of the sample data generation process implemented in BEDAWA and their relationships to each other.

The generation process of BEDAWA contains the following steps (see Figure 2): *Data source definitions* are used to identify any necessary data source for the sample data generation. They are primarily used for two purposes. First, these definitions contain the connection information for retrieving pertinent data for the definition of sample data from a data provider. Second, a data source definition has to indicate where the generated sample data has to be stored. *Dimension definitions* describe the dimension structures of the modelled sample data. Based on the tree representation of the dimension domain, the *fact effect* value of every node of the tree has to be defined. For each relationship between a dimension and a fact, one or more fact effect value set(s) can be defined. In the next step fact definitions are defined according to the statistical model. Then, the structure of the *fact tables* can be defined by specifying a list of facts and dimensions. The *distribution definition* is used to specify the distribution of the sample data that simulates real-world data. The distribution information is defined by percent numbers, and has to be provided for every node of a hierarchy level of a referenced dimension. Furthermore actors can define *constraints and post-conditions* for the data of the fact table and boundaries for other generated data. A *fact table profile definition* contains all relevant information to generate a fact table, i.e., the data source, the fact table definition, number of records. After the sample data specification the SDGE is able to generate the sample data.

The three functional areas of the sample data generation process are implemented in BEDAWA as 3-tier architecture. To provide a better accessibility for sample data consumers and to improve the performance behaviour of the sample data production, the complete functionality of the sample data generation service is separated to an extra tier. The SDGE of BEDAWA is implemented as CORBA objects, which allow sample data consumers a transparently access. Furthermore CORBA can be used to establish several instances of sample generation engines on different servers, which improves significantly the data generation performance. In the next section's 3-tier CORBA architecture of the BEDAWA tool is described in more detail.

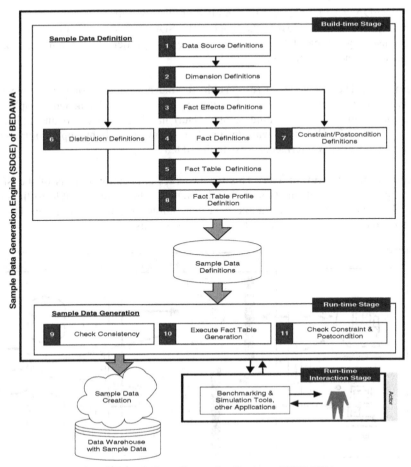

Fig. 2. The Sample Data Generation Process of BEDAWA

6   3-Tier CORBA Architecture of BEDAWA

CORBA is a specification for distributing objects across a network. CORBA provides an object distribution framework that allows you to invoke and manipulate objects on separate machines as if they were all available on the same machine. The CORBA-based development tools provide the requisite infrastructure that gives us this location transparency.

Location transparency is very interesting for the sample data generation, if the sample data generation service is supposed to be available to various sample data consumers. If sample data consumers are able to remotely specify and generate the sample data, CORBA can provide most of the logical glue that will connect them to the application objects of the SDGE.

The BEDAWA tool uses a 3-tier architecture that provides layers of abstraction to improve the reuse and access to the SDGE. The SDGE offers public interfaces, which can be used by sample data consumers. In the storage tier all relevant data for the sample data generation process is stored.

Client Tier. The client tier is responsible for the presentation of the sample data, receiving user events and controlling the user interface. The actual sample data generation logic (e.g. calculating sample data) has been moved to an application server. The BEDAWA tool was designed to allow any client application, which supports the CORBA interface, access to the SDGE.

Sample Data Generation Tier. This tier provides application objects of the SDGE that implement the full functionality of the SDGE, which are available to the client tier. This level separates the sample data generation logic to an extra tier. This tier further protects the data from direct access by the clients.

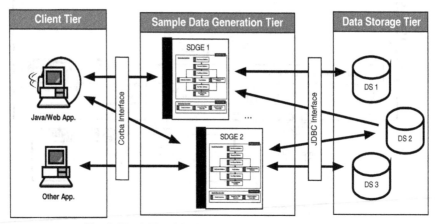

Fig. 3. The 3-Tier CORBA Architecture of BEDAWA

Data Storage Tier. This tier is responsible for the data storage. Besides the widespread relational database systems, the BEDAWA tool offers access to any data storage, which is available by JDBC. BEDAWA needs data storage to store the sample data specifications and the produced data of the sample data generation.

The 3-tier architecture of the BEDAWA tool provides following major benefits:

- Through the clear separation of user-interface-control and data presentation from the application-logic it is easier to give various clients access to the SDGE of the BEDAWA tool. The main advantage for client applications is easy and transparent access to the SDGE.
- A modification of the storage strategy won't influence the clients. RDBMS' offer a certain independence from storage details for the sample data consumers. However, cases like changing table attributes make it necessary to adapt the client's application. Even radical changes, like switching form an RDBMS to an OODBS, won't influence the sample data consumers. They can still access the

SDGE over a stable and well-designed interface, which encapsulates all the storage details.

- Application objects and data storage should be brought as close together as possible; ideally they should be physically together on the same server. This way a heavy network load is eliminated. When offering the sample data generation service over the Internet, this aspect can become crucial.
- Dynamic load balancing capabilities: if bottlenecks in terms of performance occur (e.g., generation of Terabyte data), the server process can be distributed to other servers at runtime. With CORBA architecture of BEDAWA it is simple to implement load-balancing mechanisms to distribute the generation process to several servers.
- A sample data generation process is a resource intensive service that requires authorizations from sample data consumers to avoid overloads, prohibited use of the service or denial-of-service attacks. The BEDAWA tool uses the security service of CORBA to offer authentication, authorization and encryption.

7   Results

In this section, we show evaluation results by differently generating fact tables based on the well-known *GroceryStore* star schema in [12]. We can make various evaluations on the BEDAWA tool to show its abilities to generate sample data with many levels of complexities in its structures, as well as different sizes in order to reflect a real-world situation. Because of the limitation of the paper, we focus on the abilities of the calculating model. The first evaluation is made when changing the number of dimensions (i.e., changing the right side of the formula (2)). As shown in figure 4, we generated a series of fact tables, each of which is constructed from one fact and a different number of dimensions.

Alternately, the second evaluation is made in order to show that the calculating model can be applied to the tool to capture the change in the number of facts (figure 5). Therefore, we generate in this evaluation another series of fact tables that always have three dimensions, but numbers of facts are different. To evaluate the generation performances, fact tables of four different sizes, i.e. small, medium, large, very large, are generated for each of the two evaluations.

As expected, the elapsed time of generating a fact table increases with the size of the fact table. However, the rates of growth are different for generating fact tables in the two evaluations. This indicates that the complexity of the structure also effects the time taken to synthesize fact tables. For example, the elapsed time of generating a fact table with 3 dimensions, one fact and 50000 records is smaller than the elapsed time of generating a fact table with 3 dimensions, 5 facts and 50000 records.

Although the sizes of desired sample data are considerable (100000 records), the elapsed times for generating full-featured fact tables are acceptable.

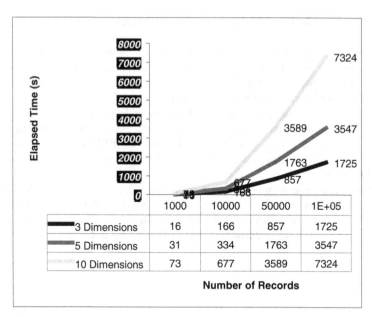

Fig. 4. Generation of Fact tables with one fact and different numbers of dimension.

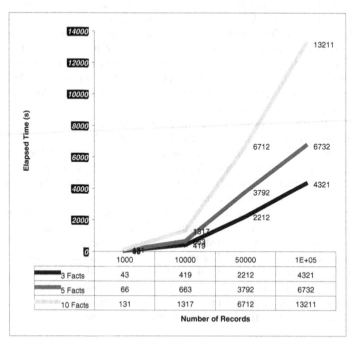

Fig. 5. Generation of fact tables with three dimension and different numbers of facts.

8   Conclusion and Future Works

In this paper, we have introduced the need of having a tool to generate sample data for data warehouses and OLAP systems. The BEDAWA tool has a sound statistical foundation by using and implementing the statistical model of an n-way classification with fixed effects. The theoretical soundness of the model provides a framework to define relationships between dimensions and facts in context of the star schema.

The paper presents a process of the generation of sample data for data warehouse environments. The tool facilitates a 3-tier CORBA architecture, in order to improve the access and performance of the sample data generation service. Furthermore, sample data is generated in the XML format to allow sample data consumers to easily transfer the data into any format.

In conclusion, the implemented prototype is able to flexibly and effectively generate sample data for specific applications. The tool has been used to generate many kinds of sample data, which have different data schemas with complexities in their structures, and has proved that the elapsed times for generating fact tables are acceptable.

In the future, we will extend the BEDAWA tool to support complex data structures for sample data such as snowflake schema. Further research will be invested for extending the adaptability of the statistical model to simulate real life data. This means that many other statistical models are needed to present various relationships in the generated sample data. To popularize the usage of the tool, we also plan to develop client tier applications that allow actors to use the SDGE of BEDAWA for example as a web service.

References

[1]   R. C.Barquin, H. A.Edelstein, Planning and Design the Data Warehouse, Prentice Hall PTR, 1997

[2]   R. Cattell, The Benchmark Handbook, 1993.

[3]   S. Chaudhuri, U. Dayal, An Overview of Data Warehousing and OLAP Technology, SIGMOD Record Volume 26, Number 1, September 1997.

[4]   R. Christensen, Plane Answers to Complex Questions: The Theory of Linear Models, Springer-Verlag, 1987.

[5]   E. F. Codd, S. B. Codd, and C. T. Salley, Providing OLAP (On-Line Analytical Processing) to user-analysts: An IT mandate, Technical report, 1993.

[6]   D. J. Dewitt, The Wisconsin Benchmark: Past, Present, and Future, The Benchmark handbook, chapter 4, 1993.

[7]   H. Gupta, V. Harimarayan, A. Jajaraman, Jeffrey D. Ullman, Index Selection for OLAP, ICDE 1997: 208-219.

[8]   D. Harkey, R. Orfali, Client/Server Programming with Java and CORBA, 2nd, John Wiley & Sons inc., 1998.

[9]   J. M. Hellerstein, The Case for Online Aggregation, SIGMOD 1997.

[10]   W.H. Inmon, Building the Data Warehouse, John Wiley, 1992.

[11]   R. Kimball, The Data Warehouse Toolkit, John Wiley & Sons inc., 1996.

[12]   R. Kimball, A Dimensional Modelling Manifesto, DBMS Magazine, 1997.

56   T.N. Huynh et al.

[13]   R. Kimball, L. Reeves, M. Ross, W. Thornthwaite, The Data Warehouse Lifecycle Toolkit, John Wiley & Sons inc., 1998
[14]   A White Paper, MicroStrategy Incorporated, The Case for Relational OLAP.
[15]   OLAP Council, APB-1 OLAP Benchmark. Release II, 1998
[16]   J. Schiefer, A. M. Tjoa, Generating Sample Data for Mining and Warehousing, Proc. Int. Conference on Business Information System, Springer-Verlag, 1999.
[17]   O. Serlin, The history of debitcredit and the TPC, The Benchmark handbook, chapter 2, 1993.
[18]   Transaction Processing Performance Council, TPC-D benchmark
[19]   C. Turbyfill, C. Orji, D. Bitton, AS^{3}AP- An ANSI SQL Standard Scalable and Portable Benchmark for Relational Database Systems, The Benchmark handbook, chapter 5, 1993.
[20]   R. Winter, P. Kostamaa, On-the-fly on the up and up, Intelligent Enterprise Magazine, November 16th, 1999.

Classification Abstraction: An Intrinsic Element in Database Systems

Elaine P. Machado[1], Caetano Traina Jr.[1], and Myrian R. B. Araujo[1]

Institute of Mathematics and Computer Science — University of São Paulo — USP
Av. Dr. Carlos Botelho, 1465 — CEP 13560-970, São Carlos — SP — Brazil
{parros,caetano,mrenata}@icmc.sc.usp.br

Abstract. The Classification Abstraction is a fundamental concept supporting the existence of database systems. However, it has been seldom supported as a database modeling element. This paper presents a technic to support classification as a semantic constructor, enabling its representation in target modeling. It is also shown that this concept can unify the commands for data definition and data manipulation in query languages, thus generating tight environments including the data model, the database management system and the software design techniques. Through classification, object types and object instances can be treated in a uniform manner, allowing the unification of concepts and commands. This approach is illustrated using the SIRIUS model and its schema editor. SIRIUS data model is a full featured, abstraction-based data model, including a semantic constructor based on the classification abstraction that supports the concepts presented here.

1 Introduction

The research on conceptual data models is an important issue in the construction of more powerful database systems. The increased availability of semi-structured data provided by the recent web explosion makes conceptual tools yet more important. Data abstractions are strong theoretical support for data models and enable creation of techniques to aid the development of software systems.

In general, conceptual data models consist of a set of semantic abstractions that represent structural and behavioral aspects of information. Each semantic constructor is backed by combining one or more abstractions. The semantic constructor adapts the concepts of each abstraction to its particular needs.

An abstraction is a relationship between two kinds of objects: abstract objects and detail objects (see Figure 1). The abstract object represents a synopsis of the details, and the set of detail objects describes the abstract object in a more detailed way. An abstraction defines properties between the objects involved. For example, the composition abstraction has composite objects as the abstract ones, and parts as details; the relationships between them are `PARTS-ARE` and `IS-PART-OF`, and properties such as existential dependencies and operations propagation hold.

T. Yakhno (Ed.): ADVIS 2000, LNCS 1909, pp. 57–70, 2000.

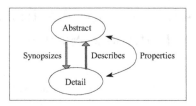

Fig. 1. Elements of an abstraction

Both abstract and detail objects may be represented in target systems, or only one kind can be represented. For example, a designer would represent just specific objects in a user interface, without explicitly representing the generic one. However, when using a semantic constructor to represent data in a schema, both the abstract and the detail objects are represented.

Semantic constructors enable systems to automatically know about the properties associated with the underlying abstraction. Classification Abstraction is essential in databases because it enables the idea of type and instances. Designers create schemas through the creation of object types, and the system instantiates these types into objects. However, database systems in general cannot instantiate objects whose instances are also types. This situation prevents the representation of the classification abstraction in the schemas.

This paper shows that the correct support for classification abstraction through a specific semantic constructor is a strong resource for a data model, completing and homogenizing diverse of its concepts. Our results are presented using the SIRIUS data model [8] [9], the central element of the SIRIUS system. However, the concept of the classification abstraction supported as an explicit semantic constructor may be extended to any other database system.

This paper is organized as follows. Section 2 reviews related works. Section 3 discusses the differences between the classification and generalization abstractions, highlighting some common misunderstandings which occur between these abstractions. Section 4 describes the classification abstraction in a way that enables its use as an intrinsic element in data models. Section 5 discusses the support of the classification semantic constructor in a DBMS. Section 6 shows a software tool aiming to model systems using the classification abstraction through a form-based Schema Editor. Section 7 summarizes the paper and gives its main contributions.

2   Related Work

Object-oriented data models support the representation of two abstractions: Association and Generalization [5] [21]. In addition, the Classification Abstraction allows the definition of common rules to sets of objects [34]. Furthermore, classification is a reusability mechanism of object-oriented database systems, since the same definition can be used to create objects with the same behavior and structure [7] [20]. Thus, Classification is the third fundamental abstraction, essential to the concept of type. However, the classification abstraction has seldom been supported in database systems as a modeling element that enables the representation of classification in the enterprise models.

The fundamental idea in classification is of elements which are members of a set, with rules to set the membership. Some works have proposed the introduction of resources in the query processing sub-systems, dealing with the manipulation of subsets summarized by the properties of another object (a object type in our framework). Resources that have been studied are tools for access and use of meta-data [1] [10] [11] [12] [25] [26] [35] and of view materialization [22], although not strong attention is paid to the representation of the properties of the set as a whole (and thus to the object type) [13]. The work of Pirotte and others [18] [28] has targeted this objective. However, although these works give special attention to the instantiation process from a well-defined type, they yet restrict the occurrence of materialization to a single level.

This paper extends these works to promote the Classification Abstraction as a full featured semantic constructor in a data model, so it can be pervasively embodied in all aspects of a database management system. As we show in the following sections, this leads to a unifying database query language, which does not need separate subsets of commands to the data definition and the data manipulation operations.

3   Classification X Generalization

Although usually not supported to be represented in data models, the classification abstraction occurs in real situations. This coerces the overload of semantic constructors based on other abstractions to represent the modeling needs. It is common to use constructors based on the generalization abstraction to fulfill these needs. However this semantic overload invariably causes loss of information that is hardly restored later during the software development process.

In fact, instantiation and specialization are closely related, and our practical experience shows that it is common for software designers to misinterpret them. The overwhelming exploration of generalization and the scantness of support for classification only contributes to hardening the distinction between both concepts. In this section we present observations aiming to help discerning between classification and generalization.

In the discussion that follows, we always consider that an *object type* is also an object, and every object has a *type*. Stopping this otherwise infinity recurrence, we admit that there is a maximal object type called Meta-Type, recognized by the system, from which "higher level" types can be instantiated.

It is common to use the IS-A relationship interchangeably between generalization and classification. The following observation distinguishes the relationships which occur in the classification and generalization abstractions.

Observation 1 - In the generalization/specialization abstraction, the relationships between specific and generic objects are IS-A and SPECIALIZED-AS, whereas in the classification/instantiation abstraction, the relationships between specific and generic objects are TYPE-IS and INSTANCE-IS.

It is also common one refers to the relationship between specific and generic objects in a generalization abstraction in both directions, as an IS-A relationship. As we always consider one relationship as distinct from its inverse, we reserve the IS-A to be used as in: "Specific IS-A generic". The opposite relationship is used as in: "Generic SPECIALIZED-AS specific". The corresponding relationships for the classification abstraction are "instance TYPE-IS type" and "type INSTANCE-IS instance". When both the abstract and the detail objects are represented, the generalization and specialization abstraction can be considered opposite from each other, so we treat both as a pair, with corresponding and opposite properties. The same holds for the classification and instantiation abstractions.

The following observation highlights the main difference between the classification and generalization abstractions.

Observation 2 — In the generalization/specialization abstraction, an object of a specific type *is the same object* of the generic type, whereas in the classification/instantiation abstraction, an object of the instance type *is different* from an object of the type.

Figure 2 illustrates this observation. In Figure 2(a), the object John Doe of type Teacher is the same object John Doe of type Person, because a teacher is a specialization of a person. In Figure 2(b), the object AA-123-05/07/2000, a real flight that took place at a certain day (e.g. 05/07/2000), is of type Flight, and the object Flight is of type Flight Number. No object of type Flight can be of type Flight Number, that is, a real plane flight cannot be a schedule because it is the real thing.

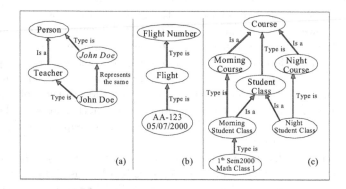

Fig. 2. Specialized and instantiated Objects

Both classification and generalization can freely interact. Figure 2(c) illustrates one situation involving both abstractions. Night Course and a Morning Course are both specializations of Course. A night edition of a course is a course instance, that is, both `Course` and `Night Course` represent the same object. On the other hand, a `Student Class` is a `Course` instance, existing classes that are instances of `Morning Course` and classes that are instances of `Night Course`, each one meeting the peculiarities of its respective type. One `Course` is not the same as a `Student Class`. The object `1stSem2000 Math Class 1` is of type `Morning Student Class`, which in turn is both an instance of `Morning Course` and a specialization of `Student Class`.

In Figure 2, the object types `Person`, `Flight Number` and `Course` are assumed to be instances of the maximal object type `Meta-Type`.

Another distinction between classification and generalization is the way used to obtain values not directly specified in the database for a given object. This leads to the following rule.

Observation 3 — In the generalization/specialization abstraction, attribute values not defined in an object would be inherited from its generic type. In the classification/instantiation abstraction, attribute values not defined in an object can be copied from the default value of its type.

Inheritance of attributes and methods is a property of the generalization abstraction. Inherited values correspond to an attribute that is not directly associated with the spotted type, but to one of its super-types. Changing the value of an attribute in a generic object should change this value in the specialized object, because the specialized object *is the same* generic object. On the other hand, when classification occurs, it is the association of default values that takes place. A default value is assigned to the object when it is instanced and it can be modified afterwards.

For example, a `Teacher` does not need to have an attribute `Age`, because it can be inherited from the `Person` who the `Teacher` is. Whenever the `Person` celebrates a birthday, so does the corresponding `Teacher`. Regarding flights, it could be defined that flight `AA-123` has a default departure time. If the `AA-123` `05/07/2000` had delay, it can be changed in this instance, without changing the default in the `AA-123`. In the same way, if `AA-123` is changed to a flight at a different time, the change will not affect the flights instanced before the change, e.g. the flights already flown.

Observation 4 — The generalization/specialization abstraction occurs between types. In the classification/instantiation abstraction, the abstract object is always a type, but the detail one can be a type or not. If the detail object is a type, it should participate in another classification occurrence assuming the role of the abstract object.

4   Classification Abstraction

The classification abstraction is the fundamental concept that allows the elements from data schemas to be instanced in an extensional database. Although many definitions and approaches to classification can be found [15] [20] [28] [29] [32], there are aspects of the classification abstraction not treated in these works. In this section we complete some of these aspects in a unified way, highlighting the ones which are important for the characterization of the classification as a pervasive abstraction in every aspect of database management systems.

Classification is used in a DBMS in a restricted way. It is usually treated as an implicit relationship between the schema and the extension. Thus, the number of levels in the classification hierarchy is always pre-determined. In some database models (such as the ER-Model [17], the EER-Model [20] or the OMT[30]), which are not meta-models, the number of levels of the classification hierarchy is one. The abstract objects (types) are in the schema and the detail objects (instances) are in the database extension. In meta-models such as the relational model and ODMG [15], classification is implicitly defined within the conceptual schemas. The correspondence between meta-type and type and between type and instance defines a two-level classification hierarchy.

Freeing the number of levels in the classification hierarchy improves the consistence of data and the completeness of the data model. A *multilevel classification hierarchy* results when classification can be explicitly represented in the schema, through a semantic constructor dedicated to represent object types and object instances. As classification is a concept already present in databases, the new constructor must seamlessly integrate with the previous usage of the classification abstraction.

The properties and methods of objects are their *attributes*. Attributes are associated to objects. Two kinds of attribute association must be distinguished to support the classification abstraction: the object attributes, which are called *instantiation attributes*, and the *classification attributes*, which characterize an object as a type and define the attributes of its instances.

Instances originate sets of structurally identical objects, whose structure is defined by the object which is their common type. In this work, we denominate *class* the set of all objects of a given type, and *type* the set of rules and characteristics that enables the instantiation of each object. Hence, a type is an object that represents the common properties to all of its instances. This is done through the classification attributes associated to the object type. An object type can be instanced in an object, which can be instanced again in another object and so on. Therefore, it is possible to define a hierarchy with many levels of classification, where the maximum number of levels is determined by the semantic of the application.

Figure 3 introduces a representation for a classification constructor, where both type and instance are homogeneously treated. An object is represented by a rectangle divided by a horizontal line. The object identifier is written in the lower part, and the object type identifier is written in the upper part. An object identifier is any of its attribute (or set of attributes) which can uniquely identify the object in its class, like a key for instances, or a user-designated name for objects whose type is the `Meta-Type`. Objects whose type is the `Meta-Type` can have its upper part left blank.

In the multilevel classification hierarchy, the instance of an object can be a type, and be instanced again. In this way, the classification attributes can be instanced not only in the immediate instance of that object type, but in any of its descendants. In the proposed notation, the instantiation attributes are placed below the rectangle, and the classification attributes are placed above the rectangle. A number preceding the classification attribute name indicates the sub-level where this attribute is instanced. Values of the attributes are placed following a colon after the name of the attribute. Both instantiation and classification attributes can have values assigned.

To exemplify a multilevel classification hierarchy, let us consider that, in the information system of a library, each *book title* may have many *editions*, and that each edition may have many *copies* in the library. This means that the concept of `Book` is instanced in `Title`, `Title` is instanced in `Edition` and `Edition` is instanced in `Copy`. As `Book` is an instance of `Meta-Type`, this defines a four-level classification hierarchy. Figure 3 illustrates occurrences in this classification hierarchy. Figure 3(a) represents `Book` as an instance of `Meta-Type`. Figure 3(b) represents the book titled `Database` as an instance of `Book`. Figure 3(c) represents the object `Databases 2ndEdition` as an instance of the book `Databases` and Figure 3(d) represents the (physical) book `2ndCopy` as an instance of the object `Databases 2ndEdition`.

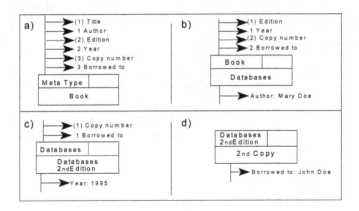

Fig. 3. Multilevel Hierarchy: (a)Book type; (b)Book; (c)Edition; and (d)Copy

The concept of multilevel classification hierarchies has an important requirement: an object of the database schema can take both roles of type and instance. For example, the object `Databases` is an instance of `Book` and it is the type of `Databases 2ndEdition`.

The classification attributes define the "template" to instantiate an object type, i.e., they define the structural and behavioral characteristics of the instances of the object type in each level of the classification hierarchy, and define the number of levels that can be created from that type. The number preceding a classification attribute indicates the level where the attribute turns into an instantiation attribute. For example, the number 2 before the attribute `Year` in object `Book` indicates that this attribute will be instanced in the second instantiation level from , i.e., `Year` receives the value 1995 as an instantiation attribute of the object `Databases 2ndEdition` and in this level it is drawn below the object. The number inside parenthesis means that the attribute is an identifier of the instances in that level. For example, the attribute `Title` of the object `Book` will be the identifier of the first instantiation level of `Book`, that is, `Databases` is the value of the attribute `Title` in such instance of `Book`.

Notice that, for example, `Databases 2ndEdition` is an instance of `Databases`, that is, `Databases` is the type of `Databases 2ndEdition`. Other books could have other `Editions`, each one the type of its respective instances. The concept of "edition" as a type is that of a *collective type*, useful to understand the classification hierarchy. Thus "editions" and "copies" represent collective types, involving every instance of books in the second and third level of the hierarchy. However, the "type" of an object instanced at the third level is an object of the second level, such as `Databases 2ndEdition` is the type for the 2ndCopy object.

When dealing with multilevel classification hierarchies, the support for *extra attributes* is an important resource. Extra attributes, which are added to an object but are not predicted in its type, allow new attributes (both instantiation and classification) to be included in any level of the classification hierarchy. All attributes associated with `Book` in Figure 3(a) are extra attributes, because the object `Meta-Type` does not have these attributes defined. Another example of extra attributes is the following: suppose the `Databases` book starts to have a companion CD beginning with its 3rd edition, and that CD can be borrowed independently from the borrowing of the book. Then, `Databases 3rdEdition` could have another attribute, `CDBorrowed to`, which does not exist in `Databases 2ndEdition`, as shown in Figure 4. Figure 4(a) shows the use of a default value, assigning the value none to a classification attribute, indicating a default value assigned to the attribute when it is instanced.

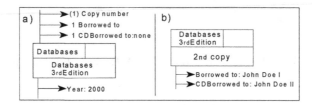

Fig. 4. (a) A new edition for the `Databases` book; (b) Copy of `Databases 3rdEdition`

Figure 5 details the example of flights (Figure 2(b)). There are three levels in the classification hierarchy: *flight number*, *flight* and *real flight*. *Flight number* stands for the routes, and each time an air-company creates a new route, the object `Flight Number` has a new instance. *Flight* is a collective type representing the set of instances of `Flight Number`. *Real flight* is a collective type representing every flight flown. In this example we are assuming that the identifier of each object is unique in its class. Thus, the object 07/05/2000 represented in Figure

5(c) is the one of type AA-123 flown at this date (its collective type is *Flight*). This particular flight had 83 passengers and departed at 10:07, although the "scheduled departure time" is 10:00.

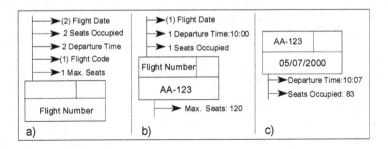

Fig. 5. A Multilevel Hierarchy for flights: (a)Flight number; (b)Flight; (c)Real flight

We claimed that classification occurs in real live, and the lack of a semantic constructor to represent it leads to semantic overload and to loss of project information. The examples presented in this section represent a real situation, and none of them could be represented using other abstractions without incurring in semantic overloads.

There is another possible modeling for the situation illustrated in Figure 5 without using classification. That is, creating two types: *Flight Number* and *Flight*, with an explicit relationship between them, e.g. "*Flight* type is *Flight Number*". However the association between instances of these two types would have to be hard-coded in the application software, because the semantic of a generic relationship is not enough to carry the information implied by the classification abstraction. Among this information are those originating in: default attributes, extra attributes in each particular *flight*, structure of instantiation (for example, in object-oriented systems a real flight could have an array of the passengers transported and a list of flight legs), and dependencies on specific categorical values.

5   Classification Semantic Constructor in a DBMS

When classification occurs in a schema, instances and types exist and can be created at runtime. Notice that, both type and instance need homogeneous treatment at the running application. Therefore, a unique set of commands is required to manipulate objects and types, unifying the commands usually separated in the DML and the DDL. There are three main issues a DBMS must satisfy to support the classification abstraction in this way: support the Meta-Type object; support extra attributes; and support classification attributes "instanceable" at more than one relative level.

The Meta-Type object represents the classification abstraction. Thus, every operation related to object instantiations is associated with this object, as well as the definition of the intrinsic definition of the classification structures. The structures pre-defined by the `Meta-Type` object enable the creation of user-defined types using insert operations in those structures. This enables the creation of types without the need of data definition commands.

Support for extra attributes enables definition of the classification attributes in user-defined types. Notice that the `Meta-Type` object has no classification attributes, so at least the classification attributes of user-defined types must be defined as extra attributes in every application schema.

The ability to represent classification attributes "instanceable" at a level other than the next one enables the use of the collective type concept. This concept is important to enable the construction of the application software and of the user interface. Support for the `Meta-Type` object and for extra attributes enable the database to deal with the classification abstraction. However, objects instanced during runtime are hard to be predicted and supported by the application software and by the user interface manager. The concept of collective type alleviates this burden, enabling the software to be prepared to accomplish at least the basic tasks of each collective type.

Support for the `Meta-Type` object and for extra attributes impacts the query language. The most remarkable changes are that no data definition commands are needed. Considering a SQL-like language, at least the `Create Table`, `Alter Table` and `Drop Table` commands are removed from the language. Their usage is supplied by the data manipulation commands in the structures (relations) which describes the `Meta-Type` object. Extra attributes require a special syntax in the `Insert` and `Update` commands, allowing definition of the extra attribute and respective value, even when the attribute is not defined in the type definition of the corresponding class (or relation), in a syntax similar to that used in object-oriented database systems [7] [19] .

To create an application schema using this approach is a hard task, as each object type must be incrementally built through many commands, each of which adding a small amount of information into the schema. This task is easier using a collection of macros, or using special commands to deal with the new concepts. As we intend to use the same commands to handle both types and instances, and yet have powerful commands, we decided to construct a schema editor to demonstrate the practical value of these concepts. This was accomplished using the SIRIUS data model.

6   Supporting the Classification Abstraction in a Schema Editor

The SIRIUS data model is based on the classification, generalization and association abstractions. The association abstraction further specializes in the aggregation and composition abstractions. These abstractions are included in the syntactic elements of the model, as objects, relationships, attributes, generalization structures, composition structures (colonies) and all their respective types [33]. The SIRIUS data model incorporates concepts from different models available in literature [3] [4] [14] [15] [16] [18] [19] [23] [24] [27] [31] [32] and the classification concepts presented in section 4.

In this section we show the schema editor of the SIRIUS model. This editor uses a graphical interface, based mainly in a set of forms to provide a coherent and complete representation of the information according to the presented concepts.

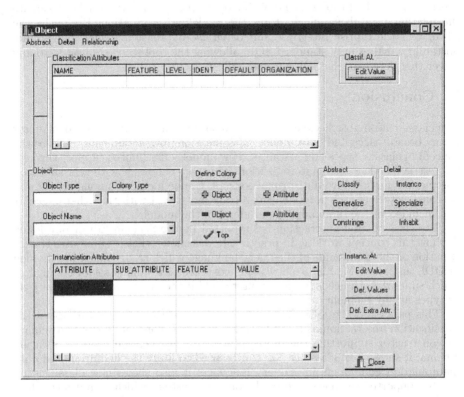

Fig. 6. The main form of the Schema Editor

Figure 6 presents the form `Object`, which is used to create objects in a SIRIUS environment. To create an object, the user chooses its type from those available in the `Object Type` list box. When a database is created, only the system-defined type `Meta-Type` exists, so the first object to be created needs to have this type. The next objects created can be of any existing type. An object is considered to be a type whenever its definition includes classification attributes. The Schema Editor provides a set of rules to enforce integrity constraints and consistency, looking for editing operations that should generate inconsistency in the database.

Unique object identifiers (OIds) are automatically generated for objects, object types and attributes. All OIds share the same domain. The operations supported by the Schema Editor use these OIds internally, so every command can be applied indistinctly to objects and object types. Therefore, the editor exhibits absolutely no difference in handling instances or types. The same editor form allows the creation and visualization of objects and object types in any level of the classification hierarchy.

The SIRIUS Object Manager is one component of the SIRIUS system. A SIRIUS Data Schema can be stored either on the SIRIUS Object Manager or in a relational database. The relation version of the editor, developed using the Inprise C++Builder tool in WindowsNT operating system, uses ODBC to create the relational database which stores the modeling information. Presently, the editor uses the Inprise InterBase and/or Oracle database manager, but aiming portability, it uses only standard SQL, allowing the tool to be connected to any server that supports the SQL through ODBC.

7   Conclusion

The classification abstraction has not been supported by a semantic constructor in any broadly accepted data model. Classification abstraction characterizes the concept of classes, types and instances. Few models offer limited support through the restricted concept of materialization. This paper shows how the classification abstraction, modeled from its primal concept, can be included among the semantic constructors of a data model, seamlessly integrated with the other concepts and included in the fundamental paradigm of the model.

The classification abstraction is supported by the Schema Editor developed to show this concept working in practice. It was shown that it eliminates the distinction of the commands of the query language which traditionally are divided in DDL and the DML. This allows the existence of a unique set of commands which support both needs, maintaining the ability to create and destroy objects or types at any moment.

This paper has the following main contributions: it presents the use of the classification abstraction as a fundamental theoretic concept to define a semantic constructor supporting the representation of classification in an application schema; it shows how classification can be used to unify the definition and data manipulation commands in a query language; it also shows how classification can be supported seamlessly through complete system, which includes the data model definition, the support in an object manager, and in the modeling methodology that permeates the software development process in all its phases.

Moreover, this paper presents the definition of the Schema Editor, showing the support for classification as a semantic constructor. The Schema Editor enables the creation and the visualization of objects and types at design time and at runtime. This homogeneous treatment is a new concept, provided that practically all object-oriented programming languages and database management systems support object types only during compilation time. Blurring the distinction between instance and type is an important step to apply database into non-structured worlds such as the Internet and the XML standard [2] [6] [10] [11] [12] [26]. It is important to stress that the concepts and techniques considered in the design and implementation of the Schema Editor are not restricted to the SIRIUS model and can be applied to every system that supports abstraction-based meta-models or in general object-oriented models that focus the basic principles of the object-oriented paradigm.

Acknowledgments. the authors would like to thank FAPESP and CNPq for the resources that made this work possible.

References

1. Abiteboul, S.: Querying Semi-Structured Data. In Proc. of ICDT '97, 6th International Conference, Afrati, F. N., Kolaitis, P. (eds.). Lecture Notes in Computer Science, Vol. 1186. Springer Verlag, Delphi, Greece (1997) 1–18
2. Abiteboul, S.,Beeri, C.: The Power of Languages for the Manipulation of Complex Values. VLDB Journal, Vol. 4. (1995) 727–794
3. Atkinsons, M.: The Object Oriented Database System Manifesto. Altaïr, GIP ALTAR in 2-INRIA-LRI, (October 1989)
4. Ayre, J., Wilkie, F. G., Hugues, J. G.: An Approach to the Design of Object Associations. Information and Software Technology, Vol. 37. 443–451
5. Beeri, C., Formica, A., Missikoff, M.: Inheritance Hierarchy Design in Object-Oriented Databases. Data & Knowledge Engineering, Vol. 30. (July 1999) 191–216
6. Beeri, C.,Milo, T.: Schemas for Integration and Translation of Structured and Semi-structured Data. In Proc. of 7th International Conference on Database Theory — ICDT '99, Vol. 1540. Springer Verlag, Jerusalem, Israel (1999) 257–276
7. Bertino, E.: Object Oriented Database Systems. Addison-Wesley (1994)
8. Biajiz, M.: Modeling Data Models Using Abstraction Parametrizations (in Portuguese). In IFSC — USP, São Carlos, São Paulo, Brazil (September 1996).
9. Biajiz, M., Traina, C., Vieira, M. T. P.: SIRIUS — An Abstraction Based Object-Oriented Data Model (in Portuguese). In Proc. of XI Brazilian Database Symposium, São Carlos, São Paulo (1996) 338–352
10. Buneman, P., Davidson, S. B., Fernandez, M. F., Suciu, D.: Adding Structure to Unstructured Data. In Proc. of 6th International Conference on Database Theory — ICDT, Afrati, F. N.,Kolaitis, P. (eds.). Lecture Notes in Computer Science, Vol. 1186. Springer Verlag, Delphi, Greece (1997) 336–350
11. Buneman, P., Fan, W., Weinstein, S.: Path Constraints in Semistructured and Structured Databases. In Proc. of 17th ACM Symposium on Principles of Database Systems, ACM Press, Seattle, Washington, USA (1998) 129–138
12. Buneman, P., Fan, W., Weinstein, S.: Interaction between Path and Type Constraints. In Proc. of 18th ACM Symposium on Principles of Database Systems, ACM Press, Philadelphia, Pennsylvania, USA (1999) 56–67
13. Bussche, J. V. d.,Waller, E.: Type Inference in the Polymorphic Relational Algebra. In Proc. of 18th ACM Symposium on Principles of Database Systems, Philadelphia, Pennsylvania, USA (1999) 80–90
14. Cattell, R. G. G.: Object Data Management. Addison-Wesley (1994)
15. Cattell, R. G. G.: The Object Database Standard: ODMG-2. Morgan Kaufmann (1997)
16. Ceri, S.,Fraternali, P.: Database Application with Objects and Rules — the IDEA Methodology. Addison-Wesley (1996)
17. Chen, P. P.: The Entity-Relationship Model — Toward a Unified View of Data. TODS, Vol. 1. (1976) 9–36
18. Dahchour, M., Pirotte, A., Zimányi, E.: Metaclass Implementation of Materialization. University of Louvain-la-Neuve, Louvain-la-Neuve, Belgium, Internal Technical Report of the EROOS Project (April 1997)
19. Deux, O., et al.: The O2 System. Communications of the ACM, Vol. 34. (1991) 34–48

20. Elmasri, R.,Navathe, S. B.: Fundamentals of Database Systems. Third ed. Benjamin Cummings (1999)
21. Formica, A.,Missikoff, M.: Correctness of ISA Hierarchies in Object-Oriented Database Schemas. In Proc. of 4th International Conference on Extending Database Technology, Jarke, M., Jr., J. A. B.,Jeffery, K. G. (eds.). Lecture Notes in Computer Science, Vol. 779. Springer Verlag, Cambridge, United Kingdom (1994) 231–244
22. Hull, R.: Managing Semantic Heterogeneity in Databases: A Theoretical Perspective. In Proc. of 16th ACM Symposium on Principles of Database Systems, ACM Press, Tucson, Arizona (1997) 51–61
23. Kim, W.,alli., e.: Composite Objects Revisited. In Proc. of ACM Conference on Data Management (1989)
24. Lamb, C., Landis, G., Orenstein, J., Weinred, D.: The ObjectStore Database Management System. Communications of the ACM, Vol. 34. (1991) 50–63
25. Milo, T.,Suciu, D.: Type Inference for Queries on Semistructured Data. In Proc. of Proceedings of the 18th ACM Symposium on Principles of Database Systems, ACM Press, Philadelphia, Pennsylvania, USA (1999) 215–226
26. Nestorov, S., Abiteboul, S., Motwani, R.: Infering Structure in Semistructured Data. SIGMOD Records, Vol. 26. (December 1997) 39–43
27. Nierstrasz, O.: A Survey of Object-Oriented Concepts, Databases, and Applications. ACM Press (1989)
28. Pirotte, A., Zimányi, E., Massart, D., Yakusheva, T.: Materialization: A Powerful and Ubiquitous Abstraction Pattern. In Proc. of 20th International Conference on Very Large Data Bases, Bocca, J. B., Jarke, M.,Zaniolo, C. (eds.). Morgan Kaufmann, Santiago de Chile, Chile (1994) 630–641
29. Pitrik, R. M.,Mylopoulos, J.: Classes and Instances. International Journal of Intelligent and Cooperative Information Systems, Vol. 1. (1992) 61–92
30. Rumbaugh, J. E.: Object-Oriented Modeling and Design. Prentice-Hall (1991)
31. Su, S. Y. W.: Modeling Integrated Manufacturing Data with SAM*. IEEE Computer, Vol. 19. (1986) 34–49
32. Traina, C.,Biajiz, M.: The use of the Classification Abstraction in Object-Oriented Data Models (in Portuguese). In Proc. of Jornadas Argentinas de Informática e Investigacion Operativa, Buenos Aires-Argentina (1996) 125–136
33. Traina, C., Ferreira, J. E., Biajiz, M.: Use of a Semantically Grained Database System for Distribution and Control within Design Environments. In Proc. of 3rd International Euro-Par Conference on Parallel Processing, Lengauer, C., Griebl, M.,Gorlatch, S. (eds.). Lecture Notes in Computer Science, Vol. 1300. Springer Verlag, Passau, Germany (1997) 1130–1135
34. Traina, C., Traina, A. J. M., Biajiz, M.: The Instantiation Abstraction Role in a Abstraction-based Meta-model for Object-Oriented Database Systems (in Portuguese). In Proc. of IX Brazilian Database Symposium, São Carlos, São Paulo (1994) 173–187
35. Tresch, M., Palmer, N., Luniewski, A.: Type Classification of Semi-Structured Documents. In Proc. of 21th International Conference on Very Large Data Bases, Dayal, U., Gray, P. M. D.,Nishio, S. (eds.). Zurich, Switzerland (1995) 263–274

Estimating Proximity of Metric Ball Regions for Multimedia Data Indexing

Giuseppe Amato[1], Fausto Rabitti[2], Pasquale Savino[1], and Pavel Zezula[3]

[1] IEI-CNR, Pisa, Italy,
{G.Amato,P.Savino}@iei.pi.cnr.it
WWW home page: http://www.iei.pi.cnr.it
[2] CNUCE-CNR, Pisa, Italy,
F.Rabitti@cnuce.cnr.it
WWW home page: http://www.cnuce.cnr.it
[3] Masaryk University, Brno, Czech Republic,
zezula@fi.muni.cz
WWW home page: http://www.fi.muni.cz

Abstract. The problem of defining and computing proximity of regions constraining objects from generic metric spaces is investigated. Approximate, computationally fast, approach is developed for pairs of metric ball regions, which covers the needs of current systems for processing data through distances. The validity and precision of proposed solution is verified by extensive simulation on three substantially different data files. The precision of obtained results is very satisfactory. Besides other possibilities, the proximity measure can be applied to improve the performance of metric trees, developed for multimedia similarity search indexing. Specific system areas concern splitting and merging of regions, pruning regions during similarity retrieval, ranking regions for best case matching, and declustering regions to achieve parallelism.

1 Introduction

As the volume and variety of digital data grow with increasing speed, huge repositories of different media data are nowadays available over computer networks. Contrary to traditional databases, where simple attribute data is used, current data is also more complex. In order to decrease the amount of operational data, reduction techniques are used producing typically high-dimensional vectors or some other objects for which nothing more than pair-wise *distances* can be measured. Such data is sometimes designated as the *distance only* data. Similar approach can be observed when dealing with the multimedia data, such as text, images, or audio. Here, the search is not performed at the level of actual (raw) multimedia objects, but on characteristic features that are extracted from these objects. In such environments, exact match has little meaning and concepts of *proximity* (*similarity, dissimilarity*) are typically used for searching.

Indexing structures of data files, designed to speedup retrieval (i.e. its efficiency or performance), have always been an important part of database technology. In order to enlarge the set of data types for which efficient search is

T. Yakhno (Ed.): ADVIS 2000, LNCS 1909, pp. 71–81, 2000.
© Springer-Verlag Berlin Heidelberg 2000

possible, researchers have also considered the case where keys are not restricted to stay in a vector space and where only the properties of the generic metric spaces are valid. This is needed by some approaches to indexing of multimedia, genome, and many other non-traditional databases, where objects are compared by measures as the *edit distance, quadratic form* distance, or the *Hausdorf* distance. This approach subsumes the case of multi-dimensional keys, which are typically compared by using an L_p metric, e.g. the Euclidean distance (L_2). Such effort has lead to generalize the notion of *similarity queries* and resulted in the design of so-called *metric trees*. Although several specific metric trees have been proposed, see for example [Ch94,Br95,BO97,CPZ97,BO99], their algorithms for partitioning and organizing objects in metric regions are based on heuristics that have no clearly defined guiding principles to rely upon. Naturally, the performance of such structures is not optimum and practical experience confirms that there is still a lot of space for improvement.

We believe that the basic reason for this state of affairs is the absence of measures able to assess or quantify properties of regions and their relative positions in generic metric spaces. In other terms: given two regions of a metric space, we need to know how "close" these regions are or what is their *proximity*. Clearly, embraced with measures for metric regions, which does not depend on specific metric space, one can address practical and important issues related to data partitioning, allocation, ranking, and searching in general.

2   Proximity of Metric Ball Regions and Its Approximation

Suppose that a *metric space* $\mathcal{M} = (\mathcal{D}, d)$ is defined by a domain of objects, $\mathcal{D}$, (i.e. the *keys* or indexed features) and by a total (distance) function, d, which satisfies for each triple of objects $O_x, O_y, O_z \in \mathcal{D}$ the following properties:

(i) $d(O_x, O_y) = d(O_y, O_x)$ (*symmetry*)
(ii) $0 < d(O_x, O_y) \leq d_m, O_x \neq O_y$ and $d(O_x, O_x) = 0$ (*non negativity*)
(iii) $d(O_x, O_y) \leq d(O_x, O_z) + d(O_z, O_y)$ (*triangle inequality*)

Notice that in our definitions, the maximum distance never exceeds d_m, that means a *bounded metric space* is considered. Given the metric $\mathcal{M}$, a ball region can be defined as

Definition 1. *A ball* $\mathcal{B}_x = \mathcal{B}_x(O_x, r_x) = \{O_i \in \mathcal{D} \mid d(O_x, O_i) \leq r_x\}$, *is the region, determined by a center* $O_x \in \mathcal{D}$ *and a radius* $r_x \geq 0$, *defined as the set of objects in* $\mathcal{D}$ *for which the distance to* O_x *is less than or equal to* r_x. □

To the best of our knowledge, ball regions are practically the only type of regions which are used in practice. Now, let us consider two regions. It is obvious that they *intersect* if there exists an object that belongs to both the regions, or alternatively, if the sum of the regions' radii is shorter than or equal to the distance between their centers. It could be correctly argued that the proximity of regions should be proportional to the amount of space shared by the two regions, and the larger their intersection, the higher the proximity of these regions. However, there is no way to compute the volume of metric ball regions, in general. That is why we define the proximity as follows.

Definition 2. *The proximity* $X(\mathcal{B}_1, \mathcal{B}_2)$ *of ball regions* $\mathcal{B}_1, \mathcal{B}_2$ *is the probability that a randomly chosen object* **O** *over the same metric space* $\mathcal{M}$ *appears in both of the regions.*

$$X(\mathcal{B}_x, \mathcal{B}_y) = \Pr\{d(\mathbf{O}, O_x) \leq r_x \wedge d(\mathbf{O}, O_y) \leq r_y\}$$

□

2.1   Computational Difficulties

The computation of proximity according to Definition 2 requires the knowledge of distance distributions with respect to regions' centers. Since any object from $\mathcal{M}$ can become ball's center, such knowledge is not realistic to obtain. However, as discussed in [CPZ98], we can assume that the distributions depend on the distance between the centers (d_{xy}), while they are independent from the centers themselves. Such assumption is realistic when the distance distributions with respect to different objects have small *discrepancies*, which was found true in [CPZ98] for many data files. Thus, we can modify our definition as:

$$X(\mathcal{B}_x, \mathcal{B}_y) \approx X_{d_{xy}}(r_x, r_y) = \Pr\{d(\mathbf{O}, \mathbf{O_x}) \leq r_x \wedge d(\mathbf{O}, \mathbf{O_y}) \leq r_y\}, \qquad (1)$$

where $\mathbf{O_x}$, $\mathbf{O_y}$, and $\mathbf{O}$ are random objects such that $d(\mathbf{O_x}, \mathbf{O_y}) = d_{xy}$.

Now, consider the way how $X_{d_{xy}}(r_x, r_y)$ can be computed. Let X, Y and D_{XY} be random variables corresponding, respectively, to the distances $d(\mathbf{O}, \mathbf{O_x})$, $d(\mathbf{O}, \mathbf{O_y})$, and $d(\mathbf{O_x}, \mathbf{O_y})$. The joint conditional density $f_{X,Y|D_{XY}}(x, y|d_{xy})$ gives the probability that distances $d(\mathbf{O}, \mathbf{O_x})$ and $d(\mathbf{O}, \mathbf{O_y})$ are, respectively, x and y, given that $d(\mathbf{O_x}, \mathbf{O_y}) = d_{xy}$. Then, $X_{d_{xy}}(r_x, r_y)$ can be computed as

$$X_{d_{xy}}(r_x, r_y) = \int_0^{r_x} \int_0^{r_y} f_{X,Y|D_{XY}}(x, y|d_{xy}) dy dx \qquad (2)$$

Notice that in general, $f_{X,Y|D_{XY}}(x, y|d_{xy}) \neq f_{XY}(x, y)$, provided that $f_{XY}(x, y)$ is the probability that the distances $d(\mathbf{O}, \mathbf{O_x})$ and $d(\mathbf{O}, \mathbf{O_y})$ are x and y, no matter what is the distance between $\mathbf{O_x}$ and $\mathbf{O_y}$. The difference between the probabilities is immediately obvious when we consider the metric space postulates. Accordingly, $f_{X,Y|D_{XY}}(x, y|d_{xy})$ is 0 if x, y, and d_{xy} do not satisfy the triangular inequality, because such distances cannot simply exist. However, $f_{XY}(x, y)$ is not restricted by such constraint, and any distance $\leq d_m$ is possible. For illustration, Figure 1 shows the distance density $f_{X,Y|D_{XY}}(x, y|d_{xy})$ for a fixed d_{xy} and the density $f_{XY}(x, y)$. They are both obtained by sampling from the same real-life dataset, but their characteristics are significantly different.

Unfortunately, an analytic form of $f_{X,Y|D_{XY}}(x, y|d_{xy})$ is unknown. In addition, computing and maintaining it as a discrete function would result in very

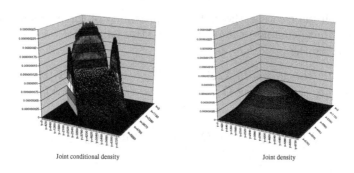

Fig. 1. Comparation of $f_{X,Y|D_{XY}}(x,y|d_{xy})$ and $f_{XY}(x,y)$

high number of values. Indeed, the function depends on three arguments so that the storage space required is $O(n^3)$, where n is the number of samples for each argument. This makes such approach of obtaining and maintaining the probabilities totally unacceptable.

An alternative is to maintain probabilities for each of the components, $f(x)$ and $f(y)$, independently and to transform them into $f_{X,Y|D_{XY}}(x,y|d_{xy})$ through a convenient function. From the storage point of view, such approach is feasible, but the problem is to find such transform. To this aim, we propose and investigate approximations that would satisfy efficiency requirements and guarantee good quality of results.

Before we proceed, we define, as reference, an approximation that is generally used in current applications.

$$X_{d_{xy}}^{trivial}(r_x, r_y) = \begin{cases} 0 & \text{if } r_x + r_y < d_{xy} \\ \frac{2 \cdot \min\{r_x, r_y\}}{2 \cdot d_m - d_{xy}} & \text{if } \max(r_x, r_y) > \min(r_x, r_y) + d_{xy} \\ \frac{r_x + r_y - d_{xy}}{2 \cdot d_m - d_{xy}} & \text{otherwise} \end{cases} \quad (3)$$

For convenience, we call this approximation *trivial*, because it completely ignores distributions of distances though, as Figure 1 demonstrates, such distributions are dramatically different.

2.2   Approximate Proximity

Given two ball regions with distance between centers d_{xy}, the space of possible x and y distances is constrained by the triangular inequality, i.e. $x + y \geq d_{xy}$, $x + d_{xy} \geq y$ and $y + d_{xy} \geq x$ – see Figure 2 for illustration. Notice that also the graph of joint conditional density in Figure 1 has values greater than zero only in this area, and that quite high values are located near the edges.

Such observation forms the basis for our heuristic functions to approximate the joint conditional density by means of the joint density. The idea can be outlined as follows:

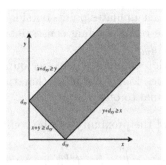

Fig. 2. distances admitted by the triangular inequality for a given d_{xy}

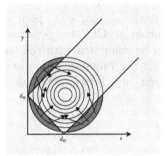

Fig. 3. Heuristic to obtain the joint conditional density by using the joint density

Move all joined density values of arguments x, y, and d_{xy} that do not satisfy the triangular inequality inside the constrained space and locate them near the edges.

Notice that such approximation preserves properties of density functions, since the integral over the whole space is 1. This is the basic assumption of any probabilistic model and that would not be the case provided the joint densities were simply trimmed out by the triangle inequality constraints. Figure 3 sketches the principle of the heuristic, where the circles represent the joint density function, and the straight lines define the constraints. The way how the densities are moved is indicated by the arrows.

In order to come out with specific solutions, we have tried four different implementations of this heuristic, varying the strategy used to move densities.

Orthogonal approximation accumulates points outside the bounded area and moves them on top of the corresponding constraint following a direction that is orthogonal to the corresponding constraint.

Parallel approximation accumulates points outside the bounded area and moves them on top of the corresponding constraint following a direction that is parallel with the axis.

Diagonal approximation accumulates points outside the bounded area and moves them on top of the corresponding constraint following a direction that always passes through d_m.

Normalized approximation eliminates densities outside the constrained space and normalizes the ones found inside so that the integral over the whole constrained space is equal to one.

Then, an approximation of the proximity can be computed as

$$X_{d_{xy}}(r_x, r_y) \approx \int_0^{r_x} \int_0^{r_y} f_{X,Y|D_{XY}}^{appr}(x, y|d_{xy})dydx$$

The fact that $f_{X,Y|D_{XY}}^{appr}(x, y|d_{xy})$ can be obtained from $f_{X,Y}(x, y)$ is also interesting from the computational complexity point of view. Since $f_{X,Y}(x, y) = f_X(x) \cdot f_Y(y)$ and, as reported in [CPZ98], $f_X(x) \approx f_Y(y)$, it is easy to show that the above integral can be computed with complexity $O(n)$, where n is the number of samples in which the function $f_X(x)$ and its distance distribution are known. It is obvious that such information can easily be kept in the main memory for quite high n.

3   Verification

In this section, we investigate the effectiveness of the proposed approaches. To this aim, we first introduce the data sets, describe the evaluation process, and define comparison metrics.

3.1   Data Sets

We have used three different datasets presenting different characteristics in order to be more confident on obtained results. We have used one synthetic data set and two real-life datasets representing color features of images. All tested datasets contained 10.000 objects.

The synthetic dataset, called UV, is a set of vectors uniformly distributed in 2-dimensional space where vectors are compared through the Euclidean (L_2) distance. Distances range in the interval [0,7000] and the graph of the overall density function $f(x)$ is shown in Figure 4-a.

The second dataset, designated as HV1, contains color features of images. Color features are represented as 9-dimensional vectors containing the average, standard deviation, and skewness of pixel values for each of the red, green, and blue channels, see [SO95]. An image is divided into five overlapping regions, each one represented by a 9-dimensional color feature vector. That results in a 45-dimensional vector as a descriptor of each image. The distance function used to compare two feature vectors is again the Eucledian (L_2) distance. Distances range in the interval [0,8000], and Figure 4-b shows the overal density function of this dataset.

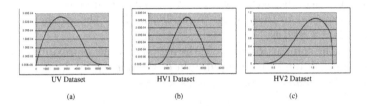

UV Dataset   HV1 Dataset   HV2 Dataset

(a)   (b)   (c)

Fig. 4. Overall density functions of the used data sets

The third dataset, called HV2, contains color histograms represented in 32-dimensions. This dataset was obtained from the UCI Knowledge Discovery in Databases Archive ([Bay99]). The color histograms were extracted from the Corel image collection as follows: the HSV space is divided into 32 subspaces (32 colors: 8 ranges of hue and 4 ranges of saturation). The value in each dimension of the vector is the density of each color in the entire image. The distance function used to compare two feature vectors is the histogram intersection implemented as L_1. Distances range in the interval [0,2]. Figure 4-c shows the overall density function of this dataset.

Notice the differences in densities for individual files: the UV dataset presents the most frequent distances on the left of the distances range, the HV1 on the center, and the HV2 on the right. In this way we have tried to cover a large spectrum of possible data.

3.2 Experiments and Comparison Metrics

In order to form the basis for comparison, we have experimentally computed the actual proximity $X_{d_{xy}}^{actual}(x, y)$ for all datasets. We have chosen several values of d_{xy} in the range of possible distances. For each of these values, we have experimentally obtained the joint conditional density using 100×100 samples for (x, y) as follows. For each d_{xy}, we have found 400 pairs of objects (O_x, O_y) such that $|d(O_x, O_y) - d_{xy}| \leq \rho$, where ρ is the smallest real that allowed to obtain at least 400 objects. For each pair, the distances x and y, respectively from O_x and O_y to all objects of the dataset, were computed to produce the joint conditional density.

The proximity for each d_{xy} was again represented using 100×100 samples for (x, y), and it was obtained by integrating the corresponding joint conditional density, produced during the previous step, for all combinations of d_{xy}, x, and y.

Notice that we did not consider distances d_{xy} of very low densities. In such cases, 400 pairs were only possible to obtain for large values of ρ, thus the actual proximity was not possible to establish with sufficient precision. Notice, however, that such situations, i.e. relative positions of regions' centers, are also not likely to occur in reality.

Having obtained the actual proximity, we have computed the approximate proximities proposed in this paper. Naturally, we have chosen the same values of variables d_{xy}, x, and y.

Given d_{xy}, the comparison between the actual and the approximate proximity is made obvious through the computation of the *average of errors*, $\epsilon'(d_{xy})$ and the *average of absolute errors*, $\epsilon''(d_{xy})$, as follows:

$$\epsilon'(d_{xy}) = E_{\mathbf{x},\mathbf{y}}(\epsilon(\mathbf{x},\mathbf{y},d_{xy}))$$

$$\epsilon''(d_{xy}) = E_{\mathbf{x},\mathbf{y}}(|\epsilon(\mathbf{x},\mathbf{y},d_{xy})|)$$

where
$$\epsilon(x,y,d_{xy}) = X_{d_{xy}}^{actual}(x,y) - X_{d_{xy}}^{approx}(x,y)$$

In similar way, we have computed the errors' variance.

$$\epsilon'_{\sigma}(d_{xy}) = Var_{\mathbf{x},\mathbf{y}}(\epsilon(\mathbf{x},\mathbf{y},d_{xy}))$$

$$\epsilon''_{\sigma}(d_{xy}) = Var_{\mathbf{x},\mathbf{y}}(|\epsilon(\mathbf{x},\mathbf{y},d_{xy})|)$$

We believe that the use of two different errors and the corresponding variance is important to understand the quality of an approximation. Small values of ϵ' are not sufficient to guarantee the quality of the approximation, because the positive and negative errors can easily be eliminated on average. However, when ϵ' and ϵ'' are significantly different, we can argue that the approximation is not stable. For instance, there can be situations where the approximation is underestimating and other situations where it is over-estimating the actual value of the proximity. In this respect, also the variance can help to correctly interpret errors. Small average errors (either ϵ' or ϵ'') but high variance means again that the errors are not stable, so it is likely that the approximate proximity may present properties that are different from the actual ones.

In fact, high errors and small variance may, sometimes, provide good approximation. To illustrate, suppose that we want to use the proximity to order (rank) a set of regions with respect to a reference region. It might happen that the results obtained through the actual and approximate proximity are completely identical even though ϵ' and ϵ'' are quite high. In fact, when the variance of the errors is very small, it means that the error is almost constant, and the approximation somehow follows the behavior of the actual proximity. In this case it is highly probable that the approximation increases (decreases) accordingly to the behavior of the actual one, guaranteeing the correct ordering.

3.3   Simulation Results

For all datasets, the actual proximity was compared with the proposed approximations, the trivial approximation, and, only for the UV dataset, with a proximity measure obtained using an analytic technique – this technique has a very high computational cost and it is not described here for brevity. The results are presented in Figures 5 and 6. It is immediate that all approximation methods

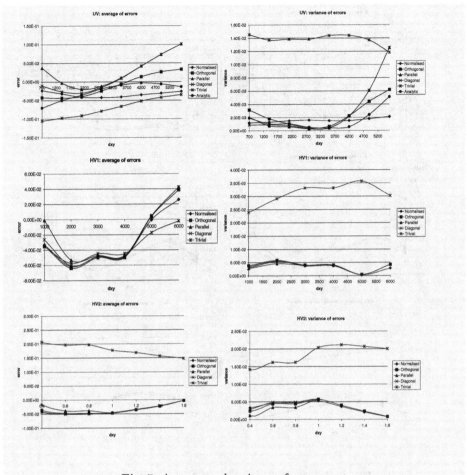

Fig. 5. Average and variance of errors

outperform the trivial one, since the error of the trivial method is even one order of magnitude higher compared to all the other methods.

For UV dataset, the trivial, normalized, and analytic methods systematically underestimate the actual proximity. However the error ϵ'' of the trivial method is much higher than ϵ'. The variance is also quite high. This implies that in specific situations, the trivial approximation may provide significantly different results with respect to the actual proximity. For what concerns the other approximations, all of them demonstrate a good and stable behavior. They demonstrate a very small variance together with small errors so can reliably be used in practice.

Tests performed on the HV1 dataset have produced errors ϵ' that are almost the same for all methods, trivial included. However, the absolute error ϵ'' is again much higher for the trivial method, revealing possible deviation of this approximation with respect to the actual proximity. This is also confirmed by

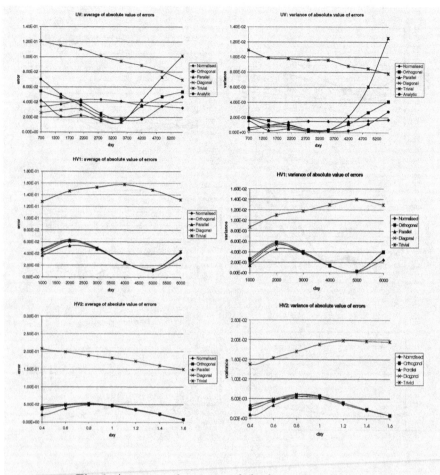

Fig. 6. Average and variance of absolute value of errors

the variance. On the other hand, all the other methods have the same stable behavior and no relevant difference between them can be noticed. Similar trends can be observed also for the HV2 dataset and support validity of the proposed approximations.

4   Conclusions

Approximation methods to quantify the proximity of metric ball regions have been proposed and evaluated. In accordance with our objectives, the proposed methods are *flexible* and do not depend on distance measure, provided it is a metric. *Accuracy* of the methods is high and depends only on proper knowledge and stability of distance distribution. The computation of proposed proximity measures is *fast*. Its computational complexity is linear, thus it is also applicable at run-time. The storage overhead of distance distribution histograms is *low*.

We are currently working on application of the method to improve the performance of metric trees. The specific problems concern the tree node *split* and *merge* functions, *ranking* of metric regions in priority queue for the *best case matching*, *declustering* of regions (partitions) to achieve parallelism, and *pruning* for approximate similarity retrieval.

Future research should concentrate on proximity measures of regions other than balls and on proximity of more than 2 regions. More effort should also be spend on developing other applications and possibly on developing new, more efficient, metric indexes.

References

[Bay99]   Bay, S. D. The UCI KDD Archive [http://kdd.ics.uci.edu]. Irvine, CA: University of California, Department of Information and Computer Science.

[BO97]   T. Bozkaya and M. Ozsoyoglu. Distance-based indexing for high-dimensional metric spaces. *ACM SIGMOD*, pp.357-368, Tucson, AZ, May1997.

[BO99]   T. Bozkaya and Ozsoyoglu. Indexing Large Metric Spaces for Similarity Search Queries. *ACM TODS*, 24(3):361-404, 1999.

[Br95]   S. Brin. Near neighbor search in large metric spaces. In *Proceedings of the 21st VLDB International Conference*, pp. 574–584, Zurich, Switzerland, September 1995.

[Ch94]   T. Chiueh. Content-based image indexing. In *Proceedings of the 20th VLDB International Conference*, pages 582–593, Santiago, Chile, September 1994.

[CPZ97]   P. Ciaccia, M. Patella, and P. Zezula. M-tree: An Efficient Access Method for Similarity Search in Metric Spaces. Proceedings of the *23rd VLDB Conference*, Athens, Greece, 1997, pp. 426-435.

[CPZ98]   P. Ciaccia, M. Patella, and P. Zezula. A Cost Model for Similarity Queries in Metric Spaces. In *Proceedings of 7th ACM SIGACT-SIGMOD-SIGART Symposium on Principles of Database Systems, PODS 1998*, Seattle, Washington, 1998, pp. 59- 68.

[SO95]   M. Stricker and M. Orengo. Similarity of Color Images. In: *Storage and Retrieval for Image and Video Databases III*, SPIE Proceedings 2420, 1995, pp. 381-392.

A Technique
for Upgrading Database Machines Online

Jiahong Wang[1], Masatoshi Miyazaki[1], and Jie Li[2]

[1] Faculty of Software and Information Science, Iwate Prefectural University,
152-52 Sugo, Takizawa, Iwate 020-0193, Japan
{wjh, miyazaki}@iwate-pu.ac.jp
[2] Institute of Information Sciences and Electronics, University of Tsukuba
Tsukuba Science City, Ibaraki 305-8573, Japan
lijie@is.tsukuba.ac.jp

Abstract. In order to improve performance of database systems, we may have to replace an existing database machine with more powerful one. Thus database migration between two machines becomes necessary. Database migration, however, generally requires taking a database off line for a long time, which is unacceptable for numerous applications. This paper addresses a very practical and important subject: how to replace a database machine on line, i.e., how to move a database from an existing machine to a new one concurrently with users' reading and writing of the database. A technique for this purpose is proposed.

1   Introduction

Current database applications require not only that database systems provide high throughputs with rapid response times, but also that they are continually available. For example, applications for reservations, finance, process control, hospitals, and police all require *highly-available databases* (databases that are to be fully available 24-hours-per-day and 7-days-per-week), and these applications cannot afford an off-line database for any significant amount of time [5].

The increased requirement for high transaction throughputs with rapid response times is generally satisfied by replacing the whole database machine or parts of it (e.g., hard disks) with more powerful ones. In fact, computing hardware technologies are developing rapidly, which suggests that the database machine should be upgraded from time to time, so that a database system provides satisfactory performance all the time. Some statistics show that uniprocessor performance doubles every 18 months, and multiprocessing system performance increases by a factor of 4 every 18 months. For example, now we can have a desktop database server with a processor of 1GMHz and a hard disk drive of 36GB, far more powerful than that of last year. For another example, on October 4, 1999, IBM announced a 72GB hard disk drive with 10,000 RPM, far more powerful than that in 1997, which was of 4GB and 7200 RPM only. All of these demonstrate the necessity of upgrading a database machine.

If we intend to upgrade a database machine except its desk drives that are connected to the other part with cables, several minutes may be enough for

T. Yakhno (Ed.): ADVIS 2000, LNCS 1909, pp. 82–91, 2000.

shutting down the system, switching the disk drive to the new environment by some plug-out and plug-in operations, and then, starting the system again. If the disk drive is to be upgraded too, however, the problem becomes very complicated. In this case, the database in the old environment is required to be moved to the new one. Moving a database, however, generally requires taking the database off line for a long time, which can be unacceptable for a highly-available database. With the aim of solving this problem, in this paper, we addresses a very practical and important subject: upgrading a database machine on line, i.e., moving a database from the old hardware environment to the new one concurrently with users' reading and writing of the database.

Although what we stated above has been seldom studied so far, it is an essential subject of the on-line database reorganization. So far as we know, two typical on-line database reorganization approaches have been proposed that can achieve the effect of upgrading a database machine on line. The one is what IBM DB2 Parallel Edition uses [9], which locks the data to be moved in exclusive mode, invalidates all transactions that involve the data, and then, moves data. This approach, however, is extremely expensive since users cannot perform any operations (e.g., update and query) on the data being moved for a long time. The other is what Tandem NonStop SQL uses [8], which is on-line, however, does not consider the index modification. In fact, for most existing systems, index modification is an important performance issue that cannot be ignored. This is because after data is moved, several days may be required for rebuilding indexes, which may degrade system performance significantly [4]. In addition, this approach needs to scan system log. The amount of log that has to be scanned, however, may be too much to make this an efficient approach, and also, the system must ensure that the relevant portion of the log is not discarded before the data migration is completed [4].

In this paper, we give a radically different solution. We propose a new technique specially for upgrading the database machine, which is on-line, takes the important index issue into account, and does not use the log. The technique is conceptually simple, but effective. We move a database by moving both its data and the corresponding indexes, and then reusing the index as is. Because users' concurrent accesses to the smallest data granule in a database is properly considered, while a database is moved, the database is continuously available to read-only transactions, and unavailable to read-write transactions for a very short time, and users can use the database normally during data migration. Our technique has adopted the idea of recycling indexes from [1], [7]. Unlike [1], [7], however, our technique is on-line.

Note that data migration is conventionally performed by moving data and then rebuilding indexes. In this paper, however, the massive database is considered that cannot fit in the main memory, therefore disk accesses are inevitable. Rebuilding indexes in such a situation would take a long time, consume a lot of system resources, and degrade system performance significantly [1], [4]. Since network transfer rate has become far greater than disk transfer rate, moving and reusing an index can be far cheaper than rebuilding an index [1].

2 Problem Model and System Model

(1) Problem Model: The problem addressed in this paper is modeled as follows. We have a highly-available database on an *old machine* that is a node (called source node and denoted by S) of a LAN. Users access this database by either building their applications on S directly, or connecting to it remotely. Now S is required to be replaced by a *new machine*, which has been connected to the same LAN (called destination node and denoted by D), so as to take advantage of the new hardware technology. Database at S is therefore needed to be moved to D. Moreover, it is required that the database is moved concurrently with users' reading and writing of the database.

It is assumed that in D, a DataBase Management System (DBMS) has been installed that is the same DBMS as that running at S, or a compatible version of it. This assumption is reasonable since what we are considering is not to change DBMS's type but to upgrade the database machine itself.

Our technique can solve the problem stated above. Its variations, however, can also be applied to such problems as that a new disk drive is added to an existing database machine, and databases on the old disk of the machine is required to be moved to the new one.

(2) System Model: The relational system is considered. Each database consists of several relation *tables*, each table contains numerous tuples, and each table has zero or more indexes. Information about databases, tables, indexes, columns, etc., is stored in the special system tables that are commonly known as system catalogs. The hierarchical index structure such as B^+-tree is considered in this paper, although variations of the proposed technique can be applied to other index structures. The storage structure is as follows, which is used by a wide range of relational DBMSes, including Oracle Rdb and nearly all IBM relational systems [1], [5]: the storage of a table (resp. an index) is divided into units called table pages (resp. index pages), and a tuple is identified using its page number and the offset of the tuple within the page. See [2] for further details.

A table and its indexes are regarded as a *table object*. A database is moved from the old machine to the new one by moving its table objects one by one. Since a table and its indexes are moved as a whole and DBMSes in both sides are compatible, no time-consuming index rebuilding work is required for the moved data.

3 Upgrading Database Machines on Line

Now we describe the technique for upgrading a database machine on line, i.e., moving databases on line from an old hardware environment to a new one. This technique is general in terms of the above system model. Here we focus on its basic idea instead of its implementation strategies. In fact, it should be implemented by considering optimization issues of the object database system. For simplicity, we restrict our description to the case that one table object (denoted by P) with a single index (denoted by $index_{P,S}$) is required to be moved from

S to D. If multiple table objects are involved, they can be moved one by one. Note that this technique can also handle the case of a table object with multiple indexes.

Because the request that changes system catalogs (e.g., adding a index to a table, dropping a table, modifying the definition of a table, etc.) may interfere with the data migration work, it is assumed that during data migration period, only normal reading and writing of database are allowed, and no requests that change system catalogs are allowed. This can be guaranteed by rejecting all such requests.

The real work of moving table objects is performed by the co-operation of *system buffer managers* at both S and D, *source redistributor* at S, and *destination redistributor* at D, which in turn, are coordinated by a *data redistributor* at S. P is moved from S to D in three steps. Firstly, copies of system catalogs are created at D. This can be done, e.g., by coping the system catalogs from S to D (see the following section). Secondly, by moving its $index_{P,S}$ in PHASE I followed by its tuple data (denoted by $Data_{P,S}$) in PHASE II (about PHASE I and II, see the following), P is eventually moved from S to D. Lastly, local users are diverted to D, and the database for describing the network is modified to divert remote user accesses to D. The former can be easily done by informing the local users that the database has been moved. The latter is slightly complicated. One solution is to quiesce all user access requests, turn off the network of both S and D, and turn on that of D with the same host name and internet address as that of S. The time requirement for quiescing user access requests can be estimated to be the average system response time multiplied by the number of user transactions running in the system, and turning a network off and on may take several seconds. Therefore the third step should be delayed until the system can afford an off-line database for a few minutes. In the meantime, S functions as a router between users and D for forwarding users' access requests and the corresponding results. The details of the technique are given below. Unless explicitly specified, P is continuously available to both read-only and read-write transactions.

PHASE I (moving index):   the following concurrent actions occur:

Action1: The source redistributor moves $index_{P,S}$ from node S to node D by:
Step1: Initialize a cursor $C_{P,S}$, and set it to the first page of $index_{P,S}$. Initialize a shared queue $Que_{P,S}$ for holding such pages that have been sent to D, but thereafter have been modified.
Step2: Repeat the following until the end of $index_{P,S}$ is reached.
(1). Send D the page pointed by $C_{P,S}$ along with its page number. Advance $C_{P,S}$ to the next page.
(2). Send D the pages in $Que_{P,S}$ (see Action2 of PHASE I) along with their page numbers.
Step3: Inform data redistributor that the end of $index_{P,S}$ has been reached, so that data redistributor can prepare to forward the users' requests for P from S to D. In the meantime, repeat the following until a response message from the data redistributor arrives (see Action4 of PHASE I). This is because

read-write transactions are still active, and $index_{P,S}$ may be modified.

(1). If $C_{P,S}$ has not reached the end of $index_{P,S}$ (this occurs in the case that new index pages are appended to $index_{P,S}$), send D the page pointed by $C_{P,S}$ along with its page number, and advance $C_{P,S}$ to the next page.

(2). If $Que_{P,S}$ (see Action2 of PHASE I) is not null, send D the pages in $Que_{P,S}$ along with their page numbers.

Step4: Start the phase of moving the tuple data.

Action2: For each page of $index_{P,S}$ that is to be forced into the disk from system buffer pool, system buffer manager at S does the following: If it is the page behind $C_{P,S}$, then force it into disk normally; else send the source redistributor a copy of this page along with the corresponding page number (this is done by queuing an element into $Que_{P,S}$), and then, force this page into disk. Source redistributor will send this page to D again (see Action1 of PHASE I), since it has been modified since it was last sent to D.

Action3: Destination redistributor initializes a storage (denoted by $index_{P,D}$) for the received pages of $index_{P,S}$ from the source redistributor, and enter each page received into $index_{P,D}$, with its page number remaining unchanged.

Action4: After initiating database migration, the data redistributor sleeps until the message from source redistributor arrives indicating that the index has been moved to D. Then it quiesces read-write transactions that access P. This quiescing blocks newly-arrived read-write transactions that access P and waits for existing read-write transactions accessing P at S to finish. When all the existing read-write transactions are finished, the newly-arrived read-only transactions that access P are forwarded to D. Then it waits for the existing read-only transactions accessing P at S to finish. When the existing read-only transactions are finished, the read-write transactions that access P are forwarded to D, and a response message is sent to the source redistributor indicating that the phase of moving tuple data can begin.

PHASE II (moving tuple data):   the following concurrent actions occur:

Action1: The source redistributor moves $Data_{P,S}$ from node S to node D by:

Step1: Initialize a cursor $C_{P,S}$, and set it to the first page of $Data_{P,S}$. Initialize a buffer $Buf_{P,S}$, in which there is an entry for each page of $Data_{P,S}$. Each entry is set to zero initially. If a page is sent to D, the corresponding entry is set to one.

Step2: Repeat the following until the end of $Data_{P,S}$ is reached:

(1). Check $Buf_{P,S}$ to see if the page pointed by $C_{P,S}$ has not been sent to D. If has not, send D the page along with its page number, and set the corresponding entry in $Buf_{P,S}$ to one.

(2). Check if there comes a page-requesting message (see Action2 of PHASE II). If there does, and the requested page has really not sent (by checking $Buf_{P,S}$), send D the requested page along with its page number, and set the corresponding entry in $Buf_{P,S}$ to one.

(3). Advance $C_{P,S}$ to the next page of $Data_{P,S}$.

Action2: The destination redistributor does the following:

Step1: Initialize a storage (denoted by $Data_{P,D}$) for the received pages of

$Data_{P,S}$ from source redistributor. Initialize a buffer $Buf_{P,D}$, in which there is an entry for each page of $Data_{P,D}$. Each entry is set to zero initially. If a page is received from S, the corresponding entry is set to one.

Step2: Repeat the following until $Data_{P,D}$ is completely received:

(1). Enter the page received into $Data_{P,D}$, with its page number remaining unchanged.

(2). Check if there comes a page request from the system buffer manager at D (see Action3 of PHASE II). If there does, send the source redistributor a page-requesting message for the requested page.

Action3: The system buffer manager at node D does the following:

— For each page access request for P, check $Buf_{P,D}$ to see if the page has been in $Data_{P,D}$. If it does, then fetch it from $Data_{P,D}$ directly, else fetch it from $Data_{P,S}$ remotely by sending a page request to the destination redistributor and waiting for the requested page.

— For each data page of P in the buffer pool that is required to be forced into the disk, send the page to $Data_{P,D}$ with its page number remaining unchanged.

4   Experimental Testbed

In order to verify and evaluate the proposed technique, we have implemented it practically (see Fig.1), which is called OLDM (On-Line Database Migration).

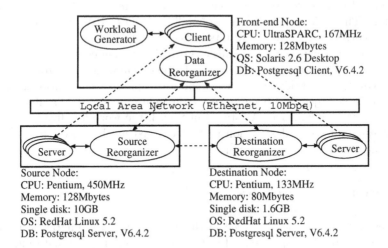

Fig. 1. The OLDM, an implementation of the proposed On-Line Database Migration technique.

For our convenience of the implementation, we let the machine with faster CPU be the source node. This has no effect on our conclusions since we are not concerned with the absolute value of, e.g., transaction throughput, but the proposed technique's ability to be on-line. In the following, we give details of the implementation.

As shown in Fig.1, a front-end node is added, and a client version of the Postgresql DBMS [6] is installed for supporting the workload generator that simulates a multi-user application environment. In order to clearly distinguish the impact of data migration on transactions, the data redistributor that coordinates source and destination redistributors is assigned to the front-end node. At either of source and destination nodes there exists a modified server version of the Postgresql, in which the proposed technique is implemented. OLDM has the process-per-client paradigm architecture. That is, for each client there exists a dedicated server process.

Such a Wisconsin benchmark relation [3] is used that is initially placed on source node, has 500000 tuples, and has a b-tree index of 10117120 bytes. Note that there is no point in having more than one relation in the experiment because the redistributor moves only one table object at a time, as stated in Sect.3.

The workload generator maintains MPL (Multi-Programming Level) clients in the system, where MPL is a system parameter that can be changed so as to change system workload. Each client starts a transaction, which is executed by the corresponding server. Each transaction accesses exactly one tuple at random. A transaction accesses one index to locate the page of the tuple, and then retrieves the page from disk and accesses the tuple. For Read-Only (RO) transactions, the content of the tuple is returned to the client. For Read-Write (RW) transactions, the tuple is updated and written back into the disk. When a transaction is completed, the corresponding client exits, and a new client is generated immediately. The ratio of occurrence probability of RO to RW transactions in the workload corresponding to the results in Sect.5 is 0.7:0.3.

Two more implementation details of OLDM, creating copies of catalogs and diverting user accesses, are as follows: (1) Before moving the table object to destination node, catalogs at source node are copied to destination node. (2) When the table object has been moved to destination node, the database migration is regarded as finished. As stated in Sect.3, the work for diverting remote user access requests to destination node is delayed until the system can afford an off-line database for a few minutes.

5   Performance Results

Two performance metrics are used in the experiments: the migration response time (denoted by RT_{migra}) defined as the time to complete the data migration, and the transaction throughput defined as the number of committed transactions per second. In addition, in order to estimate the loss of the system in terms of the number of potential transactions that could not execute due to contention caused by the data migration activity, a so-called $loss$ metric is adopted: $loss = (T_{normal} - T_{migra}) * RT_{migra}$. Here T_{normal} is the average transaction throughput in the case that no data migration occurs, and T_{migra} is the average transaction throughput within RT_{migra}. It can be seen that the loss will be high if data migration takes a long time or if the throughput during data migration is low. Interestingly, the formula for loss can be applied to the off-line data migration approaches directly, in which case the T_{migra} is 0. The loss metric thus gives a way to compare the off-line approaches and OLDM. For this purpose, the results of loss are given in the way of the normalized loss, which is defined as the loss for OLDM divided by the loss for the off-line approaches.

Figure 2 gives average transaction throughputs at different MPLs in the case that data migration is and is not performed respectively. For both cases, throug-

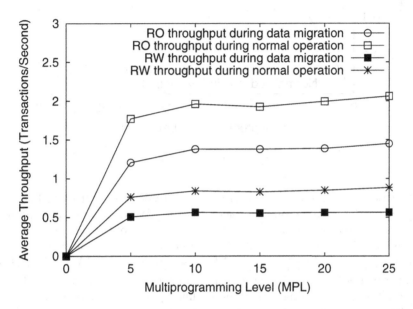

Fig. 2. Average throughput of RO and RW transactions.

hput increases as MPL increases up to 10, and then it flattens out. We think the major reason is as follows. Before MPL of 10, data contention is weak, and system resources have sufficient capacity to serve both the system's data migration request and the users' data access requests. Since increasing MPL incurs that the concurrency degree of the system is increased, throughput increases. As MPL reaches 10 and increases furthermore, however, data contention becomes so strong that transactions tend to be executed sequentially. In addition, system resources tend to be saturated. As a result, throughput remains unchanged no matter how MPL is changed. Note that for Postgresql of version 6.4.2, on which OLDM is implemented, the data contention can be very strong. This is because there is only the table-level lock, and if an RW (resp. RO) transaction is allowed to access a table, all the other RW and RO (resp. RW) transactions to access the same table are blocked.

Figure 2 demonstrates that not only RO but also RW transactions can be effectively executed during data migration period. From the RO and RW tran-

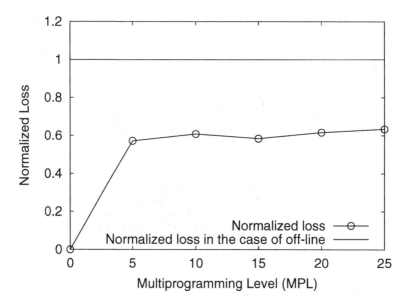

Fig. 3. Normalized loss.

saction throughput (denoted by T_{RO} and T_{RW} respectively) given in Fig.2, we can estimate the maximal off-line fraction of the RW transaction at a specified MPL (denoted by ρ_{MPL}). ρ_{MPL} is defined as the ratio of the time for quiescing RW transactions at MPL to the migration response time of the off-line approach: $\rho_{MPL} = \frac{1}{RT_{offline}} * \left(\frac{70\% * MPL}{T_{RO}} + \frac{30\% * MPL}{T_{RW}} \right)$ (see Action4 of PHASE I in Sect.3). For experiments here, ρ_{25} is the maximal among all the ρ_{MPL}s and is 16.2%. This result suggests strongly that the on-line approach should be used. Note that RO transactions are not affected by the data migration activity, and can continuously use the data (see Sect.3).

Figure 3 gives results of the normalized loss for OLDM and the off-line approach. The line identified by "Normalized loss in the case of off-line" says that, if we perform data migration off line, no transactions can be executed during data migration period, and the system losses 100% of the transactions. The off-line approach, however, provides the fastest way of moving data. On the other hand, from the line "Normalized loss" we see that, if we perform data migration on line using OLDM, 40% of performance improvement can be achieved. OLDM, however, has about two times of the migration response time of the off-line approach. In fact, if we slow down the progress of data migration, the user transactions' share of system resources increases, and more performance improvement can be achieved. For OLDM then, the following question arises: How to trade off the increase in the migration response time and the corresponding performance improvement. We have not considered this question in this paper.

6 Conclusion

We have addressed the problem of upgrading a database machine on line, and have proposed an effective technique for this purpose. Using the technique, a database machine can be upgraded concurrently with users' reading and writing of the database, so that a highly-available database need not go off line for migration. The technique has the following properties: (1) While data is moved, no locks are acquired on both data and its indexes. (2) No time-consuming log processing is required, which is necessary for the previous on-line data migration approaches. (3) Data and its indexes are just moved between machines, and re-used. The conventional time-consuming index rebuilding at new machine is not required. (4) While data and its indexes are moved, users can use the data normally. (5) Few memory and disk spaces are required.

We have implemented a prototype based on the Postgresql database system, and have demonstrated the performance benefits of the proposed technique. In this paper we did not consider the issue of recovering from failures without complete loss of work. We left it for our future work.

References

1. P.M. Aoki, Recycling Secondary Index Structures, Sequoia 2000 Tech. Rep. 95/66, Univ. of California, Berkeley, CA, July 1995.
2. J. Gray and A. Reuter, Transaction Processing: Concepts and Techniques, Morgan Kaufmann Publishers, Palo Alto, CA, 1993.
3. J. Gray, ed., The Benchmark Handbook for Database and Transaction Processing Systems, Morgan Kaufmann Publishers, Palo Alto, CA, 1993.
4. C. Mohan and I. Narang, Algorithms for creating indexes for very large tables without quiescing updates, Proc. 1992 ACM SIGMOD Conf., San Diego, California, pp.361–370, 1992.
5. G.H. Sockut, T.A. Beavin, and C.C. Chang, A method for on-line reorganization of a database, IBM Systems Journal, vol.36, no.3, pp.411–436, 1997.
6. M. Stonebraker and G. Kemnitz, The POSTGRES next-generation database management system, Comm. of the ACM, vol.34, no.10, pp.78–92, 1991.
7. M. Stonebraker, P.M. Aoki, R. Devine, W. Litwin, and M. Olson, Mariposa: a new architecture for distributed data, Proc. 10th IEEE Int. Conf. on Data Eng., Houston, TX, pp.54–65, 1994.
8. J. Troisi, NonStop availability and database configuration operations, Tandem Systems Review, vol.10, no.3, pp.18–23, July 1994.
9. DB2 Parallel Edition V1.2 Parallel Technology, IBM Corp., Apr. 1997.

Automatic Semantic Object Discovery and Mapping from Non-normalised Relational Database Systems

Sokratis Karkalas and Nigel Martin

School of Computer Science and Information Systems, Birkbeck College,
University of London, Malet Street, London WC1E 7HX, United Kingdom
{Sokratis,Nigel}@dcs.bbk.ac.uk

Abstract. In this paper we present an algorithm which automatically discovers potential semantic object structures in a relation, which may be in non-2NF. This algorithm can be utilised in reverse engineering of relational schemas, data migration from relational to object database systems and database integration. The algorithm has been implemented, and we report on issues arising from this implementation including optimization techniques incorporated.

1 Introduction

The wide use of the object-oriented paradigm in software engineering has become a major influence in the way that databases are modeled and organized. With the passage of time, traditional database management systems incorporate more OO features and native OODBMSs are becoming more mature. It is also a fact that, due to the great success of the relational approach, there exists a very large number of relational databases that need to be either enhanced with OO extensions or changed to pure OO systems in order to advance their potential. The problem becomes more serious when there are databases that may use different data models and need to be integrated either physically or virtually into a single database.

Database integration is traditionally a very labour-intensive process since it normally requires a lot of effort from experts in the domain of knowledge that the component databases belong to and database administrators that are aware of the particular way that this knowledge is organized within the databases. It is apparent that if the component databases are already in an object-oriented form, the integration process can be significantly simplified, since database objects can encapsulate more semantic information about the concepts and their inter-relationships than their equivalent relation(s). Many systems have been developed to deal with these problems using approaches that vary depending on the assumptions made by the designers and/or the degree of automation that is given to the process.

In this paper we present an algorithm that discovers potential semantic object structures from non-2NF relational schemas. Since this algorithm is designed to deal only with this particular schema configuration, it should be considered as an additional feature that can add extra functionality to existing systems and not as a stand-alone mapping mechanism.

T. Yakhno (Ed.): ADVIS 2000, LNCS 1909, pp. 92-107, 2000.

The algorithm does not require manual intervention, nor any preliminary semantic enrichment stage. We envisage that the output would provide a first-cut design for inspection and modification as necessary by a database designer.

The paper is organised as follows: In the following section we present the motivation for the algorithm presented in this paper. We also present four alternative example database models to illustrate what our algorithm aims to accomplish, given a certain input. The algorithm itself is defined in section 3 and details of a worked example are presented. Section 4 contains a presentation of the architecture of the system in which the algorithm is implemented along with a discussion on implementation issues. In section 5, a comparison with other approaches is given and finally the conclusions and suggestions for further research are given in section 6.

2   Motivation

The algorithm presented in this paper discovers patterns in a non-2NF relation, which may represent semantic objects. Many semantic object models have been defined [2],[3],[4]. A common characteristic of such models is that objects are named collections of attributes abstracting entities in the world. In this paper we adopt the model presented by [3] to illustrate our algorithm. This model uses the following terms:

Semantic object: A semantic object is a named collection of attributes that sufficiently describes a distinct identity [3]. The use of the adjective semantic here indicates that a semantic object models not only data but also semantic information about it.

Simple attribute: An atomic attribute in the sense that it represents a single non-composite concept.

Object attribute: An attribute that establishes a relationship between one semantic object and another. The value of this object must be a reference to a semantic object.

Simple object: A semantic object that contains only simple attributes.

Compound object: A semantic object that is not simple.

We believe automatic discovery techniques may be usefully applied to detect collections of attributes representing semantic objects within a non-2NF relation. This is illustrated by the following example.

2.1   Example

Consider the following alternative database models:

1st case: single-relation non-normalised database

CUSTOMERS

customer	customer_name	customer_address	customer_telephone
1	Jones	88 Bond Street	171334334
1	Jones	88 Bond Street	171334334
1	Jones	88 Bond Street	171334334
1	Jones	88 Bond Street	171334334
1	Jones	88 Bond Street	171334334
1	Jones	88 Bond Street	171334334
1	Jones	88 Bond Street	171334334
1	Jones	88 Bond Street	171334334
1	Jones	88 Bond Street	171334334
1	Jones	88 Bond Street	171334334
1	Jones	88 Bond Street	171334334
1	Jones	88 Bond Street	171334334
1	Jones	88 Bond Street	171334334

customer_credit_card	date	lunch	time	telephone	duration
334455	10/01/97	£12.00	10:00	£0.50	2
334455	10/01/97	£12.00	10:10	£0.40	1
334455	10/01/97	£12.00	10:50	£1.50	8
334455	10/01/97	£12.00	12:55	£2.60	15
334455	10/01/97	£12.00	13:15	£6.40	45
334455	10/01/97	£12.00	16:00	£7.00	50
334455	11/01/97	£8.00	09:00	£0.30	4
334455	11/01/97	£8.00	09:10	£0.60	3
334455	11/01/97	£8.00	09:55	£1.20	7
334455	11/01/97	£8.00	15:10	£1.00	10
334455	12/01/97	£7.00	11:03	£2.30	17
334455	12/01/97	£7.00	12:03	£2.80	16
334455	12/01/97	£7.00	13:10	£4.00	30

2nd case: multiple-relation normalised database

CUSTOMERS

customer_id	customer_name	customer_address	customer_telephone	customer_credit_card
1	Jones	88 Bond Street	171334334	334455

LUNCH

customer_id	date	lunch
1	10/01/97	£12.00
1	11/01/97	£8.00
1	12/01/97	£7.00

TELEPHONE_CALLS

customer_id	date	time	telephone	duration
1	10/01/97	10:00	£0.50	2
1	10/01/97	10:10	£0.40	1
1	10/01/97	10:50	£1.50	8
1	10/01/97	12:55	£2.60	15
1	10/01/97	13:15	£6.40	45
1	10/01/97	16:00	£7.00	50
1	11/01/97	09:00	£0.30	4
1	11/01/97	09:10	£0.60	3
1	11/01/97	09:55	£1.20	7
1	11/01/97	15:10	£1.00	10
1	12/01/97	11:03	£2.30	17
1	12/01/97	12:03	£2.80	16
1	12/01/97	13:10	£4.00	30

3rd case: object database

CUSTOMER

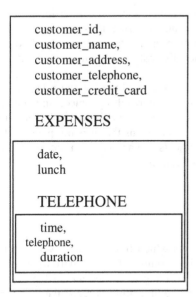

customer_id,
customer_name,
customer_address,
customer_telephone,
customer_credit_card

EXPENSES

date,
lunch

TELEPHONE

time,
telephone,
duration

4th case: object database

CUSTOMER

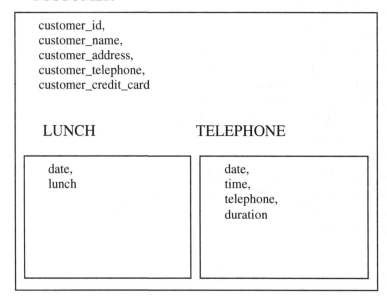

Our algorithm is aimed at identifying the 3rd case given the 1st case as input.

In the 3rd and 4th cases, we have the most semantically rich databases, in the sense that we have all the concepts and their subsumption inter-relationships identified and stored in the database schema. In our approach the 1st case is the source schema whereas the 3rd case is the target.

This algorithm can identify only the set of simple attributes in the relation that correspond to the current level of construction of the resulting object and not the object attributes. The object attributes are formed using the remaining attributes of the relation after the extraction of simple attributes. That means that it is not possible for our algorithm in its current state to derive the objects of case 4, given the relation of case 1 as input. This issue is discussed further in section 6.

3   The Algorithm

In outline, the algorithm searches for attribute sets which have common values for a significant number of tuples. This notion is captured by the definition of a "Frequency" for each candidate set of attributes. Having identified an attribute set of interest as reflecting a potential semantic object, the algorithm is recursively applied to each object to identify further potential objects nested therein.

First the algorithm is presented in section 3.1 and then in section 3.2 we present the output of each stage applied to the example relation of 2.1.

3.1 Algorithm Definition

Step 1: Identify candidate Attribute Sets

Let $A = \{A_1, A_2, A_3 .. A_n\}$, where n is the number of attributes in a relation R.

Consider a relation R (A)

We define a function `getCombinations` that given A as an argument returns the $\wp A$ (powerset of A).

   `getCombinations (A)` $= \wp A$

$\wp A$ represents the set of all the candidate sets of attributes for the relation R.

Step 2: Compute the Frequencies for each candidate set.

First of all we need to define the notion of the term Frequency in this context.

Let As $\subseteq$ A.

We define the function `executeSQL` that receives an Attribute Set and a Relation as arguments and returns a set of integer values V.

   `executeSQL (R, As)`

```
Let V be the set of values that are returned by the
following SQL statement performed on the Relation R:

            SELECT COUNT(*)
            FROM R
            GROUP BY As

            return V

End Function
```

We define F (As), the Frequency of As in R, to be the quotient of MAX (V) over STDV (V) +1 where MAX (V) is the maximum value in V and STDV (V) is the standard deviation of all the values that are members of V.

Now that we have the definition of the Frequency of an Attribute Set we can also define a function `getFrequencies` that given $\wp A$ returns the list of pairs of values P where:

   $P = <<x_1, y_1>, <x_2, y_2>, ...<x_n, y_n>> \mid \forall 1 \le i \le n, y_i \in \wp A \wedge x_i = F(y_i)$

The definition of `getFrequencies` itself uses set E to represent the Attribute Sets for which the system cannot identify a "significant" Frequency. E stands for "Exceptions". In the algorithm as it is currently implemented, a MAX(V) value greater than 1 is considered significant, though this is a parameter which could be adjusted to control the threshold at which a Frequency is considered to be significant.

$E = \{As_1, As_2, .. As_n\} \mid$
$\forall 1 \le i \le n, As_i \in \wp A \wedge MAX(executeSQL(R, As_i)) = 1$

The function getFrequencies is defined as follows.

```
getFrequencies(℘A)
          Let E=∅
          Let S=∅
          Let T=℘A-{∅}

          Loop while T≠∅
            Pick some As∈T
            T=T-{As}

            If(¬∃As' | As'∈E ∧ As'∩As≠∅)
            V=executeSQL(R,As)

              If (MAX(V)=1)
                E=E∪{As}
              Else S=S∪{<F(As),As>}

            End If

          End Loop

          //sort the elements of S to form list P
          ∀ 1≤i≤n, <xᵢ,yᵢ>∈S → <xᵢ,yᵢ>∈P ∧
          ∀ 1≤i<n, <xᵢ,yᵢ>∈P (xᵢ>xᵢ₊₁ ∨ (xᵢ=xᵢ₊₁ ∧
          card(yᵢ)≥ card(yᵢ₊₁)))

          return P
    End Function
```

The list P is in decreasing order. The first criterion that is used to maintain the order is the Frequency value and the second is the cardinality of the Attribute Sets. For every pair of values that have the same Frequency, the pair which contains the Attribute Set with the bigger number of attributes comes first in the list.

Step 3: Form the Semantic Object.

We use two functions addSimpleAttribute(O,At) and addObjectAttribute(O,Ats) that are used to add simple and object attributes to a Semantic Object O respectively. Note that At is a member of A, whereas Ats is a subset of A.

We define the function formSemanticObj(A,P) that receives the set of Attributes A that Relation R consists of and the list P as arguments and returns the semantic object that corresponds to R.

```
formSemanticObj(A,P)
          Let O be an empty semantic object
          Let M=∅

          If A is a singleton
            Let At∈A
            Call addSimpleAttribute(O,At)
            return O
          End If

          If P=<>

            Loop while A≠∅
              Let At∈A
              A=A-At
              Call addSimpleAttribute(O,At)
            End Loop

            return O
          End If
```

//Let M be the Attribute Set associated with the first
item in the list P.

$$M=y \mid <x,y>=HEAD(P)$$

```
          Loop while ∃ <x,y>|<x,y>∈P∧ y∩M≠∅ (y and M
          are not disjoint)
            Remove <x,y> from the list P
          End Loop
```

Let P' be the list with the remaining elements of P.

Let A'=A-M be the set that consists of the remaining
attributes in A.

```
          Loop while M≠∅
            Let At∈M
            M=M-{At}
            Call addSimpleAttribute(O,At)
          End Loop

          Call addObjectAttribute(O,A').
          return O
End Function
```

Step 4: Continue recursively the formation of the Semantic Object to the next level

Let R'(A') be the relation R' which now comprises the attributes A' of the object
attribute Oa, which was created in step 3.

Recursively call step 3 with formSemanticObj(A',P')

The recursive repetition of the same process for the object attribute of the resulting Semantic Object will take care of all the subsumed concepts as well as the subsumption relationships that connect them with their parent concepts.

3.2   A Worked Example

In this section we present in stages the operations that take place if we apply the algorithm on the example relation of section 2.3.

Step 1: Identify candidate Attribute Sets

```
A={customer_id, customer_name, customer_address,
customer_telephone,customer_credit_card, date, lunch,
time, telephone, duration}
```

$\wp A$ ={{customer_id}, {customer_name},
{customer_address}, {customer_telephone},
{customer_credit_card}, {date}, {lunch}, {time},
{telephone}, {duration}, {customer_id, customer_name},
{customer_id, customer_address},....
{customer_id, customer_name, customer_address,
customer_telephone,customer_credit_card, date, lunch,
time, telephone, duration}}

Step 2: Compute the Frequecies for each candidate set.

For simplicity we numbered the attributes and we use their corresponding numbers instead of their names.

```
P=<<13.0,{1,2,3,4,5}>,<13.0,{2,3,4,5}>,<13.0,{1,3,4,5}>
,<13.0,{1,2,4,5}>,<13.0,{1,2,3,5}>,<13.0,{1,2,3,4}>,
<13.0,{3,4,5}>,<13.0,{2,4,5}>,<13.0,{2,3,5}>,
<13.0,{2,3,4}>,<13.0,{1,4,5}>,<13.0,{1,3,5}>,
<13.0,{1,3,4}>,<13.0,{1,2,5}>,
```

.

.

```
<2.67,{1,2,3,4,5,6,7}>,<2.67,{2,3,4,5,6,7}>,
<2.67,{1,3,4,5,6,7}>,<2.67,{1,2,4,5,6,7}>,
```

.

.

```
<2.67,{1,7}>,<2.67,{1,6}>,
<2.67,{7}>,<2.67,{6}>>
```
Step 3: Form the Semantic Object.

We decide which attributes will be simple.

```
M={1,2,3,4,5}
```

We then reduce P by the pairs for which the Attribute Set is not disjoint with M and we get the list P'.

P'=<<2.67,{6,7}>,<2.67,{7}>,<2.67,{6}>>

And finally we create the object

CUSTOMERS

customer_id,
customer_name,
customer_address,
customer_telephone,
customer_credit_card

EXPENSES

date,
lunch,
time,
telephone,
duration

Step 4: Continue recursively the formation of the Semantic Object to the next level
We consider now the object attribute Oa as the new Relation R'.

Therefore, A'={date, lunch, time, telephone, duration}.

We skip the steps 1 and 2 for efficiency purposes and we go directly to step 3. This is part of the optimizations that have been applied on the algorithm for which a more detailed presentation is given in 4.3.

SECOND - Step 3: Form the Semantic Object.

We decide which attributes will be simple.

M={6,7}

We then reduce P by the pairs for which the Attribute Set is not disjoint with M and we get the list P'.

P'=<>

And then we continue the formation of the object.

CUSTOMERS

```
customer_id,
customer_name,
customer_address,
customer_telephone,
customer_credit_card

EXPENSES

    date,
    lunch

TELEPHONE

     time,
  telephone,
    duration
```

SECOND - Step 4: Continue recursively the formation of the Semantic Object to the next level

We consider now the object attribute `Oa` as the new Relation.

Therefore, `A'`={time, telephone, duration}.

We skip the steps 1 and 2 for efficiency purposes and we go directly to step 3.

THIRD - Step 3: Form the Semantic Object.
Since, `P=<>` we terminate.

From the above process three objects are derived.

The object CUSTOMER consists of five Simple Attributes (customer_id, customer_name, customer_address, customer_telephone, customer_credit_card) and one object attribute that references the object EXPENSES. The object EXPENSES is subsumed by CUSTOMER, with attributes concerning customer expenses (date, lunch and TELEPHONE). Finally, the object TELEPHONE is subsumed by the object EXPENSES, with attributes that are related to telephone expenses (time, telephone, duration). We believe that one-way relationship does not make sense in the semantic object model. That means that if an object of class A is related to an object of class B, then B must be related to A as well. Therefore, since it is not possible for a simple object to participate in a relationship with another object, all of the above objects are considered compound objects.

4   Implementation of the Algorithm

4.1 Architecture

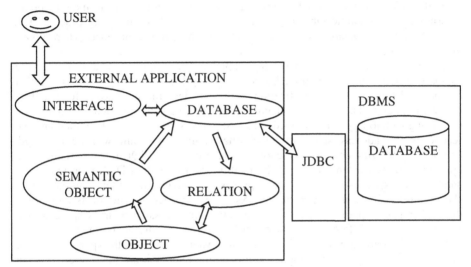

Fig. 1. Architecture of the System

The main architectural components of the system are the following:

Interface: The name of this component is self-explanatory. It is used only to facilitate the interaction between the user and the application. There are two pieces of information that need to be passed to the system by the user through this interface. First of all the application needs to know the location of the database and the user information in order to establish a connection with the database. After a connection has been successfully established and information about the database has been extracted from the DBMS, the user is required to choose which relation is going to be used by the mapping algorithm.

Database: The Database component represents the particular database that has been chosen by the user and encapsulates all the relevant information about it (Catalogs, Schemas etc.) The Database is the component that holds the connection to the DBMS and through which the application interacts with it.

Relation: The Relation component represents the relation to be mapped. This is the relation that has been chosen by the user among all the relations that belong to the Database. Part of the mining algorithm is integrated with the functional part of the Relation object. The Relation component can dynamically retrieve from the DBMS and compute all the information that is needed by the Object Engine component to make decisions concerning the mapping process.

Object Engine: This component is responsible, given all the information necessary, to decide how to decompose the given relation into smaller, less complex, structures and derive the corresponding object(s). The appropriate Relation component is passed to this component by the Database after the user's selection. Then, the Object Engine uses the information stored in the Relation component to decide how to divide it up into smaller structures.

Semantic Object: This component is the result of the mapping procedure and it represents the object-oriented equivalent of the Relation component. This component will contain, depending on the data configuration of the original relation, information about the object(s) and their subsumption relationships that derived from the mapping algorithm. The Semantic Object, after the completion of the mapping process, gets passed back to the Database component and from there to the interface, where it gets displayed in the form of a report to the user.

There are many different ways to couple a data-mining mechanism with the database system that is to be mined. According to [5] there are two main criteria that can be used for a general classification of the architectural alternatives for coupling two such systems. The first is the degree to which the coupling system can make use of existing technology that is embedded in the DBMS that needs to be mined and the second is the address space in which the mining algorithm will be executed. Considering the first criterion we chose to make use of an SQL-related architectural option that makes extensive use of the facilities that a modern DBMS provides (i.e. database indexing capabilities, query processing capabilities, SQL parallelisation, checkpointing, space management etc.) As far as the second criterion is concerned, the SQL-related options can either be executed in the same address space as the DBMS or at a different one, depending on the requirements of the specific application. There are advantages and disadvantages for both approaches. For instance, the advantages of using a loose-coupling approach are that there is greater programming flexibility and that there are no extra storage requirements for the DBMS, whereas the disadvantage is the high context-switching cost between the DBMS and the mining process. We chose to implement our algorithm using a SQL cursor interface, loosely coupled approach. Our system is written as an external Java application, that queries an ORACLE server v8 through a JDBC interface.

4.2 Testing

The example relation presented in 2.1 corresponds to one used for experimental purposes. After the algorithm was applied on the relation of case 1, it produced the object presented in case 3.

Our algorithm has also been tested with relations from real databases with bio-molecular data. The size of these relations ranged from approximately 2,000 to 10,000 rows and 2 to 20 columns. The results again were promising, since in most of the cases the derived objects had a 'meaningful' structure.

4.3 Implementation Issues

It is apparent that the computational complexity of the algorithm is an exponential function of the number of attributes that the initial relation consists of. In order to reduce the computational workload that this algorithm implies, especially with very "wide" relations, we applied a number of optimizations to it.

An important optimization that took place is related to the getFrequencies function. The getFrequencies function is the only part of the algorithm that makes extensive use of an SQL cursor interface and queries the database for all the possible combinations of attributes. It is obvious that the degree to which we can reduce the number of queries that are performed by this function can affect significantly the overall performance of the application. The first decision was to keep track of all the candidate Attribute Sets for which the maximum unique occurrence in the relation was 1 and to exclude all the Attribute Sets that were not disjoint with them from the set of Candidate Sets.

In the description of the getFrequencies function the set E which stands for Exceptions is the set that keeps track of the Attribute Sets that do not carry useful semantic information for our purposes and that should be excluded from subsequent searches.

The second optimization that took place was to develop the structure P only once in each execution of the algorithm. During the formation of the first level of the resulting object (first iteration), the system develops the structure P for all the possible combinations of Attribute Sets for the given relation. After the formation of the first level, the structure P is reduced by the Attribute Set that corresponds to the simple attributes of the newly created object and all the Attribute Sets that are not disjoint from it. In subsequent iterations the system uses the same structure P since there is no change in the underlying data and therefore there is no need to query the system again to create a new one for the remaining attributes of the initial relation.

After the optimizations took place and the algorithm was tested again on the same data sets, the results showed that there was a very significant improvement in efficiency terms. According to the figures we obtained the optimized algorithm could provide the same results on the same data using on average 60% of the time that its predecessor needed. That of course does not mean that we managed to overcome the complexity problem, which remains the most serious disadvantage of this method, since it is inherent to the nature of the algorithm.

Another important decision was the order of the <Frequency,Attribute Set> pairs in the structure P. The structure P is maintained as a bag-like singly linked list, where multiple keys with the same value are allowed. That means that a second criterion for maintaining an order between the Attribute Sets with the same key in the structure can be used. We decided this second criterion to be the size of the Attribute Set. That means that by default the system favors bigger Attribute Sets. When an attempt is made to insert in the structure an Attribute Set which is associated with a key that already exists in the list, the system checks whether the size of the Attribute Set is bigger than the size of the existing one(s) and places it accordingly. If we decided not to do that we again could end up with relatively meaningful results, but with certain data configurations we would end up with more levels of nesting that necessary. To understand why the system would behave like that you have to consider the order that is used by the system to process each individual combination of attributes. The system processes the "smaller" Attribute Sets first, which means that if we did not favor the bigger ones when constructing the linked list we would have the smaller Attribute Sets placed before the bigger ones with the same key in the list. Given the example in 2.1, it is clear that in this situation we would end up with a set of objects, each one nested into the other, representing each individual property of CUSTOMER separately which is not incorrect but not entirely sensible either.

5   Comparison with Other Approaches

The technique of searching for partitions of rows on the basis of attribute values has been used to identify functional and approximate functional dependencies in previous work [6]. In our approach, we use partitioning recursively to identify more general structures than those captured by such dependencies, representing potential semantic objects. For example, a relation with no non-key attributes and hence no non-trivial functional dependencies may nonetheless result in discovered object structures with our approach.

One other approach, which adopts a knowledge discovery technique for identifying structures within a database, is [1]. However, that work assumes that the source relational schema is at least in 2NF, which means that non-2NF based relations are excluded from the mapping. Since in real world systems it is likely that both 2NF and non-2NF relations will be found, we believe it is necessary for such a system to be equipped with the necessary tools to deal with this problem. Also other systems require manual intervention and that categorises them as the so-called semi-automatic systems [7][8][9]. Our attempt is to provide a fully automatic system to perform these tasks. The problem we face in non-2NF relations is that the data is complex. A single non-2NF relation can incorporate within it multiple concepts which may be interconnected with multiple and different types of associations. Existing algorithms for normalised schemas usually define transformation functions that can be applied on a source schema and derive the target schema. The transformations are usually collections of rules that map concepts from the source data model to the target data model [1][10]. In our approach we start from a single concept and we develop automatically (without any manual intervention) the concepts that are hidden in the relation as well as the subsumption relationships between them by mining the data.

6   Conclusions — Suggestions for Further Research

We have presented a discovery algorithm that searches for potential semantic object structures in a single non-2NF relation. We believe that the algorithm should be applied on a per-relation basis as part of a wider design approach rather than as a stand-alone mapping mechanism.

The algorithm identifies "repeating groups", not in the sense of "multi-valued" attributes, but in the sense of Attribute Sets with values that are repeated in the relation. As such, this algorithm can identify only the set of simple attributes in the relation that correspond to the current level of construction of the resulting object and not the object attributes. It would be useful to be able to identify the individual object attributes instead and to form the set of simple attributes after, using the remaining attributes in the relation. The argument that favors the second approach is that with certain data configurations it may be more logical to construct an object with two or more object attributes at the same level. At the stage that our algorithm currently is, this is not possible. Therefore, we believe that this aspect of the problem needs more investigation and experimentation with real data.

A second aspect that needs further research is the scope for further optimisation. Since the complexity of the algorithm is inherent to its nature, the more we can improve the performance of the algorithm through optimisation the more useful the algorithm can be, since it will be possible to be used for relations with bigger number of attributes.

A third aspect that also needs careful consideration is the automation and its consequences with certain data configurations. The fact that the algorithm relies solely on the data and it does not use any information provided by either the DBMS or the user could be proved dangerous with certain data configurations. The example single-relation non-normalised database presented in section 2.1 could be an example of that problem. Consider the unlikely but possible situation where the customer "Jones" was charged for every phone call he made 0.50 pounds. Having all the rest of the values in the relation fixed, we would end up with the attribute "telephone" being a Simple Attribute in the first level. That would mean that "telephone" would be considered by the system as part of the personal information about the customer, which is clearly incorrect. If the occurrence of the same value in "telephone" were not that frequent but frequent enough to be considered meaningful by the system, we would again face a similar problem as above but at a different level of nesting. In this case of course the problem is related to the size of the data. We expect problems of that magnitude to be eliminated with very large data sets because the probability that a particular Attribute Set will be distinguished because of an "accidental" difference in frequency in comparison with other Attribute Sets of the same "significance" is not great.

References

1. Ramanathan, S., Hodges, J.: Extraction of Object-Oriented Structures from Existing Relational Databases. ACM SIGMOD Record, Vol. 26 (1997) 59–64
2. Hammer, M., McLeod, D.: Database Description with SDM: A Semantic Database Model. ACM Transactions on Database Systems, Vol. 6 (1981) 351–386
3. Kroenke, D.: Database Processing. 7th edn. Prentice Hall International, Inc., USA (2000)
4. Abiteboul, S., Hull, R.: IFO: A Formal Semantic Database Model. ACM Transactions on Database Systems, Vol. 12 (1987) 525–565
5. Sarawagi, S., Thomas, S., Agrawal, R.: Integrating Association Rule Mining with Relational Database Systems: Alternatives and Implications. ACM SIGMOD Conference (1998) 343–354
6. Huhtala, Y., Kärkkäinen, J., Porkka, P., Toivenen, H.: Efficient Discovery of Functional and Approximate Dependencies Using Partitions. ICDE (1998) 392–401
7. Jahnke, J., Schafer, W., Zundorf, A.: Generic Fuzzy Reasoning Nets as a Basis for Reverse Engineering Relational Database Applications. 6th European Software Engineering Conference/5th ACM SIGSOFT Symposium of Foundations of Software Engineering (1997) 193–210
8. Behm, A., Geppert, A., Dittrich, K.: Algebraic Database Migration to Object Technology. Technical Report 99.05, Department of Computer Science, University of Zurich (1999)
9. Roddick, J., Craske, N., Richards, T.: Handling Discovered Structure in Database Systems. IEEE Transactions on Knowledge and Data Engineering, Vol. 8 (1996) 227–240
10. Akoka, J., Comyn-Wattiau, I., Lammari, N.: Relational Database Reverse Engineering, Elicitation of Generalization Hierarchies. International Workshop on Reverse Engineering in Information Systems (1999) 173–185

Estimation of Uncertain Relations between Indeterminate Temporal Points

Vladimir Ryabov and Seppo Puuronen

Department of Computer Science and Information Systems,
University of Jyväskylä, P.O.Box 35, FIN-40351, Jyväskylä, Finland
{vlad, sepi}@jytko.jyu.fi

Abstract. Many database applications need to manage temporal information and sometimes to estimate relations between indeterminate temporal points. Indeterminacy means that we do not know exactly when a particular event happened. In this case, temporal points can be defined within some temporal intervals. Measurements of these intervals are not necessarily based on exactly synchronized clocks, and, therefore, possible measurement errors need to be taken into account when estimating the temporal relation between two indeterminate points. This paper presents an approach to calculate the probabilities of the basic relations (before, at the same time, and after) between any two indeterminate temporal points. We assume that the probability mass functions of the values of these two temporal points and the maximum measurement difference between these values are known. Using these assumptions we derive formulas for the lower and the upper bounds of the probabilities of the basic relations between two indeterminate temporal points.

1 Introduction

Time is important in modeling dynamic aspects of the world. Temporal formalisms are applied, for example, in natural language understanding, planning, process control, and temporal databases, i.e. in the areas, where the time course of events plays an important role. Even though representation and reasoning about temporal knowledge has already achieved significant results, there still exist topics in the field of temporal databases, which require and deserve further research attention.

Temporal events are often represented using temporal points as ontological primitives as it is in temporal databases, where each event is associated with a timestamp that indicates when the event occurred. Many of the published articles have been based on the assumption that complete and accurate temporal information is available. Generally, the proposed approaches give limited support for situations in which imperfect temporal information exists.

One kind of imperfect temporal information is indeterminacy [5], which means that we do not know exactly when a particular event happened. Indeterminacy can arise from different sources, as mentioned in [5]:

- granularity mismatch (when an event is measured in one granularity, and is recorded in a system with a finer granularity);
- clock measurements (every clock measurement includes some imprecision);

T. Yakhno (Ed.): ADVIS 2000, LNCS 1909, pp. 108-116, 2000.

- dating techniques (many dating techniques are inherently imprecise);
- unknown or imprecise event times (in general, occurrence times could be unknown or imprecise);
- uncertainty in planning (projected dates are often inexactly specified).

Motro [7] suggests that imperfection can also result from an unreliable source, a faulty sensor, an input error, or the inappropriate choice of representation. Kwan et al. [6] also mention a number of sources of uncertainty and incompleteness in databases used in scientific applications. For example, some data is recorded statistically and so is inherently uncertain, whilst other data is deliberately made uncertain for reasons of security, and some data can not be measured accurately, due perhaps to some mechanical effect.

Often the exact value of a temporal point is unknown, but the interval of possible values is given. Such a temporal point is referred to as an indeterminate temporal point in this paper. We assume that the probability mass function (p.m.f.) and thus the probability distribution of the values of the indeterminate temporal point is known. We also assume that the measurement of the values of the indeterminate temporal point can include some imprecision, and the value of the maximum measurement error is known.

Often there is a need to know the relation between two temporal points, for example, during query processing when we need to estimate the temporal relation between two events. When the points are indeterminate it is in general case impossible to derive a certain relation between them. Based on the above assumptions we present formulas, which calculate the lower and the upper bounds of the probabilities of the basic relations: "<" (before), "=" (at the same time), and ">" (after) between any two indeterminate temporal points

In Section 2 we define the basic concepts used in the paper. Section 3 presents the formulas for the basic relation "at the same time". In Section 4 we present the formulas for the other two basic temporal relations. Section 5 presents an example, Section 6 includes a discussion about related research, and in Section 7 we make conclusions.

2 Basic Concepts

In this Section we present the main concepts used in the paper.

The various models of time are often classified as discrete, dense, and continuous models. We use the discrete model, which is commonly used in the field of temporal databases. In the discrete time model temporal point is the main ontological primitive. Temporal points are isomorphic to natural numbers, i.e. there is the notion that every point has a unique successor. The time between two points is known as a *temporal interval*. A *chronon* is an indivisible time interval of some fixed duration, which depends on the application. A time line is represented by a sequence of chronons of identical duration. Ontologically, a temporal point is located during one chronon.

Definition 1. A temporal point is *determinate* when it is exactly known during which particular chronon it is located. An *indeterminate* temporal point $\mathbf{a}$ is such that $\mathbf{a} \in [\mathbf{a}^l, \mathbf{a}^u]$, where $\mathbf{a}^l$ (the lower bound) is the first chronon of the interval $[\mathbf{a}^l, \mathbf{a}^u]$, $\mathbf{a}^u$ (the upper bound) is the last chronon of the interval, and $a^l \leq a^u$. An indeterminate temporal point **a** is attached with *the probability mass function* (p.m.f.) $f(x)$ so, that

$f(x)=0$ when $x<a^l$ or $x>a^u$; $f(x)\in[0,1)$ and $\displaystyle\sum_{x=a^l}^{a^u} f(x) = 1$ when $x\in[a^l,a^u]$, $f(a^l)>0$, and $f(a^u)>0$.

The requirement that the sum of the probabilities $f(x)$ is equal to 1 results from the definition of our time ontology, according to which, a temporal point is located during exactly one particular chronon.

Generally, the p.m.f. stems from the sources of indeterminacy, such as granularity mismatch, dating and measurements techniques, etc. When the granularity mismatch is the source of indeterminacy the uniform distribution is commonly assumed. For example, if an event is known in the granularity of one hour then in a system with the granularity of one second the temporal point is indeterminate, and we have no reason to favor any one second over another. Some measurement techniques or instruments can have fixed trends in measurements, for example, the normal distribution of the measurement error. In some situations, the analysis of past data can provide a hint to define the p.m.f. For example, we may know that a particular type of event in a particular situation tends to occur during the last chronons of the interval.

Several other means of determining the p.m.f. were suggested by Dey and Sarkar [2]. Also Dyreson and Snodgrass [5] point out that in some cases a user may not know the underlying mass function because that information is unavailable. In such cases the distribution can be specified as missing, which represents a complete lack of knowledge about the distribution and then, one of the above mentioned means of estimating the p.m.f. can be applied.

Definition 2. The temporal relation between two determinate temporal points is exactly one of the basic relations: "before" ($<$), "at the same time" ($=$), and "after" ($>$). In general case, more than one of the basic relations can hold between two indeterminate temporal points. This kind of relation is called an *uncertain temporal relation* between two points. Let an uncertain relation between two indeterminate temporal points **a** and **b** be represented by a vector $(e^<,e^=,e^>)$, where the value $e^<$ is the probability that **a**<**b**, the value $e^=$ is the probability that **a**=**b**, and the value $e^>$ is the probability that **a**>**b**. The sum of $e^<$, $e^=$, and $e^>$ is equal to 1, since these values represent all the possible basic relations that can hold between **a** and **b**.

In some situations the measurements of the values of the indeterminate temporal points are not based on exactly synchronized clocks, and in this case it is impossible to estimate the probabilities of the basic relations precisely.

Definition 3. Let ε_{max} be the maximum measurement difference, expressed as a number of chronons, between the measurements of the values of two temporal points. Then, the error of estimation $\varepsilon\in[0,\varepsilon_{max}]$.

Taking into account different values of ε we can calculate the maximum and the minimum values of the probabilities of the basic relations. In the next section we propose formulas for the lower and the upper bounds of the temporal relation "=" when the measurement error is taken into account.

3   Lower and Upper Probabilities of the Temporal Relation "At the Same Time"

In this section we present the lower and the upper probabilities of the temporal relation "at the same time" between two indeterminate temporal points $a \in [a^l, a^u]$ and $b \in [b^l, b^u]$, attached with the p.m.f.s $f_1(a)$ and $f_2(b)$ correspondingly. Let ε_{max} be the maximum measurement difference between the values of a and b expressed as a number of chronons. The probability of a certain pair of temporal points a and b from the intervals $[a^l, a^u]$ and $[b^l, b^u]$ correspondingly is $f_1(a) \times f_2(b)$. Each pair of the values of the temporal points contributes to one of the three basic relations depending on the values of a, b, and ε. The temporal relation "at the same time" is supported by all the pairs of a and b that can be located during the same chronon. This means that the upper limit of the probability is achieved when the probabilities of all the possible equal pairs taking the maximum measurement error into account are summed up as it is shown by formula (1):

$$ e^{=} = \sum_{a=a^l}^{a^u} \left\{ f_1(a) \times \sum_{b=a-\varepsilon_{max}}^{a+\varepsilon_{max}} f_2(b) \right\}. \tag{1} $$

Formula (1) gives in general case too high upper limit, which can not be achieved. More sharp upper limit is achieved taking into account that, in reality, two time measurements must have a fixed temporal measurement difference between their values during a certain period of time. This temporal difference, expressed as a number of chronons, between the values is exactly one value within the interval $[-\varepsilon_{max}, \varepsilon_{max}]$, but this value is unknown.

Therefore, the sharp upper and lower limits can be calculated using formulas (2) and (3) correspondingly.

$$ e^{=}_{max} = \max\left(\sum_{a=a^l}^{a^u} \left\{ f_1(a) \times f_2(a+t) \right\} \right), t \in \left[-\varepsilon_{max}, \varepsilon_{max} \right]. \tag{2} $$

$$ e^{=}_{min} = \min\left(\sum_{a=a^l}^{a^u} \left\{ f_1(a) \times f_2(a+t) \right\} \right), t \in \left[-\varepsilon_{max}, \varepsilon_{max} \right]. \tag{3} $$

In the next section we present the upper and the lower probabilities of the other two basic relations between points a and b.

4   Lower and Upper Probabilities of the Temporal Relations "before" and "after"

In this section we derive formulas for the lower and the upper probabilities of the basic relations "before" ($<$) and "after" ($>$) between points a and b.

To derive the lower and the upper probabilities of the relations "before" and "after" we need to consider different values of the measurement error, as it was done in the previous section. The upper probability of the relation "before" is achieved when the sum, defined by formula (4) includes the maximum number of probability values from the interval $[\mathbf{b'},\mathbf{b''}]$. This happens when the summation takes into account the maximal possible error plus one chronon:

$$e_{max}^{<} = \sum_{a=a'}^{a''}\left\{\mathbf{f}_1(\mathbf{a}) \times \sum_{b=a-\varepsilon_{max}+1}^{b''}\mathbf{f}_2(\mathbf{b})\right\}. \tag{4}$$

Similarly, we can obtain the upper probability of the relation "after" defined by formula (5):

$$e_{max}^{>} = \sum_{b=b'}^{b''}\left\{\mathbf{f}_2(\mathbf{b}) \times \sum_{a=b-\varepsilon_{max}+1}^{a''}\mathbf{f}_1(\mathbf{a})\right\}. \tag{5}$$

The lower probability of the relation "before" is achieved when the second sum includes as fewer probability values as possible taking the maximum measurement error into account as it is shown by formula (6):

$$e_{min}^{<} = \sum_{a=a'}^{a''}\left\{\mathbf{f}_1(\mathbf{a}) \times \sum_{b=a+\varepsilon_{max}+1}^{b''}\mathbf{f}_2(\mathbf{b})\right\}. \tag{6}$$

Similarly, we can obtain the lower probability of the relation "after", defined by formula (7):

$$e_{min}^{>} = \sum_{b=b'}^{b''}\left\{\mathbf{f}_2(\mathbf{b}) \times \sum_{a=b+\varepsilon_{max}+1}^{a''}\mathbf{f}_1(\mathbf{a})\right\}. \tag{7}$$

In the next section we present an example, which shows how the proposed formulas can be used.

5 Example

In this section we consider an example of using the proposed formulas to estimate the uncertain relation between two indeterminate temporal points.

Let us suppose that a tube manufacturing plant keeps records about the production process, and that they are included in the temporal databases, fragments of which we consider in this example. The granularity of the databases is assumed to be one hour. We further assume that tubes are normally made of steel blanks that are delivered to the plant by warehouses. Usually, tubes are made in series, and for each series the production record includes: "Series ID" - the identity number of the produced series of tubes, "Start of production" - the temporal interval defining the start of the production of the particular series, "End of production" - the temporal interval defining the end of the production, and "Defective tubes" - the percentage of defective tubes achieved for each produced series (Table 1).

Table 1. Production of tubes by a tube manufacturing plant

Series ID	Start of production	End of production	Defective tubes
#T1	4 - 11	25 - 31	11
#T2	22 - 35	45 - 55	2
#T3	35 - 50	60 - 70	3

It is supposed that the start of the production is a temporal point from the interval "Start of production". For each series the percentage of defective tubes is achieved, which normally should be less than 10.

Table 2 presents a fragment of the database that keeps records about the delivery of steel blanks to the plant by warehouses.

Table 2. Delivery of steel blanks

Series ID	Delivery
#B1	1 - 10
#B2	12 - 16

Each series of steel blanks has a unique "Series ID" number included in the first column of Table 2. The second column specifies the delivery period, during which, the series of blanks arrived to the tube manufacturing plant.

If the percentage of defective tubes for a produced series is above 10, then we are interested in which series of blanks was used in the production of that series. To answer the question we need to estimate the temporal relation between the start of the production of the series of tubes and the delivery interval for a series of blanks. We suppose that steel blanks could be used in the production of tubes if they arrived to the plant exactly at the day when the production of tubes started. The basic temporal relation "at the same time" supports this assumption, and it's probability is the probability that the particular series of blanks could be used.

The series of tubes #T1 has a percentage of defective tubes equal to 11. Let us estimate the probability that the series of blanks #B1 was used in the production of #T1. Let us denote the time when the series of blanks #B1 arrived to the plant as a temporal point $a[1,10]$, and the start of the production of the series of tubes #T1 as a temporal point $b[4,11]$.

We assume that all chronons within the intervals $a[1,10]$ and $b[4,11]$ are equally probable, which means that $f_1(a)=0.1$ when $a\in[1,10]$, and $f_1(a)=0$ when $a\notin[1,10]$. Similarly, $f_2(b)= \frac{1}{8}$ when $b\in[4,11]$, and $f_1(b)=0$ when $b\notin[4,11]$. Let the value of the maximal measurement error ε_{max} be equal to 2. Our goal is to estimate the minimal and the maximal probabilities that the indeterminate temporal point a is "at the same time" with the indeterminate temporal point b.

The upper probability $e_{max}^=$ of the relation "at the same time" is calculated by formula (1) with $\varepsilon_{max}=2$, and in this case $e_{max}^=$ is equal to 0.425. Using the sharp formula (2) $e_{max}^=$ is equal to 0.1 both with $\varepsilon=2$ and $\varepsilon=1$. The lower probability $e_{min}^=$ of the relation "at the same time" is calculated by formula (3) and in this case, $e_{min}^=$ is equal to 0.0625. Using formulas (4), (5), (6), and (7) from Section 4 the lower and the upper probabilities of the relations "before" and "after" can also be calculated. So, the lower probability of "before" is equal to 0.45, the upper probability of "before" is equal to 0.8125, the lower probability of "after" is equal to 0.125, and the upper probability of "after" is equal to 0.45 (Figure 1).

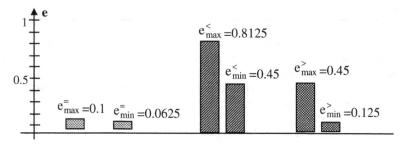

Fig. 1. Extreme values of the probabilities of the basic relations between points **a** and **b**

Now we can answer the question what is the probability that the series of blanks #B1 was used in the production of the series of tubes #T1. The minimum value of this probability is 0.0625, and the maximum value is 0.1 when the estimation error is equal to 2.

6   Related Research

Many researchers deal with imperfect information and various approaches are mentioned in the bibliography on uncertainty management by Dyreson [4], in the surveys by Parsons [8], and by Parsons and Hunter [9], although not many of them consider temporal imperfection. Formalisms intended for dealing with imperfect information are often distinguished as symbolic and numerical. Among the numerical approaches the most well known are: probability theory, Dempster-Shafer's theory of evidence [11], possibility theory [3], and certainty factors [12]. We have selected a probabilistic approach in this paper.

van Beek [14] and by van Beek and Cohen [13] discussed representation and reasoning about temporal relations between points, and introduced the notion of indefinite temporal relation, which is a disjunction of the three basic relations, similarly as we do in this paper. They proposed also algorithms for reasoning about such relations and studied the complexity. However, no numerical means were included in their representation compared to the representation used in our paper, and temporal points were supposed to be determinate only.

The probabilistic representation of uncertain relations was already studied in Ryabov et al. [10], where an algebra for reasoning about uncertain relations was proposed. The algebra included negation, composition, and addition operations, which make it possible to derive unknown relations in a relational network combining known uncertain relations. That approach can be used together with the one proposed in this paper, that derives uncertain relations between indeterminate temporal points.

Barbara et al. [1] proposed a Probabilistic Data Model (PDM) intended to represent in a database entities whose properties can not be deterministically classified. The approach, however, is applied to relational databases and does not discuss explicitly the management of imperfect temporal information. Dyreson and Snodgrass [5] proposed a mechanism supporting valid-time indeterminacy in temporal databases, which can be seen as an extension of PDM of Barbara et al. [1]. They represent indeterminate temporal points similarly as we do in this paper, although their main concern was the development of query language.

7 Conclusions

Many database applications need to manage temporal information, which is often imperfect. Sometimes there is a need to estimate relations between temporal points, which can be indeterminate. In this paper we assume that an indeterminate temporal point is defined within the interval of possible values, and is attached with the probability mass function of the values in that interval. We further assume that the measurement error between the values of the indeterminate points might exist, and the maximum value of this measurement error is known. We proposed formulas for the probabilities of the basic relations (before, at the same time, and after) between any two indeterminate temporal points.

The proposed formulas can easily be applied in chronon-based temporal contexts as we have also shown by the example. Further research is needed to export the approach into other temporal contexts, and the benefits of the approach should be evaluated with other real word situations as well.

References

1. Barbara, D., Garcia-Molina, H., Porter, D.: The Management of Probabilistic Data. IEEE Transactions on Knowledge and Data Engineering **4**(5) (1992) 487-502
2. Dey, D., Sarkar, S.: A Probabilistic Relational Model and Algebra. ACM Transactions on Database Systems **21**(3) (1996) 339-369
3. Dubois, D., Prade, H.: Possibility Theory: An Approach to the Computerized Processing of Uncertainty. Plenum Press, New York (1988)
4. Dyreson, C.: A Bibliography on Uncertainty Management in Information Systems. In: Motro, A., Smets, P. (eds.): Uncertainty Management in Information Systems: From Needs to Solutions. Kluwer Academic Publishers, Boston (1996) 415-458
5. Dyreson, C., Snodgrass, R.: Supporting Valid-Time Indeterminacy. ACM Transactions on Database Systems **23**(1) (1998) 1-57
6. Kwan, S., Olken, F., Rotem, D.: Uncertain, Incomplete and Inconsistent Data in Scientific and Statistical Databases. In: Second Workshop on Uncertainty Management and Information Systems: From Needs to Solutions, Catalina, USA (1993)
7. Motro, A.: Sources of Uncertainty in Information Systems. In: Second Workshop on Uncertainty Management and Information Systems: From Needs to Solutions, Catalina, USA (1993)
8. Parsons, S.: Current Approaches to Handling Imperfect Information in Data and Knowledge Bases. IEEE Transactions on Knowledge and Data Engineering **8**(3) (1996) 353-372
9. Parsons, S., Hunter, A.: A Review of Uncertainty Handling Formalisms. In: Hunter, A., Parsons, S. (eds.): Applications of Uncertainty Formalisms. Lecture Notes in Artificial Intelligence, Vol. 1455. Springer-Verlag (1998) 8-37

10. Ryabov, V., Puuronen, S., Terziyan, V.: Representation and Reasoning with Uncertain Temporal Relations. In: Kumar, A., Russel, I. (eds.): Twelfth International Florida AI Research Society Conference. AAAI Press, Orlando, Florida, USA (1999) 449-453
11. Shafer, G.: A Mathematical Theory of Evidence. Princeton University Press (1976)
12. Shortliffe, E.: Computer-Based Medical Consultations: MYCIN. Elsevier, New York (1976)
13. van Beek, P.: Temporal Query Processing with Indefinite Information. Artificial Intelligence in Medicine **3** (1991) 325-339
14. van Beek, P., Cohen, R.: Exact and Approximate Reasoning about Temporal Relations. Computational Intelligence **6** (1990) 132-144

Comparison of Genetic and Tabu Search Algorithms in Multiquery Optimization in Advanced Database Systems*

Zbyszko Królikowski, Tadeusz Morzy, and Bartosz Bębel

Institute of Computing Science, Poznań University of Technology
60-965 Poznań, Piotrowo 3A, Poland
E-mail: Bartosz.Bebel@cs.put.poznan.pl

Abstract: In several database applications sets of related queries are submitted together to be processed as a single unit. In all these cases the queries usually have some degree of overlap, i.e. may have common subqueries. Therefore a significant performance improvement can be obtained by optimizing and executing the entire group of queries as a whole, thus avoiding to duplicate the optimization and processing effort for common parts. This has suggested an approach, termed *multiquery optimization (MQO)* that has been proposed and studied by several authors. In this paper we suggest a new approach to multiple-query optimization based on *Genetic* and *Tabu Search* algorithms that ensure the tractability of the problem even for very large size of the queries. To analyze the performance of the algorithms, we have run a set of experiments that allow to understand how the different approaches are sensitive to the main workload parameters.

1 Introduction

The importance of databases in non-traditional applications, such as decision support systems, is well understood and needs no explanation. Specific feature of such applications is that complex queries are becoming commonplace. These complex queries often have a lot of common sub-expressions. The scope for finding common sub-expressions increases greatly if we consider a set of queries executed as a batch. For example, SQL-3 stored procedures may invoke several queries, which can be executed as a batch [7]. Data analysis and aggregate computation in data warehouse systems often also requires a batch of queries to be executed [1, 3].

In this paper, we address the problem of optimizing sets of queries, which may have common sub-expressions. This problem is termed *multiquery optimization.*

The idea behind such a way of query optimization in advanced database systems is to minimize the total optimization and execution cost avoiding to duplicate the processing of common sub-expressions of queries. Taking advantage of these common tasks, mainly by avoiding redundant access of data, may prove to have a considerable effect on execution time.

---

* This research is supported in part by Polish State Committee for Scientific Research, Grant 8 T11C043-15

T. Yakhno (Ed.): ADVIS 2000, LNCS 1909, pp. 117-126, 2000.

A new alternative approach to multiquery optimization, that we present in this paper, is to use some heuristic methods and *Genetic* and *Tabu Search* algorithms, that ensure the tractability of the problem even for very complex queries.

The paper is organized as follows. In the next Section, we will formulate multiquery optimization problem. In Section 3 we will introduce three heuristic multiquery optimization methods that we propose to use in advanced database systems. In Section 4 and 5 we will present implementation of *Genetic* and *Tabu Search* algorithms that we propose to use for searching of the space of the possible query execution plans. In Section 6 we present performance evaluation of the algorithms implemented in our simulation experiments. Finally conclusions are given in Section 7.

2 Multiquery Optimization Problem Formulation

We assume that a query Q is represented by a query graph $G_Q=(V_Q, E_Q)$, where the vertices $V_Q = \{ R_i \}$ represent base relations in the database, and each edge $c_{ij} \in E_Q$ represent a join condition between the relations R_i and R_j.

As in most of the literature we restrict our work to queries typical in advanced database systems, that means, involving mainly join operations [2, 6, 8]. The execution of a query Q can be expressed by a *query execution plan - QEP_Q* which is a tree representing the order in which join operations are performed during the query evaluation process. The leaves of QEP_Q are the base relations in the query graph and the internal nodes represent join operations together with the specification of the way the joins are performed (e.g. join method, argument order, use of indices etc.). Any subtree of QEP_Q represents a partial execution of Q and will be called a *Partial Query Execution Plan (PQEP)*. As most systems implement the join operation as a 2-way join, we shall assume that QEP_Q is a binary tree. The execution cost of a query, e.g. time necessary to process the query, depends on the QEP_Q that is selected to evaluate the query. For a given query there might be many different *QEPs*. The general goal of query optimization is to find *QEP* with the lowest cost.

Assume now that a set of queries $S=\{Q_1, ... , Q_m\}$ that should be processed together is given. We shall call it a *Multiple-Query Set (MQS)}*. In order to reduce the execution cost, the common parts of queries are outlined and executed only once in an integrated process. Therefore, the execution of a *MQS* may be represented by a *Multiple-Query Execution Plan - MQEP = {C_1, ... ,C_k, R_1, ... ,R_m}*, where $C_1, ... ,C_k$ are *PQEP* representing the execution of common parts of queries, and $R_1, ... , R_m$ are *PQEP* representing the residual parts of execution of the queries in the *MQS*. More precisely, each C_i and each R_i are *PQEPs* having as leaves either base relations or the results of the execution of some C_j. Therefore, in the *MQEP* a partial order is defined among the $C_1, ... ,C_k$ and $R_1, ... ,R_m$.

The cost function associates an execution cost to each *MQEP*, that means, an execution cost corresponding to the sum of costs of all component *PQEPs* in following way:

$$cost \, (MQEP) = \sum_{i=1}^{k} cost(Ci) + \sum_{i=1}^{m} cost(Ri)$$

Given the cost function defined above, the *Multiquery Optimization (MQO)* problem can now be formulated as selecting among all possible *MQEPs* for a given multiple-query set *S* the one with the minimum cost.

3 Heuristic Methods Proposed for Multiquery Optimization

In this section, as a solution of the *MQO* problem in advanced database applications, we propose three heuristic multiquery optimization methods.

Compared to the single query optimization (*SQO*) problem, the major problem to be addressed in *MQO* is dealing with the large size of the search space of *MQEPs* that results from considering all possible overlapping among single query execution plans. Thus, it is necessary to introduce some simplifications to make the *MQO* problem tractable. Such simplifications we consider in the optimization methods presented below. The first one is based on the decomposition of the problem into a set of single query optimization problems; the second and third one are based on the integration of the query graphs and on solving the single query optimization problem for the integrated query graph.

The first approach that we shall call *Divide&Merge (D&M)* [4] consists of the following steps:

- Solve separately the *SQO* problem for every query Q_i in the query set to get the optimal execution plan.
- Determine the common parts $C_1, \dots C_k$ by intersecting all individual optimal query execution plans - *QEPs*.
- Build the multiquery execution plans - *MQEP* from $C_1, \dots C_k$, and the residual parts $R_1, \dots, R_m$ that are determined from the individual optimal *QEPs*.

The second approach that we shall call *Integrate&Split (I&S)* [4] consists of the following steps:

- Merge the query graphs of all the queries Q_i to get a *multiquery query graph MQG(MQS)* with maximum overlapping for *n* queries.
- Solve the *SQO* problem for the *MQG* graph to get an optimal *QEP*, which is a forest of trees with a separate root r_i for every query in the query set.
- Build the *MQEP* with $C_1, \dots C_k$ being the subtrees in optimal *MQEP* common to more than one of the individual *QEPs*, and the residual parts $R_1, \dots, R_m$ being the rest of the query trees.

The third approach that we shall call *Extended_Integrate&Split (EI&S)* consists of the following steps:

- Merge the query graphs of all the queries Q_i to get a *multiquery query graph MQG(MQS)* (that actually is a forest in a general case).
- Find common parts: $C_1^n, \dots, C_k^n$ for *n* queries in *MQG*; $C_1^{n-1}, \dots, C_k^{n-1}$ for *n-1* queries in *MQG*, $C_1^{n-2}, \dots, C_k^{n-2}$ for *n-2* queries in *MQG*; ...; $C_1^2, \dots, C_k^2$ for 2 queries in *MQG*, where *n* is the number of queries in *MQS*.
- Optimize the common parts of queries determined in previous step, first for *n* queries, than for *n-1* queries, etc.

- Optimize the residual parts of queries $R_1, \ldots, R_m$ in the same way as in previous step.
- Build *MQEP(MQS)* consists *PQEP*: $QEP(C_1^n), \ldots, QEP(C_k^n)$ - common for n queries $Q_i \in MQS$; $QEP(C_1^{n-1}), \ldots, QEP(C_k^{n-1})$ - common for $n-1$ queries $Q_i \in MQS$; etc.; and *PQEP* of the residual parts of queries $QEP(R_1), \ldots, QEP(R_m)$.

The advantage of the *Divide and Merge* approach is clearly connected to the combinatorial structure of the search space, since it allows to reduce the large *MQO* problem to several *SQO* problems with considerably smaller search space sizes. The advantage of the *Integrate&Split* and *Extended_Integrate&Split* approach is due to the preliminary integration of the query graphs that allows to determine all the common subqueries before performing the optimization process and hence always leads to a set of execution plans where maximum overlapping is ensured. A further advantage is given especially when the amount of overlapping among query graphs is large, by the possibility of concentrating the optimization effort on the common part.

As a search engines in the exploitation of the *MQEP* space to find optimal multiquery execution plan we propose to use some *Genetic algorithm* (*GA*) and combinatorial *Tabu Search* (*TS*) algorithm. In other words, we assume that *GA* and *TS* algorithms are integral parts of heuristics presented above.

4   Implementation of *Genetic* Algorithm in Multiquery Optimization Methods

Genetic algorithm (*GA*) differs from traditional optimization methods in a few aspects [5]. First of all, *GA* applies strategies adapted from the natural process of evolution, namely, natural selection – "survival of the fittest" and inheritation. Such special approach results in better proof (compromise between efficiency and effectiveness) in comparison to other combinatorial techniques that are commonly in use.

Genetic algorithm can be outlined as follows:

- Create a random, start population of the chromosomes.
- Evaluate the fitness of each individual in the population.
- Create the next generation by applying crossover and mutation operators to the individuals of the population until the new population is complete.
- Select parent – the best chromosome.
- Stop when a termination criterion (time, number of generations) is met.

Genetic algorithm starts from initialization, what means the creating some initial population, made by the random choice of a number of chromosomes (in the case of query optimization problem, that means random choice of some number of *QEP* or *MQEP*). In next step *GA* evaluate the fitness of the each chromosome by calculate fitness function (in the case of query optimization problem – cost function *cost (MQEP)* or *cost (QEP)* – lower value of the cost function is corresponding to the better fitness). Further, termination criterion is evaluated, which can be time slice expired or number of generation is met. If termination criterion is met then genetic algorithm returns the best chromosome, otherwise selection, recombination and fitness evaluation are made. Selection operator materializes the concept of "survival of the fittest". Based on the relative fitness of the individuals, it selects parents from the population, on which crossover and mutation operators described below are applied to produce the next generation. Selection strategies can be following: roulette wheel selection and tournament selection. The crossover is a binary recombination operator: two input chromosomes are transformed to two the new output chromosomes. The mutation is a unary operator that applies a small change on an individual. The new individual thus produced brings new genetic material to its generation. Crossover and mutation are applied to chosen chromosomes, what leads to the new population. For the new population evaluation of the fitness and termination criterion are then applied.

More formally genetic algorithm can be shown in following way:

```
Population_S = newStartPop(popsize); /* generate start
population */
minS = Evaluate(Population_S); /* at the moment it is
the best population */
    while (not StopCond) /* do it until termination
    criterion is false */
    {
    ParentPop_S = Selection(Population_S); /* select a
    temporary parent population */
    Population_S' = Recombination(ParentPop_S /* new
    generation is achieved by using crossover and
    mutation operators */
    minS = Evaluate(Population_S'); /* population with
    the best fitness function */
    }
return(minS);
```

In the context of multiquery optimization problem we propose following implementation of genetic algorithm described in details below. This implementation is destined for multiquery environment and it is an extended and modified version of the genetic algorithm implementation proposed in [9].

```
/* parameters given: optimization_time, popsize */
start_time = time();
switch(optimization)
    {query Qi:  for (i = 0; i<popsize; i++)
               { qep = GenQEP(Qi);
                 population[i] = qep;
               }
    set_of_queries MQS:  for(i = 0; i<popsize; i++)
               { qep = GenMQEP(MQS);
                 population[i] = qep;
               }
    }
qepcost = Evaluate(population);
min_cost = qepcost.min_cost;
passed_time = time() - start_time;
```

```
while (passed_time < optimization_time)
    {
    generation++;
    Selection(Population, ParentPop);
    Recombination(ParentPop, NewPop);
    qepcost = Evaluate(NewPop);
    min_cost = qepcost.min_cost;
    passed_time = time() - start_time;   }
    return(min_cost);
```

Declared optimization time is the termination criterion in this case. Fitness function called *Evaluate(population)* is following.

```
for(i = 0; i<popsize; i++)
    {
    qep = population[i];
    cost = QEPCost(qep);
    if (i == 0)
    {
        qepcost.min_cost = cost;
        qepcost.best_qep = qep;
    }

        if (cost < qepcost.min_cost)   {
        qepcost.min_cost =cost;
        qepcost.best_qep = qep;        }
        return(qepcost);        }
```

Presented above implementation of genetic algorithm consists following functions and procedures.

- *GenQEP(Q$_i$)* – function that generates *QEP$_Q$* based on query graph.
- *GenMQEP(MQS)* – function which generate *MQEP(MQS)*.
- *QEPCost(qep)* – cost evaluation function, which compute query or set of queries execution costs (parameter *qep* of the function can be *QEP$_Q$* or *MQEP(MQS)*.
- *Evaluate(population)* – function that is calling procedure *QEPCost(qep)* in purpose to evaluate the survival of the fittest of the chromosome (*QEP$_Q$* or *MQEP(MQS)*) in a population – returned value: the best plan and its cost.
- *Selection(population1, population2)* – procedure which is implementing selection operator; based on „*population1*" parent population „*population2*" is created.
- *Recombination(population1, population2)* – procedure which is implementing crossover and mutation operators. Based on „*population1*" new population „*population2*" is generated.

In our proposition of the genetic algorithm implementation we are using following data structures.

- *qep* – structure which stores currently processed *QEP$_Q$* or *MQEP(MQS)*;
- *population* – matrix of the size „*popsize*" in which all plans of a given population are stored;
- *qepcost* – record stored plan identifier and its execution cost;
- *optimization* - parameter, which determines what kind of optimization should be done: *SQO* or *MQO*.

5   Implementation of *Tabu Search* Algorithm in Multiquery Optimization Methods

The main idea behind *Tabu Search* (*TS*) consists in performing an aggressive exploration of the *MQEPs* space that seeks to make at each step the best possible move, with the restriction that the move has to satisfy certain constraints. These constraints are contained in a cyclic list *T* (called *Tabu list*), and are used to prevent the reversal, or sometimes repetition, of certain moves. The primary goal of the *tabu* restrictions is to allow the method to go beyond states of local optimality while still making high quality moves at each step. Without such restrictions, the method could take a best move away from a local optimum (making an upward move), and then at the next step fall back into the local optimum. In general, the *tabu* restrictions are intended to prevent such cycling behavior and more broadly to induce the search to follow a new trajectory if cycling in a narrower sense occurs.

```
S=initialize(); minS=S;    /* Set an initial solution */
T=0 (emptyset)             /* Set the Tabu list */
while not ("stopping condition satisfied") do
{
   generate the set V* ⊆ N(S) -T;
   choose the best solution S* ∈ V*;
   S = S*;
   T=(T - (oldest)) ∪ {S}; /* update the Tabu list T */
   if cost(S) < cost(minS) then
      minS=S;
}
return (minS);
```

The *Tabu Search* algorithm sketched above starts from a randomly generated initial state, and repeatedly performs moves from a state to a neighbor one. The critical step is choosing the best move from a state *S* to another state. In the context of the query optimization problem, state *S* is a single *QEP* or *MQEP*. At each iteration the procedure generates a subset V^* of the set $N(S)$ of the neighbors the current state *S*, from among the set of neighbors. A best solution S^* in V^* is determined and a move from *S* to S^* is made. The *TS* accepts both downward as well as upward moves, i.e. accepts moving from *S* to S^*, even if $cost(S) > cost(S^*)$. At each move, in order to avoid cycling or at least to reduce the probability, a list *T* of already visited states (the *Tabu list*) is updated. This forbids moves that should bring back to a previously (recently) visited state. Therefore, when the subset V^* of neighbor states has to be generated, we check that a candidate for membership of V^* is not in *T*. The procedure stops ("stopping condition satisfied") if no improvement of the best solution *minS* found so far has been made during a given number of iterations, or if there is no way of leaving state *S* because computing V^* with *the tabu* restrictions gives the empty set.

6 Performance Study

Due to the complexity of the *MQO* problem, it is not feasible, at least for non-trivial values of *N* (number of joins involved in a given query), to compute the global optimum, and hence to compare it with a solution given by a combinatorial or genetic algorithm. Therefore it is not possible to make an absolute statement on how good a given algorithm is. The only way to carry on the evaluation is to run a large set of experiments, and to perform a parametric analysis to compare the solutions given by different algorithms, in a large range of situations. To run our experiments we implemented *GA* and *TS* algorithms as a part of *D&M., I&S and EI&S* heuristics in our query optimization simulator. More specifically in setting up the experiments we concentrated on a few parameters: *S* - number of queries in multiquery set, *N* - number of joins in each query (called latter - query size) ranging from 10 to 30, ω - size of the common part in the multiquery set and optimization time ranging from 10 seconds up to 3 minutes. For any set of parameter values 50 different set of queries (with S=3) were generated. Each set were then optimized using all three *D&M., I&S and EI&S* heuristics and *GA* and *TS* algorithms.

In our experiments we restricted to bushy queries with equality joins. Joining relations were randomly generated according to database profile, and their joining attributes randomly selected. We made the usual assumption about uniform distribution and independence of joining attribute values. We model a distribution of the relation cardinalities with predominance of medium size relations (between 1000 and 10000 tuples).

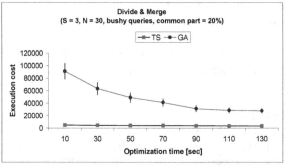

Fig. 1. Average execution cost versus optimization time for *Divide&Merge* method

A first series of our experiments shows how the quality of the *MQEPs* improves with the optimization time (Fig. 1-4), either for *SQO* method (Fig. 2) and *MQO* heuristic methods proposed in the paper (Fig. 1, 3-4). In Figure 1 we consider a large queries (N = 30) with bushy structure. In the range between 10 and 70 seconds *GA* improves very much, then it stabilizes around 90 seconds, but *TS* still performs about 8 times better than *GA*, in case of *MQO* method – *Divide&Merge*. In all the cases we analyzed for large queries *TS* seems to have a faster convergence, and therefore its relative performance improves when the optimization time (i.e. the length of the search) is reduced.

In Figures 2–4 we consider a relatively small queries (N = 10) with bushy structure. In case of the small queries the differences between algorithms are not such a large, like in case presented above. In the range between 10 and 70 seconds all heuristic methods with *TS* and *GA* algorithms investigated, improve results very much, then they stabilize around 70–80 seconds. In all investigated cases, *TS* outperforms *GA* in 80 %–200 %.

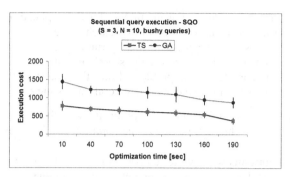

Fig. 2 Average execution cost versus optimization time for sequential query execution

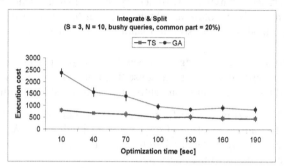

Fig. 3 Average execution cost versus optimization time for *Integrate&Split* method

Comparison of the Figures 2 and 3, 4 shows that for *Divide&Merge, Integrate&Split* and *Extended Integrate&Split* methods with *GA* and *TS* algorithms, common part of queries equal to 20 % is quite enough to generate most effective solutions in comparison to these generated by the *SQO* approach.

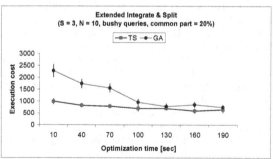

Fig. 4 Average execution cost versus optimization time for *Extended Integrate&Split* method

7 Conclusions

In this paper we have studied the application of *Genetic* and *Tabu Search* algorithms to the multiquery optimization problem. This is a hard combinatorial problem with very large search space, and higher complexity than the classical single query optimization problem. Therefore it is well suitable for the application of *GA* and *TS* algorithms to perform a partial exploration of the search space.

The parametric analysis has been performed by running a set of experiments in a wide range of situations to understand how the search algorithms are sensitive to the main parameters that characterize the problem. From the results of the experiments, presented in the previous section, we can draw several interesting conclusions:

- The relative performance of multiquery optimization methods clearly depend on the overlapping, i.e. on the relative size of the common part between the queries in the query set.
- All the search algorithms we have tested perform reasonably well on the problem. Of course we can make no absolute statement since computing an optimal solution to use as a reference is anyway precluded by the size of the search space, but the spread between the best solutions obtained by different algorithm is relatively limited, and the quality of the solution stabilizes after a fairly short optimization time.
- *Tabu Search* in a very large range of cases has a better performance. It shows a very stable and robust behavior, in the sense that it is fairly insensitive to variations of the workload parameters, and constantly approaches faster the solution.

As usual when analyzing the performance of randomized algorithms, it is very difficult to give a formal explanation of their behavior in all cases. We believe that an explanation can be found by understanding the structure of the search space, since this in turn depends on the shape of the query tree.

References

1. Chaudhuri S., Dayal U., An overview of data warehousing and OLAP technology, in: ACM SIGMOD Record (26), 1997, pp. 65-75.
2. Graefe G., Research Problems in Database Query Optimization, in: Proc. of the Intern. Workshop on Database Query Optimization, Portland, Oregon, 1989.
3. Hellerstein J.M., Haas J.P., Wang H.J., On-line Aggregation, in: Proc. of ACM-SIGMOD Conference on Management of Data, 1997, pp. 171 – 183.
4. Królikowski Z., Matysiak M., Morzy M., Salza S., A Combinatorial approach to the multiple-query optimization problem, in: Proc. COMAD' 94, 1994, Bangalore.
5. Michalewicz Z., Genetic Algorithms + Data Structures = Evolution Programs, Spring Verlag (Second Edition), 1994.
6. Park J., Segev A., Using Common Subexpressions to Optimize Multiple Queries, in: Proc. of the 4th Intern. Conf. On Data Eng., Los Angeles, 1988, pp. 311 - 319.
7. Roy P., et al., Efficient and Extensible Algorithms for Multi Query Optimization, in: Proc. of the SIGMOD' 2000 Intern. Conf., Dallas, 2000.
8. Sellis T., Multiple-Query Optimization, ACM TODS, 13(1), 1988, pp.23-52.
9. Stillger M., Spiliopoulou M., Genetic Programming in Database Query Optimization, Research Rep. – ICS, Humboldt Univ. of Berlin, 1997.

Interval Optimization of Correlated Data Items in Data Broadcasting*

Etsuko Yajima[1], Takahiro Hara[2], Masahiko Tsukamoto[2], and Shojiro Nishio[2]

[1] Sales Department, Tokyo Office, FM Osaka Co., Ltd.
[2] Dept. of Information Systems Eng., Grad. Sch. of Engineering, Osaka University

Abstract. The server strategy of repeatedly broadcasting data items can result in higher throughput than sending the requested data items to individual clients. Various methods have been studied to reduce the average response time in such systems. In this paper, we introduce a strategy to determine the optimal broadcast interval between two correlated data items. Based on these estimated optimal intervals, we propose a new scheduling strategy of a broadcast program to accommodate an environment where a large number of correlated data items exist in the broadcast program.

keywords: data broadcast, data correlation, scheduling strategy, broadcast interval

1 Introduction

Recently, there has been increasing interest in "push-based" data delivery mechanisms in which a server repetitively broadcasts various data to clients using a broad bandwidth. In a system which uses such kind of data delivery mechanism, in order to access a particular piece of data, each client has to wait for the data broadcast period as shown in Figure 1. In contrast, in the case where conventional "pull-based" mechanisms are used, a server delivers data by responding separately to every request message sent by each client. One remarkable advantage of the "push-based" delivery mechanism is that throughput is higher for data access in distributed systems with a large number of clients. This is made possible because the absence of communication contention among the clients (no request transmitted from the clients to the server) enables them to efficiently share the bandwidth.

Information broadcast systems are good examples of "push-based" systems. In these systems, the server's clients are typically portable computers, desktop computers, or electrical household appliances. The following are some scenarios in which information broadcast systems could be useful:

---

* This research was supported in part by Research for the Future Program of Japan Society for the Promotion of Science under the Project "Advanced Multimedia Content Processing (JSPS-RFTF97P00501)" and Grant-in-Aid for Scientific Research numbered 12480095 from the Ministry of Education, Science, Sports and Culture of Japan.

T. Yakhno (Ed.): ADVIS 2000, LNCS 1909, pp. 127–136, 2000.

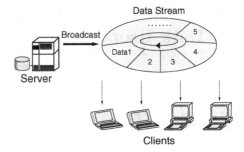

Fig. 1. Information system based on message broadcast

- In a shopping center, the store directory could be broadcast to enable customers with portable computers to locate shops of interest.
- At home, desktop computers or electrical household appliances could automatically receive digital data of various formats (image, audio, text) via satellite, ground communication, or cable TV. The data could include stock quotes, hit charts, commercial advertisements, sports news, and automatic software updates.

Several strategies have been proposed to improve the performance of these information systems. These strategies are categorized into four groups: scheduling strategies at servers[1,5,8,9,11,13], caching strategies at clients[1,2], a combination of "push-based" and "pull-based" data delivery mechanisms[4,11], and dissemination of updates[3,6,7,12].

Although correlation generally exists among broadcast data (e.g., clients collectively access certain sets of data), the conventional strategies do not take this into account. When clients access a set of correlated data frequently, the scheduling strategies which take the correlation into account can reduce the response time for data access.

In this paper, it is assumed that the client accesses a set of correlated data by submitting multiple access requests with some intervals between the requests. We discuss a method to determine the optimal broadcast interval between two correlated data. Moreover, assuming an environment where a large number of correlated data exist in a broadcast program, we propose a scheduling strategy of the broadcast program based on the estimated optimal broadcast intervals.

The following system environment is assumed:

- The system has a single server.
- Data is handled in clusters called *data items*.
- All data items are the same size.
- The server creates a broadcast program consisting of data items (identifiers: $1, \cdots, M$). The program is repeatedly broadcast. Each data item appears only once in the program cycle, and it takes L units of time to broadcast one item (Broadcast period is $T = ML$)

- Neither the data items nor the broadcast program is updated.
- Each client does not have a cache.
- Data item i can be accessed when i is first broadcast after an access request to i is issued.

The remainder of the paper is organized as follows: In section 2, correlation among broadcast data item is described. In section 3, a method to determine the optimal broadcast interval is discussed. In section 4, a scheduling algorithm of the broadcast program based on the estimated optimal broadcast intervals is proposed. Finally, in section 5, we summarize the paper.

2   Correlation among Data Items

There generally exists some correlation among the data items which a server broadcasts. For example, if a server broadcasts HTML files of various home pages, a client often collectively accesses several pages which are linked to each other. The probability that the client collectively accesses a set of data items represents the strength of the correlation among those data items. The stronger the correlation, the higher the probability that a particular set of data items will be accessed together. The probability differs for each set of data items.

When a client collectively accesses correlated data items, there are two possible cases: (i) the client accesses a set of correlated data items by submitting multiple access requests at the same time, and (ii) the client accesses a set of correlated data by submitting access requests with some intervals between these requests. In this paper, we assume the latter case, since this case is more general in a real environment. For example, if HTML files of homepages are broadcast, a user often refers to a certain page for a while and then refers to one of the linked pages from the first page. Moreover, if binary files of various tools are broadcast, a user often uses a word processor for a while in order to produce a document and then uses a drawing tool in order to create some figures in the document.

The probability that an access request to item j is issued with interval t after i is accessed is represented by a probability-density function, $F_{i,j}(t)$. The function $F_{i,j}(t)$ differs for each i and j depending on the contents of the broadcast items and the property of each client. By scheduling a broadcast program based on $F_{i,j}(t)$, the average response time for access request is expected to be shorter.

3   Optimal Broadcast Interval between Two Correlated Data Items

In this section, we analyze the optimal broadcast interval between two data items, i and j, when the probability-density function, $F_{i,j}(t)$, is given. Let us assume that the broadcast time of item i is given during one cycle of program, and the access request to item j is issued after the access to item i based on the probability-density function $F_{i,j}(t)$. Figure 2 shows an example of $F_{i,j}(t)$. For the purpose of simplicity, we assume that the access request to j is issued

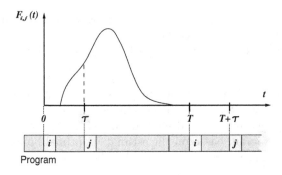

Fig. 2. Example of $F_{i,j}(t)$

between the access time to i and the moment when i is broadcast in the next cycle. Thus, $F_{i,j}(t) = 0$ while $t < L$ or $t > T$.

Suppose that item j is broadcast with interval τ from the broadcast period of i ($t = 0$). The response time for the request to j which is issued at the time t is $\tau - t$ if $L \leq t \leq \tau$, and $(T + \tau) - t$ if $\tau < t \leq T$. Thus, the average response time, $avg_{i,j}$, for requesting j is expressed by the following equation:

$$avg_{i,j} = \int_0^\tau F_{i,j}(t) \cdot (\tau - t)dt + \int_\tau^T F_{i,j}(t) \cdot \{(T + \tau) - t\}dt. \qquad (1)$$

In the following, we evaluate the optimal values of τ for several typical $F_{i,j}(t)$.

Example 1:

$$F_{i,j}(t) = q \qquad (k \leq t \leq h).$$

Figure 3(a) shows the graph of $F_{i,j}(t)$. Here, $q = 2/(h-k)$ since $\int_0^T F_{i,j}(t)dt = 1$. One example of this $F_{i,j}(t)$ is the case that item i is a publicity page of lottery, item j is its application page, and the validated period of application is from k to h. In this case, access requests may be uniformly issued in the validated period.

The value of $avg_{i,j}(\tau)$ where $0 < \tau < k$ or $h < \tau < T$ is always larger than the value where $\tau = h$. Therefore, it is enough to discuss the case where $k \leq \tau \leq h$ in order to find the minimum value of $avg_{i,j}(\tau)$.

For $k \leq \tau \leq h$, $avg_{i,j}(\tau)$ is expressed by the following equation:

$$avg_{i,j}(\tau) = q\{(h - k) - T\}\tau - q\left(\frac{h^2 - k^2}{2} - Thq\right) = Tq(h - \tau) + \frac{1}{2q}.$$

$avg_{i,j}$ takes the minimum value $1/2q$ when $\tau = h$. Therefore, the optimal value of τ is h.

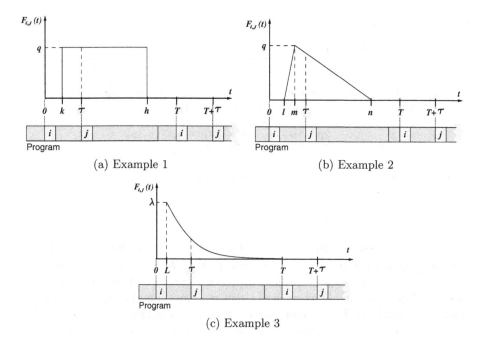

Fig. 3. Graphs of $F_{i,j}(t)$ in three examples

Example 2:

$$F_{i,j}(t) = \begin{cases} \dfrac{q}{m-l}(t-l) & (l \le t \le m) \\ \dfrac{q}{m-n}(t-n) & (m < t \le n). \end{cases}$$

Figure 3(b) shows the graph of $F_{i,j}(t)$. Here, $q = 1/(n-l)$ since $\int_0^T F_{i,j}(t)dt = 1$. This is a typical case that the distribution of access requests has a peak. One example of this $F_{i,j}(t)$ is the case of an automatic switching page, i.e., the system automatically issues an access request to page j at time m if there is no operation by a user during the time $0 \sim m$ while page i is referred. However, the point of time at which the processing is done may be vary between l and n due to the difference in computational power among clients.

The value of $avg_{i,j}(\tau)$ where $0 < \tau < l$ or $n < \tau < T$ is always larger than the value where $\tau = l$. Therefore, it is enough to discuss the cases where $l \le \tau \le m$ and $m \le \tau \le n$ in order to find the minimum value of $avg_{i,j}(\tau)$.

(1) Case 1: $l \le \tau \le m$

$$avg_{i,j}(\tau) = \frac{Tq(m^2 - \tau^2)}{2(m-l)} + \left(\frac{Tlq}{m-l} + 1 \right)\tau - \frac{n+m+l}{3} + Tq\left\{ \frac{-lm}{m-l} + \frac{n-m}{2} \right\}.$$

$avg_{i,j}(\tau)$ takes the following minimum value when $\tau = m$:

$$\frac{Tq}{2}(n-m) - \frac{n-2m+l}{3}.$$

(2) Case 2: $m \leq \tau \leq n$

$$avg_{i,j}(\tau) = \frac{Tq(\tau^2 + n^2)}{2(n-m)} - q\left(\frac{Tn}{n-m} - \frac{n-l}{2}\right)\tau - \frac{q(n-l)(n+m+l)}{6}.$$

$avg_{i,j}(\tau)$ takes the maximum value when $\tau = m$. On the contrary, $avg_{i,j}(\tau)$ takes the following minimum value when $\tau = n - (n-l)(n-m)/2T$:

$$\frac{2n-m-l}{3} - \frac{(n-l)(n-m)}{4T}.$$

From the above calculations, the optimal value of τ is $n - (n-l)(n-m)/2T$.

Example 3:
$$F_{i,j}(t) = \lambda e^{-\lambda(t-L)} \qquad (L \leq t \leq T).$$

$F_{i,j}(t)$ is the Poisson-distribution where the mean is $1/\lambda$ (λ is the arrival rate). For the purpose of simplicity, it is assumed that $\int_L^T F_{i,j}(t)dt \simeq 1$ is satisfied:

Figure 3(c) shows the graph of $F_{i,j}(t)$. One example of this $F_{i,j}(t)$ is the case that item i is a publicity page of ticket reservation and item j is a page of the reservation form. In this case, since a reservation is processed according to the access order, access requests to j are considered to be issued with highest probability just after i is accessed.

Here, $avg_{i,j}(\tau)$ is expressed by the following equation where $L \leq \tau \leq T$:

$$avg_{i,j}(\tau) \simeq \tau + Te^{-\lambda(\tau-L)} - L - \frac{1}{\lambda}.$$

$avg_{i,j}(\tau)$ takes the minimum value $(\log(T\lambda))/\lambda$ when $\tau = L + (\log(T\lambda))/\lambda$. Therefore, the optimal value of τ is $L + (\log(T\lambda))/\lambda$.

Although three examples do not express every case of correlation among data items, several typical cases are represented. As well, some other cases can be represented by combinations of the examples.

4   Discussions

In section 3, we have calculated the optimal broadcast interval between two items. However, it is not easy to schedule a broadcast program even if $avg_{i,j}(\tau)$ is given for every combination of two data items. For example, focusing two data items i and j in the broadcast program, while $\tau = (d_{i,j}+1)L$ gives the minimum $avg_{i,j}(\tau)$, $\tau = (d_{j,i}+1)L$ $(d_{j,i} = M - 2 - d_{i,j})$ does not always give the minimum $avg_{j,i}(\tau)$. Moreover, even if we assume the simplest case that access requests to two correlated data items are always issued at the same time, it is very difficult to schedule the optimal broadcast program[10].

Therefore, in this paper, a heuristic scheduling algorithm is proposed. This algorithm uses the calculated optimal broadcast interval and the access probability of each combination of two items to compose a broadcast program so that data items with strong correlation are assigned suitable intervals in a program.

4.1   Scheduling Algorithm

We assume that a frame of reference consists of discrete M points with distance L. For arbitrary z in this frame, the following equation is satisfied:

$$z = kML + z \quad (k = 0, \pm 1, \pm 2, \cdots).$$

Scheduling a broadcast program is equivalent to assigning a point in this frame to each data item so that each data item is assigned to just one point. In this section, we use the following notations:

E: A set of permutations of two data items. Initially, all permutations belong to E.

S_i $(0 \leq i \leq M)$: Sets of data items. Initially, S_i is $\emptyset$ for every i. No two sets S_i and S_j $(i \neq j)$ share the same data item, i.e., one data item belongs to only one S_i.

G: A set of permutations of two data items. This is the union set of all S_i. In the algorithm proposed later, permutations selected from E belong to G.

R: A set of data items. Initially, this set is empty. In the algorithm proposed later, items that are assigned a point belong to R.

N: A set of data items. Initially, all data items belong to this set.

In order to schedule a program, union-find operations are used:

union(A, B): Operation to create a new set A as the union of A and B, and create a new set B as an empty set.

find(i): Operation to return the name of S_j that contains element i.

Here, we propose the scheduling algorithm, *UFL algorithm*. Let $v_{i,j}$ denote the optimal broadcast interval of items i and j which gives the minimum $avg_{i,j}(\tau)$, and z_i denote a point assigned to data item i.

UFL (Union and Find Layout) Algorithm

First, the optimal broadcast interval, $v_{i,j}$, is calculated for each permutation $\langle i, j \rangle$ in E. Here, $v_{i,j}$ is calculated as a multiple of L. Then, z_i of each item i is determined as follows.

Let $P_{i,j}$ denote the probability that the access request to data item j is issued after i is accessed $(\sum_{j=1}^{M} \sum_{i=1}^{M} P_{i,j} = 1)$. A permutation $\langle \alpha, \beta \rangle$ whose $P_{\alpha,\beta}$ is the largest in all permutations $(P_{\alpha,\beta} \geq P_{i,j}$ for arbitrary $\langle i, j \rangle)$ is deleted from E, and set S_α is made as $\{\alpha, \beta\}$. z_α is set to 0 and z_β is set to $v_{\alpha,\beta}$. The items α and β are deleted from set N and added to set R.

Then, the following operations are executed repetitively till $N = \emptyset$ is satisfied.

1. A permutation $\langle i, j \rangle$ whose $P_{i,j}$ is the largest in E is added to G.
2. The permutations $\langle i, j \rangle$ and $\langle j, i \rangle$ are deleted from E.
3. Based on the results of operations find(i) and find(j), one of the following operations is executed. During the operation execution, when a point is assigned to an item, the item is deleted from N and added to R.

- If find(i)= none and find(j)= none, then: New set S_i is created as $\{i, j\}$.
- If find(i)= none and find(j)= S_a, then: union(S_a, $\{i\}$) is executed. If $S_a \cap R \neq \emptyset$, z_i is set to $z_j - v_{i,j}$. (z_j has already been determined.)
- If find(i)= S_a and find(j)= none, then: union(S_a, $\{j\}$) is executed. If $S_a \cap R \neq \emptyset$, z_j is set to $z_i + v_{i,j}$. (z_i has already been determined.)
- If find(i)=find(j)= S_a, then: No operation is executed.
- If find(i)= S_a and find(j)= S_b, then: union(S_a, S_b) is executed. If $S_a \cap R \neq \emptyset$, one of the following operations is repetitively executed on items $i' \in N$ and $j' \in R$ which belong to S_a till $S_a \cap N = \emptyset$ is satisfied. The permutation $\langle i', j' \rangle$ or $\langle j', i' \rangle$ in G is selected by access probability order.
  - If permutation $\langle i', j' \rangle$ is in G, then: $z_{i'}$ is set to $z_{j'} - v_{i',j'}$.
  - If permutation $\langle j', i' \rangle$ is in G, then: $z_{i'}$ is set to $z_{j'} + v_{j',i'}$.

In this algorithm, when $S_k \cap R \neq \emptyset$ is satisfied for set S_k, all data items in S_k are assigned points. Moreover, if union(S_k,S_h) is executed when $S_k \cap R \neq \emptyset$ and $S_h \cap R = \emptyset$, all data items in S_h are also assigned points. In the case that a point which should be assigned to item j has been already assigned to other item i, z_j is set to the nearest free point from z_i.

In this way, all data items are assigned points, and data items are broadcast in ascending order of the z_i value. The UFL algorithm can reduce the average response time for data access because data items with stronger correlation are broadcast with more suitable intervals. The time complexity of this algorithm is $O(M^2 \log M)$.

4.2   Simulation

We show the simulation results regarding the performance evaluation of the UFL algorithm. The following assumptions were made for the simulation experiments:

- There exist 120 data items ($M = 120$). It takes 10 units of time to broadcast one item in the program. (Although this value proportionally affects the response time, it has relatively no effect on the simulation results, so we can choose this value arbitrarily.)
- The probability that a client issues consecutive requests to two items is 0.1 at each unit of time. This value does not affect the response time, since the system's access frequency does not affect our proposed scheduling algorithm. Hence, this value can also be chosen arbitrarily.
- The probability that two items i and j will be requested continuously, i.e., the strength of correlation, is expressed by an i-j element, $P_{i,j}$ ($P_{i,j} \neq P_{j,i}$), in the access probability matrix (a 120 dimensional matrix). For the purpose of simplicity, it is assumed that $P_{i,i} = 0$ for each item i. Either 0 or a constant positive value is assigned to the remaining elements. The percentage of the elements which are not assigned is changed in the simulation experiments. As the percentage of the non-zero elements gets lower, the correlation between each two items becomes stronger.
- $F_{i,j}(t)$ in Example 1 where $k = 0$ and $0 \leq h \leq ML$ is used as the probability-density function.

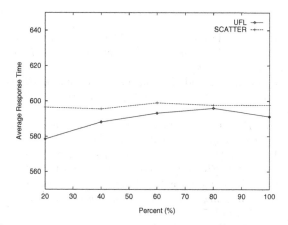

Fig. 4. Simulation result

In the simulation experiments, we evaluate the interval between the responses to two access requests which are issued continuously. Evaluating the interval between two responses is sufficient because the response time to the first access does not depend on the employed scheduling strategy. Here, we call this interval the 'response time' for simplicity. For comparison, we also evaluate the average response time of the random scheduling strategy which does not consider the correlation among data items.

Figure 4 shows a set of simulation results. The horizontal axis indicates the percentage of elements which are not 0. In the graph, the random scheduling algorithm is shown as 'SCATTER'. These results show that the UFL algorithm gives shorter response time than the random scheduling algorithm. The difference in response time between the UFL algorithm and the random scheduling algorithm is not very large because the probability-density function whose distribution is uniform is used as Example 1. If the probability-density function whose distribution includes a clear peak such as those of Example 2, 3, and 4 is used, it is expected that the superiority of our proposed algorithm becomes more evident.

5   Conclusion

In this paper, the optimal broadcast interval between two data items were calculated, assuming that the clients continuously issue access requests to two correlated data items with some intervals. Moreover, we proposed the scheduling algorithm of broadcast program, based on the calculated optimal intervals for combinations of every two data items in the program. In addition, the performance of the proposed algorithm was evaluated assuming the simplest case.

As a part of our future work, we are planning to evaluate our proposed algorithm in various cases such as those discussed in this paper. We are also planning to consider the caching strategies which take the correlation among data items into account.

References

1. Acharya, S., Alonso, R., Franklin, M., and Zdonik, S.: "Broadcast Disks: Data Management for Asymmetric Communication Environments," *Proc. ACM SIGMOD Conference*, pp.199–210 (1995).
2. Acharya, S., Franklin, M., and Zdonik, S.: "Dissemination-Based Data Delivery Using Broadcast Disks," *IEEE Personal Communications*, Vol.2, No.6, pp.50–60 (1995).
3. Acharya, S., Franklin, M., and Zdonik, S.: "Disseminating Updates on Broadcast Disks," *Proc. VLDB Conference*, pp.354–365 (1996).
4. Acharya, S., Franklin, M., and Zdonik, S.: "Balancing Push and Pull for Data Broadcast," *Proc. ACM SIGMOD Conference*, pp.183–194 (1997).
5. Chen, M.S., Yu, P.S., and Wu., K.L.: "Indexed Sequential Data Broadcasting in Wireless Mobile Computing," *Proc. Int'l Conf. on Distributed Computing Systems (ICDCS)*, pp.124–131 (1997).
6. Dao, S. and Perry, B.: "Information Dissemination in Hybrid Satellite/Terrestrial Networks," *Proc. Int'l Conf. on Data Engineering*, Vol.19, No.3, pp. 12–18 (1996).
7. Franklin, M. and Zdonik, S.: "Dissemination-Based Information Systems," *Proc. Int'l Conf. on Data Engineering*, Vol.19, No.3, pp.20–30 (1996).
8. Hameed, S. and Vaidya, N.H.: "Log-time Algorithms for Scheduling Single and Multiple Channel Data Broadcast," *Proc. MOBICOM 97*, pp.90–99 (1997).
9. Hameed, S. and Vaidya, N.H.: "Efficient Algorithm for Scheduling Data Broadcast," *Wireless Networks*, Vol.5, No.3, pp.183–193 (1999).
10. Hara, T., Yajima, E., Tsukamoto, M., Nishio, S., and Okui, J.: "A Scheduling Strategy of a Broadcast Program for Correlative Data," *Proc. ISCA Int'l Conf. on Computer Applications in Industry and Engineering*, pp.141–145 (Nov. 1998).
11. Imielinski, T., Viswanathan, S., and Badrinath, B.R.: "Data on Air: Organization and Access," *IEEE Transaction on Knowledge and Data Engineering*, Vol.9, No.3, pp.353–372 (1997).
12. Stathatos, K., Roussopoulos, N., and Baras, J.: "Adaptive Data Broadcast in Hybrid Networks," *Proc. VLDB Conference*, pp.326–335 (1997).
13. Su, C.J., Tassiulas, L., and Tsotras, V.J.: "Broadcast Scheduling for Information Distribution," *Wireless Networks*, Vol.5, No.2, pp.137–147 (1999).

Store and Flood: A Packet Routing Protocol for Frequently Changing Topology with Ad-Hoc Networks

Hiroaki Hagino, Takahiro Hara, Masahiko Tsukamoto, and Shojiro Nishio

Dept. of Information Systems Eng., Graduate School of Engineering, Osaka University
2-1 Yamadaoka, Suita, Osaka 565-0871, Japan
TEL: +81-6-6879-7820, FAX: +81-6-6879-7815
Email: {hagino,hara,tuka,nishio}@ise.eng.osaka-u.ac.jp

Abstract. In ad-hoc networks, packet flooding is used to discover a destination terminal of communication. However, since, in ad-hoc networks, disconnection and reconnection of terminals frequently occur, it often occurs that the destination terminal cannot often be discovered. Moreover, since the route from a source terminal to a destination terminal frequently changes, the route information obtained by packet flooding may become unavailable in a short period of time. In this paper, to solve these problems, we propose the *store-and-flood protocol* to improve the packet reachability in ad-hoc networks. In this protocol, the source terminal does not flood route request packets to discover destination mobile terminal but blindly floods data packets. Then, terminals which relay the flooded packets store the packets and later flood the packets when new other terminals connect to them. We show the protocol behavior and the packet format of our proposed protocol. Moreover, the performance and the overhead of the proposed protocol are shown by simulation experiments.

keywords: ad-hoc networks, routing protocol, mobile computing environments

1   Introduction

Recent advances of network and computer technologies have led to development of mobile computing environments. As one of the research fields in mobile computing environments, there has been increasing interest in *ad-hoc networks* which are constructed by only mobile terminals. In ad-hoc networks, to realize multihop communication, mobile terminals have the function to relay packets from the source terminal to the destination one.

In ad-hoc networks, mobility of mobile terminals causes changes of network topology. Since route information in routing table may become unavailable in a short period of time, a new routing technique different from the conventional one used in fixed networks is required in ad-hoc networks. Based on this idea, packet flooding is generally used for searching an appropriate route of packets in ad-hoc networks. A typical procedure is as follows: A source mobile terminal

T. Yakhno (Ed.): ADVIS 2000, LNCS 1909, pp. 137–146, 2000.

floods a route request packet to neighbor terminals. Then, mobile terminals which receive the packets send the packet to their neighbor terminals. If the destination mobile terminal receives the packet, it informs the source terminal of the route information through which the packet has been sent. As a result, the source terminal can know the route to the destination terminal and starts to send data packets to the destination. If the route becomes unavailable during the communication, the source terminal floods a route request packet again. In this way, packet flooding plays a very important role in communication in ad-hoc networks.

However, packet flooding is not always efficient for all situations in ad-hoc networks. For instance, if mobile terminals move fast, a route may be already unavailable after the route is found by flooding. As a result, it is difficult to achieve high packet reachability using such a simple flooding method.

In this paper, we propose the *store-and-flood protocol* (SFP for short) to improve the packet reachability in ad-hoc networks. In SFP, a source mobile terminal does not flood route request packets but blindly sends data packets to the all neighbors. Furthermore, the received terminals store the packets and later flood the packets when new terminals connect to them. Flooding data packets improves the communication performance when mobile terminals move fast and network topology frequently changes. Moreover, even though there is not a route between two mobile terminals which want to communicate with each other, data packets may reach the destination terminal by storing and reflooding the packets. We also show the protocol behavior and the packet format of our proposed protocol. The performance and the overhead of SFP are shown by simulation experiments. From the results, we show that our proposed protocol drastically improve the reachability in ad-hoc networks.

This paper is organized as follows. The next section discusses some related works. Section 3 describes the proposed SFP. The performance evaluation of SFP is presented in section 4. Some considerations are given in section 5. Finally, section 6 presents conclusions and future work.

2   Related Works

IETF (Internet Engineering Task Force) proposed several routing protocols for ad-hoc networks. DSDV (Destination-Sequenced Distance Vector)[11,12] is one of the protocols which do not use packet flooding. DSDV uses a routing method similar to RIP. Namely, in DSDV, each mobile terminal has a routing table consisting of route information to all other terminals and manages it by periodically exchanging the latest distance vector information among terminals. Since all terminals have route information of the entire network, they do not need packet flooding to discover a destination terminal. However, to guarantee correctness of route information, it is necessary for mobile terminals to exchange route information more frequently than changes of network topology. Therefore, DSDV is not suitable for ad-hoc networks in which mobile terminals move fast. On the other hand, AODV (Ad-hoc On Demand Distance Vector)[13] and DSR

(Dynamic Source Routing)[2,6] use only flooding of route request packets for communication. However, these protocols have problems described in section 1. Moreover, if mobile terminals frequently communicate with each other, flooding of route request packets causes large traffic.

In ZRP (Zone Routing Protocol)[4] and CBRP (Cluster Based Routing Protocol)[5], mobile terminals are classified into several groups, and route information only inner groups are managed. When a mobile terminal communicates with other one included in a different group, the source terminal floods a route request packet. ULSR (Uni-directional Link State Routing)[9] uses a similar approach.

In all the above protocols, a mobile terminal cannot communicate with other one if a route between the two terminals does not exist. On the other hand, in our proposed protocol, a mobile terminal may be able to communicate with other one in such a situation.

3   Store-and-Flood Protocol (SFP)

In this section, we explain our proposed SFP. SFP is an extension of simple packet flooding. Thus, SFP can be applied to other conventional protocols which use packet flooding for communication.

3.1   Outline

SFP has two notable features compared with the conventional protocols which flood route request packets:

- **Storing and reflooding of packets:** Each mobile terminal has a buffer and stores packets which are flooded by other mobile terminals. When the terminal which stores flooded packets connects with new other mobile terminal, it restart flooding of the stored packets. By doing so, the packet reachability is expected to be improved.
- **Flooding of data packets:** As described in section 1, if there are many mobile terminals which move fast, flooding of route request packets does not work well. Moreover, in SFP, packets may reach the destination mobile terminal even if there is not a route to the terminal. Thus, route information obtained by flooding of route request packets cannot be reused. From these reasons, a mobile terminal does not flood route request packets but data packets.

3.2   Details of SFP

In this subsection, we describe the details of our proposed SFP. In SFP, periodical exchange of route information is not performed. It only stores and floods data packets for communication between mobile terminals. Here, it is assumed that each mobile terminal can detect other mobile hosts which are directly connected to the terminal by periodical exchange of hello packets.

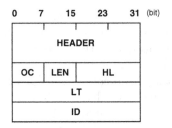

Fig. 1. The header format

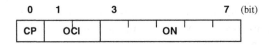

Fig. 2. The option code field format

SFP is realized by using option field of IP packet. In the following, we explain the header format of data packets in SFP and the behavior of SFP.

Header format: The header format of data packets in SFP is shown in Figure 1. Each field in the header is as follows:

- **HEADER:** Normal IP header.
- **OC:** Option code. 1byte.
- **LEN:** Length of the option field. 1byte.
- **HL:** Hop limit. 2bytes.
- **LT:** Life time of the packet. 4bytes. HL field and LT field of the packet are used to check whether the packet is valid or not.
- **ID:** Packet identifier. Each mobile terminal sets this value independently of other mobile terminals. Mobile terminals can identify a receiving packet from a combination of the source address described in HEADER field and the value of ID field.

As shown in Figure 2, OC field includes the following three subfields:

- **CP:** Copy flag. 1bit. The value of this field is fixed to 1.
- **OCl:** Option class. 2bits. The value of this field is fixed to 2.
- **ON:** Option number. 5bits. The value of this field is fixed to 18.

Behavior of SFP: We show the behavior of SFP when a mobile terminal wants to communicate with other one.

– **Behavior of a mobile terminal which sends data packets:**
  A mobile terminal which wants to communicate with other one prepares data packets and sets the values of HL field, LT field, ID field and other fields in normal IP header. Then, it starts to flood packets and also stores the packets. MAC address of the packets are set to the broadcast address.
– **Behavior of a mobile terminal which receives flooded packets:**
  Each mobile terminal that receives a flooded packet behaves according to the following algorithm:
  1. The mobile terminal checks the destination address of the received packet. If the terminal is the destination of the packet, the terminal accepts the packet, and the algorithm finishes. If not, it decrements the value of HL field, and the algorithm goes to the next step.
  2. If the value of HL is more than 0 and the value of LT does not expire, the mobile terminal continues to flood the packet and stores the packet. Then, the algorithm goes to the next step. If not, it discards the packet, and the algorithm finishes.
  3. If a buffer space of the mobile terminal is full of packets, the mobile terminal checks LT fields of the stored packets. If there are packets whose values of LT fields expire, the packets are discarded. Then, the new received packet is stored in the buffer, and the algorithm finishes. If not, a packet whose value of LT field is minimum is replaced by the new received packet.
– **Behavior of a mobile terminal which stores flooded packets:**
  When a mobile terminal which stores packets detects an other mobile terminal newly connected to the terminal, the terminal first checks LT field of all packets stored in its buffer. Then, packets whose values of LT fields do not expire are reflooded. Other packets are not flooded and are discarded.

4   Simulation Experiments

In this section, we show results of simulation experiments regarding the evaluation of our proposed protocol. In the simulation experiments, our proposed protocol is compared with a protocol which uses flooding of route request packets to discover the destination terminal.

4.1   Simulation Models

In the simulation experiments, the network consists of 5 mobile terminals, and they randomly move in all directions in 10×10 square. The moving speed is determined based on exponential distribution with mean 5. The number of packets created by each mobile terminal is determined based on exponential distribution with mean COM. The radio communication range of each mobile terminal is a circle with the radius of 2. The values of LT and HL in each packets in SFP are set to 100 and 10, respectively. Each mobile terminal has a buffer with size of BUF packets.

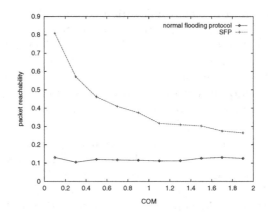

Fig. 3. Relationship between the communication frequency and the packet reachability

4.2   Simulation Results

Our simulation experiments are done based on the simulation models described in section 4.1. First, we examine the relationship between the communication frequency and the packet reachability, where BUF is fixed to 10. Here, the packet reachability represents the ratio of the number of packets accepted by destination terminals to the number of all packets generated by source terminals. The result of this simulation experiment is shown in Figure 3. In this figure, a solid line represents the reachability in SFP and a dotted line represents that in the protocol using the normal flooding. It is also shown that the result in Figure 3 shows that SFP drastically improves the packet reachability. Especially, when the communication frequency is very small, the packet reachability in SFP is 6 times as mush as that in the normal flooding protocol. The communication frequency has a great influence on the packet reachability in SFP, while it does not in the normal flooding protocol. This is because the buffer size on each mobile terminal is not sufficient, and thus flooded packets are discarded before reaching the destination terminals. This shows that sufficient buffer size is necessary to guarantee high packet reachability.

Figure 4 shows the relationship between the buffer size and the packet reachability where $COM = 5$. From the result, the buffer size does not have influence on the packet reachability in the normal flooding protocol. On the other hand, larger buffer size becomes, higher packet reachability becomes in SFP. This result shows that sufficient buffer size improves the packet reachability.

From the above simulation results, the buffer size is a factor to improve the packet reachability in SFP. The last simulation experiment shows the influence of the value of BUF on the network load. Here, the network load is represented by the total number of packets per unit time received by all mobile terminals. The result of this simulation experiment is shown in Figure 5. If the buffer size is small, the network load in SFP is almost equal to that in the normal flooding

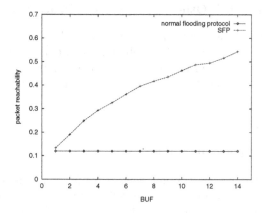

Fig. 4. Relationship between the buffer size and the packet reachability

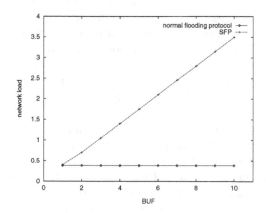

Fig. 5. Relationship between the buffer size and the number of packets

protocol. However, the difference in network load between SFP and the normal flooding protocol gets larger as the buffer size gets larger. Especially, when the buffer size is 10, the network load is 9 times as much as that in the normal flooding protocol. High network load causes the degradation of packet processing capacity, and thus the communication performance is also degraded. From the result, it is shown that a tradeoff between the communication performance and the packet reachability is a important issue in SFP.

5 Discussions

In this section, we give some considerations to SFP including the protocol overhead.

5.1   Relationship with Other Conventional Protocols

SFP is an extension of the normal packet flooding. Therefore, it is easy to apply SFP to other conventional protocols which make use of packet flooding. However, in most of the conventional protocols, since mobile terminals flood route request packets, it is necessary to extend them to flood data packets.

5.2   Applications

Some application examples of SFP are as follows:

- SFP is very effective for communication in disaster regions. Of course, in general, ad-hoc networks are considered to be effective in disaster regions since they do not assume network infrastructures such as the fixed networks or telephone lines which are often broken in a disaster. Since SFP improves the packet reachability in ad-hoc networks, it is more effective in a disaster. For instance, when a user looks for his/her friends of whom he/she has lost sight, SFP makes higher chance to discover them.
- Let us assume that there are some fixed terminals which have radio communication devices and that they are not connected to a fixed network. Let us also assume that a fixed terminal which is far from them wants to send data to them. In this case, the data can be sent from the source terminal to other terminals by a user who can physically carry the data and intentionally moves among the destination terminals.
- The application shown first is an example of non-intentional use of SFP, while that shown second is an example of intentional use. Now, we give an example of an intermediate case of the two. Let us assume that two users with mobile terminals exist along a road and far from each other. If many cars, which have communicatable computers, run along the road, the two terminals can expect to communicate with each other since the cars physically carry the packets.

5.3   Network Load

In SFP, the number of floodings caused by sending one data packet is more than that in protocols which flood route request packets. This causes waste of battery power in mobile terminals. Moreover, since flooded packets in SFP exist in the network for a long time, resources of mobile terminals are also occupied for a long time.

A possible solution of these problems is a usage of geographical information given by GPS. Ko and Vaidya [7] propose a strategy to reduce the number of packets by using geographical information. In this strategy, when a mobile terminal wants to communicate with an other terminal, the terminal predicts the current location of the destination terminal according to the past location information obtained in the previous communication. Then, the source terminal floods a route request packet. Each mobile terminal that receives the packet discards it if the terminal is further from the predicted location of the destination terminal than a mobile terminal which relayed the packet previously. In another paper of the same authors [8], this approach is used for multicast in ad-hoc networks. SFP can also be improved by applying this approach. In SFP, since a mobile terminal physically carries packets, not only location information but also velocity information of mobile terminals are important. For instance, a mobile terminal which exists further from the destination terminal than a terminal which previously relayed packets may move closer to the destination terminal. Moreover, we consider another approach in which the time duration to store packets on mobile terminals is varied based on location information.

6   Conclusion

In this paper, we have proposed the Store-and-Flood Protocol (SFP) to improve the packet reachability in ad-hoc networks. In this protocol, a mobile terminal which relays a flooded packet stores the packet and later floods the packet when the mobile terminal detects new mobile terminals. As a result, our proposed protocol allows a mobile terminal to communicate with other mobile terminal to which there is not a route. Moreover, by simulation experiments, we show that our proposed protocol improves the packet reachability in ad-hoc networks.

As part of our future work, we are planning to implement the proposed protocol and to evaluate it on a practical platform.

References

1. Broch, J., Maltz, D. A., Johnson, D. B., Hu, Y. C., and Jetcheva, J.: "A Performance Comparison of Multi-Hop Wireless Ad Hoc Network Routing Protocols," *Proc. Mobicom'98*, pp.159–164, 1998.
2. Broch, J., Johnson, D. B. and Maltz, D. A.: "Dynamic Source Routing in Ad Hoc Wireless Networks," *Internet Draft*, draft-ietf-manet-dsr-00.txt, 1998.
3. Chen, T. W. and Gerla, M.: "Global State Routing: A New Routing Scheme for Ad-hoc Wireless Networks," *Proc. IEEE ICC'98*, 1998.
4. Haas, Z. and Pearlman, M.: "The Zone Routing Protocol (ZRP) for Ad Hoc Networks," *Internet Draft*, draft-ietf-manet-zone-zrp-01.txt, 1998.
5. Jiang, M., Li, J., and Tay, Y. C.: "Cluster Based Routing Protocol (CBRP) Functional Specification," *Internet Draft*, draft-ietf-manet-cbrp-spec-00.txt, 1998.
6. Johnson, D. B.: "Routing in Ad Hoc Networks of Mobile Hosts," *Proc. the IEEE Workshop on Mobile Computing Systems and Applications*, pp.158–163, 1994.
7. Ko, Y. B. and Vaidya, N. H.: "Location-Aided Routing (LAR) in Mobile Ad Hoc Networks," *Proc. MOBICOM'98*, 1998.
8. Ko, Y. B. and Vaidya, N. H.: "Geocasting in Mobile Ad Hoc Networks: Location-Based Multicast Algorithms," *Proc. IEEE Workshop on Mobile Computing Systems and Apprications (WMCSA'99)*, 1999.
9. Nishizawa, M., Hagino, H., Hara, T., Tsukamoto, M., and Nishio, S.: "A Routing Method Using Uni-directional link in Ad Hoc Networks," *Proc. of 7th International Conference on Advanced Computing and Comuunications (ADCOM'99)*, pp.78-82, 1999.

10. Park, V. and Corson, S.: "A Highly Adaptive Distributed Routing Algorithm for Mobile Wireless Networks," *Proc. INFOCOM'97*, 1997.
11. Perkins, C. and Bhagwat, P.: "Highly Dynamic Destination-Sequenced Distance-Vector Routing (DSDV) for Mobile Computers," *Proc. SIGCOMM*, pp.234–244, 1994.
12. Perkins, C. and Bhagwat, P.: "Destination-Sequenced Distance-Vector," *Internet Draft*, draft-ietf-manet-dsdv-00.txt, 1998.
13. Perkins, C. and Royer, E.: "Ad Hoc On Demand Distance Vector (AODV) Routing," *Internet Draft*, draft-ietf-manet-aodv-02.txt, 1998.

Management of Networks with End-to-End Differentiated Service QoS Capabilities

Lisandro Zambenedetti Granville, Rodrigo Uzun Fleischmann,
Liane Margarida Rockenbach Tarouco, and Maria Janilce Almeida

Federal University of Rio Grande do Sul – UFRGS,
Computer Science Institute,
Bento Gonçalves Avenue – Block IV
Porto Alegre – Rio Grande do Sul, Brazil
{granville, uzun, liane, janilce}@inf.ufrgs.br

Abstract. Offering QoS on TCP/IP networks is the object of intense research. New applications, such as telemedicine, distance learning, videoconference and others can only be implemented on environments that ensure QoS for the existing services. Usually, routers are the network resources that undergo changes in order to implement it – hosts have little or no influence. This paper presents an implementation of QoS structures offered directly from the host. It also presents a marking process that is run on the end systems and can be remotely managed via SNMP. A signaling application is also described in order to allow QoS requests to inherited applications that do not perceive QoS service on the networks.

1 Introduction

The best-effort paradigm currently found on TCP/IP networks is able to offer enough services to support the great number of applications spread on the Internet. However, an important set of applications cannot operate properly on best-effort services only. Applications such as videoconference, distance learning and telemedicine, for instance, will only operate on environments where services are guaranteed. Given this setting, one of the great current challenges is to offer QoS [1] on TCP/IP networks.

One of the apparently promising propositions is the use of differentiated services architecture [2] of IETF [3]. Its importance is clear since the Internet2 project [4] has elected differentiated services as the solution for QoS supply. In addition, several manufacturers have supported the solution, and software houses are making support tools available on initial versions (for example, Bandwidth Brokers).

The main solution to offer QoS is through network equipment. Currently, hosts perform little or no role in it, leaving the routers with most of the work to be done. On the other hand, chances of success increase with the distribution of QoS tasks.

Therefore, it is reasonable to think that involving hosts is also important to run the services properly.

This paper presents an implementation of QoS structures based on host implementation. The packet marking process is now run directly on the end systems, separately from the applications. The network manager can program all hosts

T. Yakhno (Ed.): ADVIS 2000, LNCS 1909, pp. 147-158, 2000.

remotely and control the marking process globally, exchanging SNMP (Simple Network Management Protocol) messages [5] with the hosts. A Marking MIB (Management Information Base) [6] has been implemented so that the manager can program the hosts.

QoS management on the hosts can also be automated to the extent of replacing the management station with a Bandwidth Broker (BB) [7]. Users can request QoS parameters to the BB. This request uses a signaling tool. Resources can be obtained immediately or scheduled for future use, according to the capability of the BB. The request communication also uses SNMP and a QoS MIB.

This work is organized as follows. Section 2 presents the packet marking process on the hosts. Such remote marking management uses the Marking MIB and the SNMP marking support shown in section 3. Section 4 presents considerations on user resource requests. Section 5 describes the QoS MIB that implements the host/BB signaling. Section 6 presents management and QoS supply scenarios that may benefit from this work. Finally, section 7 wraps it up and presents guidelines for future studies.

2   Packet Marking on End Systems

The process of marking IP packets through the DS Field allows for the differentiation of data flow on leaf routers and differentiation of aggregate flow on other routers. Marking is based on service classes that represent a set of previously defined QoS parameters, which does not mean such classes are static.

The packet marking process is performed normally on the first routers close to the flow sources (leaf routers). On the other hand, aiming at providing data flow QoS throughout the whole transmission path (from data transmitting host to receiving host), this work suggests that the marking process be performed on the hosts. As a consequence, leaf router activity is minimized, which frees the routers to process other functions.

Also, since this host marking process offers marked packets from their source, the first flow segment can make use of level three switches that run traffic conformation and prioritization functions. This enables QoS as early as on the local network, which can be relevant for a context of applications that require QoS but do not have flow beyond the first router.

2.1   Applications Should Not Mark

Who in the host is to mark the packets? Some solutions suggest that applications assign a service class to the flows they generate. This could be done by marking the IP packets that the application directly generates. We understand this should not happen for two main reasons: architecture and management.

The DS Field belongs to the IP packet (network level). As such, only the IP layer of the protocol hierarchy should access it. If the application could access this field, hierarchy would be broken, since the transport level (between the application and network levels) would not be used. Besides that, as a rule, applications do not have direct access to IP packets. Rather, applications use API interfaces so that programmers are free from network implementation details.

However, the worst problem with application packet marking is related to management. Let's examine what could happen during the development and testing of network applications. A programmer develops an application with no concern about the priority of application flows and notes that the performance is not satisfactory. In an attempt to optimize communication and having the ability to mark, the application will start generating high priority packets and performance will increase. All developers want their applications to have optimal network data flow performance. If all applications generate high priority flows, the result will be similar to best-effort because several flows will fight for the same high priority treatment. This is certainly not advisable. Packet marking control should not be left to applications, but to a neutral, resource management entity capable to determine if the network can deal with a new flow. Therefore, the marking process must be implemented on a different context than the application.

2.2   Independent Applications

Nowadays, there are several applications that use network resources: plant production line monitors – called supervisors –, financial transaction applications, e-mail clients, web browsers and so on. Some of these use more and others use fewer network resources. This work does not intend to ignore these existing applications. On the contrary, one of the key points was not to cause changes to the applications because of the model, or at least minimize the adaptation effort to fit to the proposed architecture. Several architectures have been considered so that applications run regardless of QoS on the network. A first distinction can be made between two large groups: networks that implement QoS and networks that do not.

The first group can be further divided into hosts that implement QoS and hosts that do not. The former can be subdivided again into applications that perceive QoS and applications that do not. In every case, it is desirable that the application does not have to change its source code. In all situations, we will show that the source codes will not need changing.

2.3   The Marking Process and the Protocol Stack

Since the marking process that differentiates data flows will not be implemented by the application, we believe the best way to do so is to implement the marking process on the host protocol stack. It is important to notice that the packet is only marked after the network layer has generated the IP packet and before the data-link layer has received the data (Fig. 1). The implementation adds a module that captures the packet sent by the network layer to the data-link layer in a way that neither perceives the marking process.

3 Marking Management on End Systems

The packet marking process on the hosts is performed by looking up a local database that indicates how the packets generated by the host should receive the DS Field values. Programming this base allows changing packet marks and new flows can be identified and receive appropriate DS Field values. Packets that do not have a matching rule on the database have a programmable default handling, usually best-effort (setting zero to the DS Field).

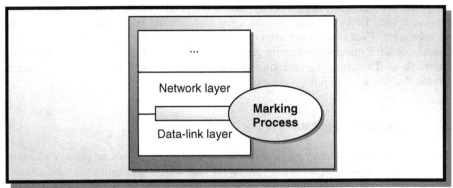

Fig. 1. Protocol Stack and Marking Process Architecture

The database can be programmed on the specific corresponding host. However, that programming requires the network manager's presence, which can be impossible where environments have a lot of equipment. Thus, we propose a solution for programming and managing the host marking process by accessing a Marking MIB.

3.1 Objectives

On a network management environment, the manager must access resources remotely. This access allows for listing the configuration of equipment and detecting faults. According to the information obtained from the resources, the network utilization ought to be optimized by performance management. The resources used can be equipment (routers, switches, hosts) as well as services (web servers, databases, news servers). Considering the packet marking process on a host as a service available on that host, one must implement mechanisms to manage this service.

Once these mechanisms are present, a sequence of management forms can be implemented. Initially, central management can statistically determine the programming of every host with a marking process. Where hosts have inadequate flow marking, the manager can remotely reconfigure hosts through a management platform. In more dynamic environments, a Bandwidth Broker (BB) substitutes for the management platform [7]. Host programming becomes automatic through the use of management policies on the BBs.

Finally, remote access to host programming should be such as to integrate management and other network features. The network manager should be able to control and monitor equipment, services and the marking process by using the same management environment. In order to do so, there must be a standard method to access hosts. Our solution uses the SNMP protocol to communicate manager and host marking process, and a Marking MIB to define what information the manager can access and program.

3.2   The Marking Process and the SNMP Protocol

In order to program the marking process remotely, it is necessary to support the management protocol chosen – SNMP. This can be achieved through direct or indirect integration.

Direct integration involves the programming of the marking process to install SNMP support. The marking process observes its own internal configuration and returns the requested information to the requesting manager. Where there is an attribution, the marking process starts marking packets according to the performed programming and updates the database so that programming is consistent (i.e., not lost next time the host is restarted).

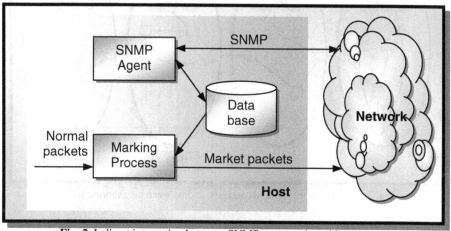

Fig. 2. Indirect integration between SNMP agent and marking process

Indirect integration does not need any changes in the marking process. However, there must be a new process to support SNMP exclusively. Integration is indirect because the marking process communicates with the SNMP support through the database. Upon request, an SNMP agent returns the values found on the base. When it is programmed, the SNMP agent changes the existing data values (Fig. 2). The marking process perceives these changes every time a new packet is marked on the host. Because the marking directly sees the base alteration, there is little or no performance loss in indirect integration. Our solution uses indirect marking so that packet marking and database programming via SNMP remain independent.

4 User-Based Signaling

QoS requires a signaling process that is used to reserve network resources. This reservation ensures that the requested QoS parameters can be effectively supplied. Signaling can be done in different ways, depending on the QoS architecture under use. For instance, ATM network signaling requires resources to be reserved in each node between source and destination of a flow. In the differentiated services architecture, signaling is performed by the BBs of the domains located between source and destination.

Signaling must start at the source of the flow and continue to the destination reserving resources. At the source, the application that perceive QoS services on the network have to use programming APIs to start the request. One of the problems is how to let applications that do not perceive QoS benefit from the network services without having their code changed.

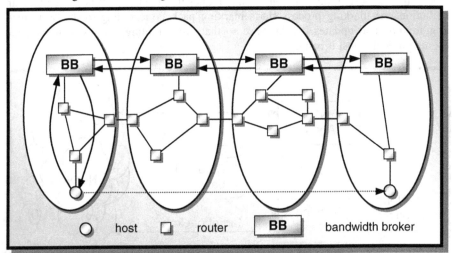

Fig. 3. Signaling between hosts and BB, and between neighboring BBs.

4.1   Signaling and BBs

The differentiated services architecture includes a resource negotiation through communication with the BBs. Request messages are sent to a domain's BB, which verifies the availability of the requested resources. The received requests can originate in neighboring domain BBs or hosts of the domain itself. Therefore, there must be a signaling protocol to enable communication between neighboring BBs, and between hosts and a BB within the same domain (Fig. 3).

Proprietary protocols can be built on TCP or UDP and used for this task. However, they may not be interesting because the communication between neighboring BBs supplied by different manufacturers is limited since BBs are not likely to interact. The use of open standards increases the chances of interactivity between BBs. IETF [3] has workgroups that conduct research on such protocols. From a technical point of view, COPS [8] is the most promising because it was developed aiming at reserving resources on a QoS environment. On the other hand, RSVP [9] is the most spread and supported signaling protocol. Another possible option is to use SNMP [5] as a protocol for requesting resources.

4.2   Are the BBs Management Systems?

The role of a BB is to receive requests, verify the resources of the local domain and make new requests towards the destination of a flow. If all resources requested to a destination can be allocated, each BB shall proceed network programming on its domain. This programming involves accessing information on the routers and configuring them so that they correctly handle the new data flow to be generated from the source.

Accessing and configuring routers are typical functions run by network management systems on best-effort environments. In this sense, one can say a BB is a network manager with a specific task: QoS management. The main difference is that whereas the management system typically needs human intervention to program the routers, a BB will do so automatically. Thus, we can say that, given this context, a BB is a management system specialized in configuring routers automatically.

A common issue to BBs and management systems is interactivity. Two management systems on different domains currently cannot communicate except on proprietary solutions. IETF faced this problem when it started SNMPv2 work [10]. The definitions of new protocol functionalities planned for manager communication using InformRequest messages. The problem was not solved because the final protocol adopted by the industry was SNMPv2c [11], which does not have such messages.

The interactivity between different domain BBs is similar. Since there is not a standard protocol to exchange information between the BBs, proprietary solutions are used in spite of their incompatibility. Although it is a serious problem for both environments (management systems and BBs), it is more urgent to solve communication between BBs. Unless there is communication, it is not possible to ensure QoS between source and target located on different domains. We believe that solving the problem of communication between managers will eventually solve the problem of communication between BBs, which in its turn will define the communication standard between hosts and the BB of the associated domain. For the time being, our work present a communication between host and BB based on SNMP, as shown below.

4.3   Motivation

Our solution needs to look up on two main problems: offering a communication mechanism between host and BB and ensuring that applications that do not perceive QoS on the network services can use them anyway.

One way to offer QoS to applications that do not perceive this possibility is by having the manager identify flows statically. Imagine a critical database application such that can be programmed for the flow generated by a replication between 6 and 7 p.m. to have priority over the HTTP flow. On one hand, the network manager is not always able to determine which flows generated by a host are the most important. In this situation, it is the user who knows it and should decide the conditions of each flow. Thus, the most appropriate way to determine QoS parameters for each flow is to obtain this information from the user.

Item 2 presented a method to identify flows from their source, by marking packets on the hosts. It is also necessary to supply a mechanism to request resources on the hosts. This mechanism must be such that applications do not need changing. Supplying such mechanisms focuses on the host a considerable part of the differentiated services logic.

5   Intra-domain Signaling Protocol

Our solution uses SNMP protocol to request QoS parameters from the host to the BB. We believe SNMP can also be used for neighboring domain BBs, but this is not within the scope of this study.

Host requests parameters by accessing a QoS MIB on the BB. Parameters are sent as a sequence of objects. When the ASK object (an integer) of the QoS MIB receives value 1, the request analysis process is started on the BB. When ASK is altered to 2, the analysis process has been completed successfully, which means the request has been accepted and the network programmed to support the new flow described. If ASK is 0, the BB has denied the request and the host can look up what parameters could be accepted by observing the new status of other objects of the QoS MIB.

5.1   Users Ask – Applications Get

It is the host that must access the QoS MIB described above. Applications that perceive QoS must proceed with SNMP request and monitor the MIB objects on the BB to check for the request return. If the request is accepted, the flows can rely on the network services since they will be programmed to support the flow described on the request.

A signaling tool has been developed for applications that do not perceive QoS (most of them). The host user uses the application to describe QoS parameters for applications that do not perceive such services on the network. Once parameters are described for each application, the user can proceed and request services to the BB. If the application started by the user is accepted, the network is programmed and the application can use the services. In short, the user defines which applications are important and which QoS parameters are most appropriate. Finally, the user sends the requests to the BB and the applications use the programming run on the network.

5.2   Implementation and Operations

The request tool is built as a standard application that uses SNMP as an information exchange protocol. Internally, the tool knows the QoS MIB on the BB. The QoS parameter requesting process takes the following steps:

1. **Selecting an application profile**. Some application profiles are previously supplied for more usual applications. For instance, a profile for POP3 clients describes an application that may have high delay and irrelevant jitter. Therefore, the discard priority can be high, since a late e-mail message does not threaten this application. On the other hand, the profile for a videoconferencing application requires low delay, controlled jitter, high flow and low discard priority.
2. If none of the existing profiles satisfactorily describes the requisites of an application, **the user can define new profiles** with appropriate parameters. The parameters of existing profiles can also be changed. It is important to notice that it is the user who decides on the appropriate QoS requisites. Pre-recorded profiles are only a starting point for a more precise definition.
3. **Once the most appropriate profile for an application is determined**, the user must schedule the utilization time of the profile. If immediate use is wanted, only the interval is to be set. As an option, the user can also assign the selected profile a process number to be monitored. If the process is ended before the profile interval finishes, it may be disabled as soon as this situation is detected.
4. **Request the QoS parameters stored on the profile to the BB**. This step is only taken on the application if the user has not scheduled a request and wishes to do it immediately. The tool will communicate with the BB and map user-supplied information onto objects of the QoS MIB. The tool will monitor the BB while waiting for a reply to the request.

When a request is scheduled, it will be run by the tool at the programmed time. Where requests are scheduled but not accepted by the BB, there is an event log that directs to the failure on the reservation process. In fact, scheduling can be carried out in two different ways, depending on the capacity of the BB:

a) **Immediate confirmation scheduling**. When a user requests scheduling, the tool checks the BB to find out whether the necessary resources will be available at the requested time. Upon affirmative response from the BB, the scheduled request will certainly be successfully run. In this case, the BB must be able to store information about reserved resources internally in order to determine whether resources will be available at a given moment. BBs that do not implement this solution are in the category presented below.
b) **Delayed confirmation scheduling**. If the BB can not immediately determine whether a scheduling can be run, the tool will request resources to the BB at the scheduled time. Denied requests will be informed through entries on a log file that the user can study.

6 Putting All Together

The local domain must have a BB that informs about the requested resources so that the signaling tool is used effectively. The BB implements the QoS MIB, which lets the signaling tool request and monitor resource requests.

Currently, no BB works according to our definitions. Therefore, there are two options to offer QoS on this environment: creating a proxy for the existing BB or creating a small testing BB. We have chosen the latter because it could be more promptly implemented in addition to providing our team with the experience of programming a BB. Our BB is simple because it implements the necessary QoS MIB, but it does not communicate with neighboring domain BBs. This communication is part of future developments and projects.

This BB has a resource scheduling feature that can be disabled. Scheduling allows hosts to mark resources that will only be used later. We have decided to allow disabling the functionality so that the host-based scheduling could be tested. In this case, the BB does not store any previous information about the resources, but the host waits and makes the request at the time set by the user (section 5.2).

Figure 4 presents a general diagram of our solution. The structures developed are enough to offer QoS from the hosts on different environments, as follows.

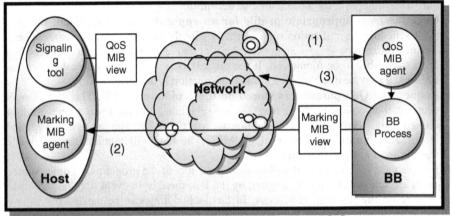

Fig. 4. Management architecture of host-based QoS

6.1   Static Flow Priority and Human Network Manager

On a network environment where there is only one management station, the human manager is responsible for accessing the hosts that will generate the many network flows. Since there is not a BB, signaling between host and BB is not possible. The users know their needs and contact the network manager personally to inform their priority flows.

Upon receipt of user requests, the network manager determines the priority flows, where they come from and their destination. The most critical hosts (web servers, SMTP and news servers) will certainly be more important. Strategic users (company president, financial manager, and so on) also have priority over common users. Finally, critical applications (data base updating, e-commerce and videoconferencing) are chosen.

Once flows are qualified, priorities are attributed. Then, the manager starts programming hosts by accessing the Marking MIB (Fig. 4, number 2). The computer network is also statically programmed to meet the specifications of the qualified flows (Fig. 4, number 3)

6.2   Dynamic Flow Priority with Network Manager Scheduling

If the managing station has an application to automate network host programming, it is possible to optimize the environment described above. The manager can delimit critical periods when special flows should have differentiated priority. For example, the managing station tool can program the host that keeps the database so that it has maximum priority when base backup is run.

6.3   Dynamic Flow Priority with Signaling Scheduling

The most complete environment is the one in which the managing station is replaced by a BB. Users request QoS parameters for their applications directly to the BB (Fig. 4, number 1). The BB checks whether the requested resources will be available at the set time. If this is not possible the signaling tool informs the user of the network capabilities available on the BB. Then, the user can proceed with a new request.

When possible, the BB will program the host (Fig. 4, number 2) at the scheduled time so that the requested flows take the appropriate QoS parameters. Network programming is also carried out (Fig. 4, number 3) so that the new flows can be effectively supported. On this last environment, QoS negotiation is dynamic and does not require intervention of a human manager, who keeps the usual management tasks.

The three environments described here show the possibilities of using the solution in different situations. There can be several other configurations depending on the necessity of each environment. The architecture of the solution allows the generation of differentiated traffic to be run from its source through the Marking MIB. The other structures (reduced BB and signaling tool) attempt to make the configuration of the host marking process easy and automatic.

7   Conclusions and Future Works

QoS management on hosts that are placed on a differentiated services network allows applications to operate more appropriately according to each user's specification. To do so, we have presented a host management architecture based on SNMP protocol and divided into two main elements: a flow marker and a signaling tool.

The flow marker lets the network manager program the marking process on the host, which involves attributing a value to the DS Field. That requires exchanging messages between the source and target IP protocol addresses, the transport protocol used (usually TCP or UDP) and the source and target transport protocol ports. Usually, the marking process is run by a router on the entry interface. Our solution improves the performance of leaf routers because packets are marked at their source (i.e., the hosts), which frees routers from this task. Because the marking process responds to SNMP messages, marking can be remotely controlled by a network management system (NMS) that accesses the Marking MIB on the host.

The second element is a signaling tool. The network manager can program the machines to prioritize certain flows, but it is the user who knows the needs of the applications on the hosts. In order to supply the manager with more precise information about how to deal with the several flows, the signaling tool is used. Users identify an application profile on their hosts and schedule a resource reservation. Also by using an SNMP protocol, the signaling tool requests the manager a special set of resources. Such request is obtained by accessing the QoS MIB that has been developed.

The management system that is used can automate the host marking process based on the requests of the signaling tool. This makes a human network manager unnecessary. Such automation can also be obtained by using Bandwidth Brokers (BBs), which receive requests from the signaling tools and negotiate resources with the neighboring BBs towards the destination of a flow. If the requested resources can be allocated, the BB also programs the source host to mark the packets generated.

Future projects to be developed to manage QoS-perceiving hosts are currently related to two issues: security and applications. Security attempts to ensure that requests from a signaling tool are not monitored by sniffers. Also, it is important to certify the user who accesses the QoS MIB in order to ensure that only valid users can access information on the BB. The plan is to use the SNMPv3 protocol [12] to solve such matters, since it can encrypt messages and because its verification system is more sophisticated than the simple use of community strings found in SNMPv1 and v2.

Application-related matters aim at supplying communication mechanisms between new applications and the signaling tool. This will allow applications to request QoS parameters, but will not need the user to know the corresponding new parameters. Communication between new applications and the signaling tool can be done in several ways and encapsulated in API programming functions still to be developed. This will require the creation of communication libraries and the alteration of the signaling tool to allow external communication.

References

1. Campbell, A., Coulson, G., Garcia, F., Hutchison, D., Leopold, H.: Integrated Quality of Service for Multimedia Communications. Proc. IEEE INFOCOM'93, pp. 732-739, San Francisco, USA (1993)
2. DiffServ Workgroup Homepage. URL: http://www.ietf.org/html.charters/diffserv-charter. html (2000)
3. IETF Homepage. URL: http://www.ietf.org (2000)
4. Internet2 Homepage. URL: http://www.internet2.org (2000)
5. Case, J., Fedor, M., Schoffstall, M., Davin, J.: A Simple Network Management Protocol (SNMP). RFC1157 (1992)
6. Perkins, D., McGinnis, E.: Understanding SNMP MIBS. Prentice Hall (1996)
7. Nichols, K., Jacobson, V., Zhang, L.: A Two-Bit Differential Services Architecture for the Internet. Internet Draft <draft-nichols-diff-svc-arch-00.txt> (1997)
8. Reichmeyer, F., Chan, K. H., Durham, D., Yavatkar, R., Gai, S., McGloughrie, K., Herzog, S., Smith, A.: COPS Usage for Policy Provisioning. <draft-sgai-cops-provisioning-00.txt>, Work in Progress (1999).
9. Braden, R., Clark, D., Shenker, S.: Resource ReSerVation Protocol (RSVP) – Version 1 Functional Specification. RFC 2205 (1997)
10. Stallings, W.: SNMP, SNMPv2 and CMIP. Addison Wesley (1993)
11. Miller, M. A.: Managing Internetworks with SNMP. 2nd edition. M&T Book (1995)
12. Stallings, W.: SNMPv3: A security Enhancement for SNMP. http://www.comsoc.org/ surveys (2000)

Throughput Stability of Reliable Multicast Protocols[*]

Öznur Özkasap[1] and Kenneth P. Birman[2]

[1] Ege University, Department of Computer Engineering, 35100 Bornova, Izmir, Turkey
ozkasap@bornova.ege.edu.tr
[2] Cornell University, Department of Computer Science, 14853 Ithaca, NY, USA
ken@cs.cornell.edu

Abstract. Traditional reliable multicast protocols depend on assumptions about flow control and reliability mechanisms, and they suffer from a kind of interference between these mechanisms. This in turn affects the overall performance, throughput and scalability of group applications utilizing these protocols. However, there exists a substantial class of distributed applications for which the throughput stability guarantee is indispensable. Pbcast protocol is a new option in scalable reliable multicast protocols that offers throughput stability, scalability and a bimodal delivery guarantee as the key features. In this paper, we focus on the throughput stability of reliable multicast protocols. We describe an experimental model developed for Pbcast and virtually synchronous protocols on a real system. We then give the analysis results of our study.

1 Introduction

Several distributed applications require reliable delivery of messages or data to all participants. Example applications include electronic stock exchanges, air traffic control systems, health care systems, and factory automation systems. Multicast is an efficient communication paradigm and a reliable multicast protocol is the basic building block of such applications. Communication properties and the degree of reliability guarantees required by such applications differ from one setting to another. Thus, reliable multicast protocols can be broadly divided into two classes, based on the reliability guarantees they provide. One class of protocols offers strong reliability guarantees such as atomicity, delivery ordering, virtual synchrony, real-time support, security properties and network-partitioning support. The other class offers support for best-effort reliability in large-scale settings.

For large-scale applications such as Internet media distribution, electronic stock exchange and distribution of radar and flight track data in air traffic control systems, the throughput stability guarantee is extremely important. This property entails the steady delivery of multicast data stream to correct destinations. For instance, Internet media distribution applications, that transmit media such as TV and radio, or teleconferencing data over the Internet, disseminate media with a steady rate. An important requirement is the steady delivery of media to all correct participants in spite of possible failures in the system. Another application group is electronic stock exchange and trading environments like the Swiss Exchange Trading System (SWX)

---

[*] This work was partially supported by a TÜBİTAK-NATO A2 grant.

T. Yakhno (Ed.): ADVIS 2000, LNCS 1909, pp. 159-169, 2000.
© Springer-Verlag Berlin Heidelberg 2000

[1]. Similarly, such applications use multicast communication protocols to disseminate trading information to all participants at the same time and with minimal delay. Throughput instability problem applies to both classes of reliable multicast protocols that we mentioned.

In this study, we focus on a new option in reliable multicast protocols. We call this protocol Bimodal Multicast, or Pbcast (probabilistic multicast) for short [2]. Pbcast offers throughput stability, scalability and a bimodal delivery guarantee. The protocol is based on an epidemic loss recovery mechanism. It exhibits stable throughput under failure scenarios that are common on real large-scale networks. In contrast, this kind of behavior can cause other reliable multicast protocols to exhibit unstable throughput.

In this study, we develop an experimental model for Pbcast protocol and virtually synchronous reliable multicast protocols offering strong reliability guarantees. We construct several group communication applications using these protocols on a real system. The aim is to investigate protocol properties, especially the throughput stability and scalability guarantees, in practice. The work has been performed on the IBM SP2 Supercomputer of Cornell Theory Center that offers an isolated network behavior. We use emulation methods to model process and link failures. Ensemble system has been ported on SP2, and a detailed analysis study of Pbcast protocol and its comparison with Ensemble's virtual synchrony protocols has been accomplished.

The paper is organized as follows: Section 2 describes reliability properties offered by two broad classes of multicast protocols, and Pbcast protocol. In section 3, we describe the throughput stability requirement and causes of the instability problem. Section 4 presents our experimental model and settings. Section 5 includes analysis and results of the study. Section 6 concludes the paper.

2   Reliability Properties of Multicast Protocols

Reliability guarantees provided by multicast protocols split them into two broad classes: *Strong reliability* and *best-effort reliability*. There is a great deal of work on communication tools offering protocols with strong reliability guarantees. Example systems include Isis [3,4], Horus [5,6], Totem [7], Transis [8] and Ensemble [9]. The other class of protocols offers support for best-effort reliability in large-scale settings. Example systems are Internet Muse protocol for network news distribution [10], the Scalable Reliable Multicast (SRM) protocol [11], the Pragmatic General Multicast (PGM) protocol [12], and the Reliable Message Transfer Protocol (RMTP) [13,14]. A new option in the spectrum of reliable multicast protocols is the Pbcast protocol [2]. We now describe basic properties offered by these multicast protocols.

Among the key properties provided as strong reliability guarantees are atomicity, ordered message delivery, real-time support and virtual synchrony. *Atomicity* means that a multicast message is either received by all destinations that do not fail or by none of them. Atomicity, which is also called all-or-nothing delivery, is a useful property, because a process that delivers an atomic multicast knows that all the operational destinations will also deliver the same message. This guarantees consistency with the actions taken by group members [15]. Some applications also require *ordered message delivery*. Ordered multicast protocols ensure that the order of messages delivered is the same on each operational destination. Different forms of ordering are possible such as FIFO, causal and total ordering. The strongest form among these is the total order guarantee that ensures that multicast messages reach all of the members in the same order [16]. Distributed real-time and control applications need *real-time support* in reliable multicast protocols. In these systems, multicast messages must be delivered at each destination by their deadlines. The *virtual synchrony* model [17] was introduced in the Isis system. In addition to message ordering, this model guarantees that membership changes are observed in the same order by all the members of a group. In addition, membership changes are totally ordered with respect to all regular messages. The model ensures that failures do not cause incomplete delivery of multicast messages. If two group members proceed from one view of membership to the next, they deliver the same set of messages in the first view. The virtual synchrony model has been adopted by various group communication systems. Examples include Transis [8], and Totem [7].

The other category includes scalable reliable multicast protocols that focus on best-effort reliability in large-scale systems. Basic properties offered are: *best-effort delivery*, *scalability* as the number of participants increases, *minimal delivery latency* of multicast messages. This class of protocols overcomes message loss and failures, but they do not guarantee end-to-end reliability. For instance, group members may not have a consistent knowledge of group membership, or a member may leave the group without informing the others.

Pbcast, which is a new option in reliable multicast protocols, is constructed using a novel gossip based transport layer. The transport layer employs random behavior to overcome scalability and stability problems. Higher level mechanisms implementing stronger protocol properties such as message ordering and security can be layered over the gossip mechanisms. In this paper, we do not go into details of the protocol. Detailed information on Pbcast is given in [2]. The protocol has the following properties:

Bimodal delivery: The atomicity property of Pbcast has a slightly different meaning than the traditional 'all-or-nothing' guarantee offered by reliable multicast protocols. Atomicity is in the form of 'almost all or almost none', which is called bimodal delivery guarantee.

Message ordering: Each participant in the group delivers Pbcast messages in FIFO order. In other words, multicasts originated from a sender are delivered by each member in the order of generation. As mentioned in [18], stronger forms of ordering like total order can be provided by the protocol. [19] includes a similar protocol providing total ordering.

Scalability: As the network and group size increase, overheads of the protocol remain almost constant or grow slowly compared to other reliable multicast protocols. In addition, throughput variation grows slowly with the log of the group size.

Throughput stability: Throughput variation observed at the participants of a group is low when compared to multicast rates. This leads to steady delivery of multicast messages at the correct processes.

Multicast stability detection: Pbcast protocol detects the stability of multicast messages. This means that the bimodal delivery guarantee has been achieved. If a message is detected as stable, it can be safely garbage collected. If needed, the application can be informed as well. Although some reliable multicast protocols like SRM do not provide stability detection, virtual synchrony protocols like the ones offered in Ensemble communication toolkit include stability detection mechanisms.

Loss detection: Because of process and link failures, there is a small probability that some multicast messages will not be delivered by some processes. The message loss is common at faulty processes. If such an event occurs, processes that do not receive a message are informed via an up-call.

3 Throughput Stability

For large-scale distributed applications that motivate our work, the throughput stability guarantee is extremely important. This property entails the steady delivery of multicast data stream to correct destinations.

Traditional reliable multicast protocols depend on assumptions about response delay, failure detection and flow control mechanisms. Low-probability events caused by these mechanisms, such as random delay fluctuations in the form of scheduling or paging delays, emerge as an obstacle to scalability in reliable multicast protocols. For example, in a virtual synchrony reliability model, a less responsive member exposing such events can impact the throughput of the other healthy members in the group. The reason is as follows. For the reliability purposes, such a protocol requires the sender to buffer messages until all members acknowledge receipt. Since the perturbed member is less responsive, the flow control mechanism begins to limit the transmission bandwidth of the sender. This in turn affects the overall performance and throughput of the multicast group. In effect, these protocols suffer from a kind of interference between reliability and flow control mechanisms. Moreover, as the system size is scaled up, the frequency of these events rises, and this situation can cause unstable throughput.

An observation on the throughput instability problem of reliable multicast protocols offering strong reliability is mentioned in the Swiss Exchange Trading System (SWX) [1]. SWX developers observed some shortcomings that they attribute to the multicast protocols (and strong reliability guarantees) provided by Isis. For instance, one slow client could affect the entire system, especially under peak load. Also, multicast throughput was found to degrade linearly as the number of clients increased.

Throughput instability problem does not only apply to the traditional protocols using virtually synchronous reliability model. Scalable protocols based on best-effort reliability exhibit the same problem. As an example, recent studies [20,21] have shown that, for the SRM protocol, random packet loss can trigger high rates of request and retransmission messages. In addition, this overhead grows with the size of the system.

4 Experimental Model

A theoretical analysis of Pbcast is given in [2]. A simulation model and analysis of the protocol, and also its comparison with a best-effort protocol are discussed in [22]. In this paper, we describe our experimental study for the protocol and Ensemble's virtually synchronous protocols. The main focus is to investigate and analyze protocol properties, giving attention to stability and scalability, in practice.

The experimental work has been performed on the SP2 system of Cornell Theory Center that offers an isolated network behavior. SP2 consists of nodes connected by an ethernet and a switch. A node is a processor with associated memory and disk. Cornell Theory Center's SP2 system has total 160 nodes that share data via message passing over a high performance two-level cross bar switch.

In this work, as shown in fig. 1, Ensemble group communication system has been ported on SP2, and many process group applications utilizing Pbcast and Ensemble's traditional reliable multicast protocols have been designed. Emulation methods have been used to model process and link failures. A detailed experimental study and analysis of Pbcast, and also its comparison with Ensemble's reliable multicast protocols has been accomplished.

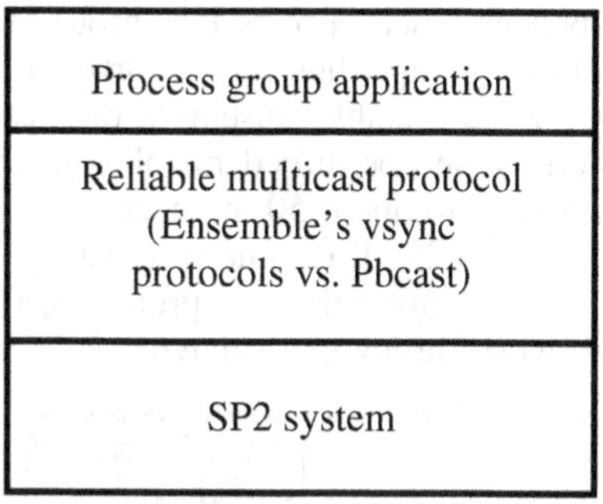

Process group application
Reliable multicast protocol (Ensemble's vsync protocols vs. Pbcast)
SP2 system

Fig. 1. Block diagram of the model

Our interest is in the performance of the protocols in the case of soft process failures. We emulate a process failure, such as a slow or overloaded member, by forcing the process to sleep with varied probabilities. We call a group member subject to such a failure as '*perturbed*', and the probability of failure that impacts the process as '*perturb rate*'. We have constructed process group applications on Ensemble toolkit for various group sizes starting from 8-member case up to 128-member process groups. There exists one sender process that disseminates 200 multicast messages per second to the group participants. During the execution of group application, some members were perturbed, that is forced to sleep during 100 millisecond intervals with varied perturb rates. First, we designed experiments so that one member is perturbed for various group sizes. Then, we increased the percentage of perturbed members up to 25% of the group size. In other words, we arranged the application so that, one or more group members would occasionally pause, allowing incoming buffers to fill and eventually overflow, but then resume computing and communication before the background failure detection used by the system have detected the situation. This behavior is common in the real world, where multicast applications often share platforms with other applications.

5 Analysis and Results

Based on the results of process group executions described above, we first examine the throughput behavior of Ensemble's virtually synchronous protocols. We then focus on the throughput behavior of Pbcast and also protocol overhead associated with soft failure recovery. During our experiments, we varied a number of operating parameters. These are; n: size of process group (8 to 128), f: number of perturbed processes (1 to n/4), p: degree of perturbation (0.1 to 0.9).

5.1   Throughput Behavior of Virtually Synchronous Protocols

In this part, process group applications utilize Ensemble's traditional and scalable multicast protocols based on virtual synchrony reliability model. We investigate and analyze the throughput behavior of these protocols. We varied operating parameters n, f and p. We measure throughput at the unperturbed or correct group members. The data points in the analysis correspond to values measured during 500 millisecond intervals. Fig. 2 shows some analysis results for 32 and 96-member process groups. Graphs show the superimposed data for cases f=1 and f=n/4. We see that even a single perturbed group member impacts the throughput of unperturbed members negatively. The problem becomes worse as the group size (n), percentage of perturbed members (f), and perturb rate (p) grow. If we focus on the data points for a single perturb rate, we see that the number of perturbed members affects the throughput degradation. For instance, in fig. 2, for a 96-member group when the perturb rate is 0.1, the throughput on non-perturbed members for the scalable Ensemble multicast protocol is about 90 messages/second when there is one perturbed member in the group. The throughput for the same protocol decreases to about 50 messages/second when the number of perturbed members is increased to 24. The same observation is valid for the traditional Ensemble multicast protocol. Among the two protocols, the traditional Ensemble multicast protocol shows the worst throughput behavior.

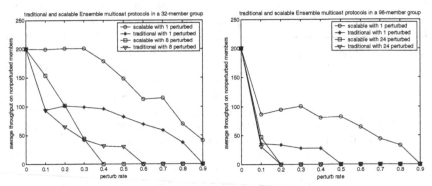

Fig. 2. Throughput performance of Ensemble's reliable multicast protocols

Fig. 3 shows the impact of an increase of group size on the throughput behavior clearly, when f=1. In the next section, we show that, under the same conditions, Pbcast achieves the ideal output rate even with high percentage of perturbed members.

5.2   Throughput Behavior of Pbcast Protocol

In this part, process group applications utilize Pbcast protocol. We investigate the scalability and stability properties of Pbcast. We mainly focus on the following analysis cases:

   a) Throughput as a function of perturb rate for various group sizes
   b) Throughput as a function of proportion of perturbed members
   c) Protocol overhead associated with soft failure recovery as a function of group size.

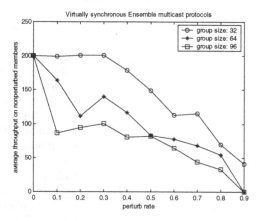

Fig. 3. Throughput behavior as a function of group size

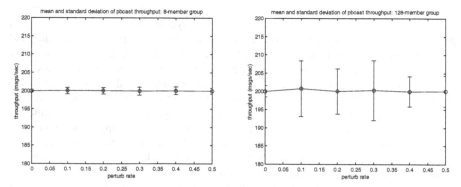

Fig. 4. Variation of pbcast throughput

We varied operating parameters n, f and p. We measure throughput at the unperturbed or correct group members. The data points in the analysis correspond to values measured during 500 millisecond intervals. Since the throughput was steady, we also computed the variance of these data points. Fig. 4 shows variation of throughput measured at a typical receiver as the perturb rate and group size increase. The group size is 8 and 128 respectively. These sample results are for the experiments where f=n/4. We can conclude that as we scale a process group, throughput can be maintained even if we perturb some group members. The throughput behavior remains stable as we scale the process group size even with high rates of failures. During these runs no message loss at all was observed at unperturbed members. On the other hand, the variance does grow as a function of group size. Fig. 5 shows throughput variance as group size increases. Although the scale of our experiments was insufficient to test the log-growth predictions of computational results for Pbcast [2], the data is consistent with those predictions. As we saw in the previous section, the same conditions provoke degraded throughput for traditional virtually synchronous protocols.

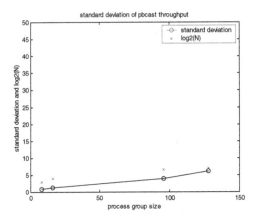

Fig. 5. Throughput variance of Pbcast as a function of group size

We can conclude that Pbcast is more stable and scalable compared to the traditional multicast protocols. As the perturbed process begins to sleep for long enough to significantly impact Ensemble's flow control and windowed acknowledgement, the fragility of the traditional multicast protocols becomes evident very quickly,. Furthermore, in such a condition, high data dissemination rates can quickly fill up message buffers of receivers, and hence can cause message losses due to buffer overflows.

In the case of virtually synchronous protocols, a perturbed process is particularly difficult to manage. Since the process is sending and receiving messages, it is not considered to have failed. But, it is slow and may experience high message loss rates, especially in the case of buffer overflows. The sender and correct receivers keep copies of unacknowledged messages until all members deliver them. It causes available buffer spaces to fill up quickly, and activates background flow control mechanisms. Setting failure detection parameters more aggressively has been proposed as a solution [1]. But, doing so increases the risk of erroneous failure detection approximately as the square of the group size in the worst-case. Because, all group members monitor one another and every member can mistakenly classify all the other (n-1) members as faulty where n is the group size. Then, the whole group has n*(n-1) chances to make a mistake during failure detection. Since the failure detection parameters are set aggressively in such an approach, it is more likely that randomized events such as paging and scheduling delays will be interpreted as a member's crash. As group size increases, failure detection accuracy becomes a significant problem. Most success scenarios with virtual synchrony use fairly small groups, sometimes structured hierarchically. In addition, the largest systems have performance demands that are typically limited to short bursts of multicast.

In this study, we analyzed protocol overhead associated with soft failure recovery, as well. For this purpose, retransmission behavior at a correct member was investigated. Fig. 6 shows overhead as perturb rate increases, for 8 and 128-member groups, respectively. For these graphs f=n/4, and each region in the graphs illustrates data points measured during 500 msec intervals for a certain perturb rate. For instance, the first region contains data points for p=0.1, second one is for p=0.2, and so on. Fig. 7.a superimposes the data for n= 8, 16, 64 and 128, and shows the percentage of messages retransmitted as p increases for various n.

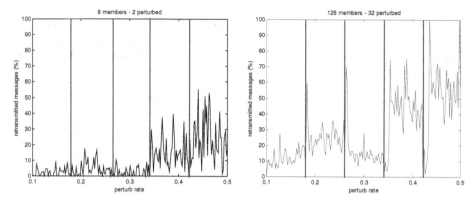

Fig. 6. Pbcast overhead associated with soft failure recovery

For these experiments, we also compute the theoretical worst-case bounds for retransmission behavior at a correct member (fig. 7.b). Assume r is the number of multicast data messages per second disseminated to the group by the sender, and p is the perturb rate. In every 100 msec (which is the duration of a gossip round in the experiments), at most $((r/10)*p)$ messages are missed by a faulty member, and a correct member gossips to two randomly selected group members. In the worst-case, if these two members are faulty and they lack all $((r/10)*p)$ data messages, they request retransmissions of these messages from the correct member. Then, the correct member retransmits at most $2*((r/10)*p) = (r*p)/5$ messages in every 100 msec. In our experiments, we measured data points during 500 msec intervals, and computed the percentage of retransmitted messages to the multicast data messages disseminated by the sender during each interval. If we compute theoretical values for 500 msec intervals, the correct member retransmits at most $5*(r*p)/5 = r*p$ messages, and the sender disseminates r/2 messages during every 500 msec interval. Then, the bound for the percentage of retransmitted messages would be $(r*p)/(r/2) = 2*p$ in the worst-case. Fig. 7.b shows the computed theoretical worst-case bounds. Note that, our experimental results are below the theoretical bound, and the results confirm that overhead on the correct processes is bounded as the size of process group increases. As the group size increases, we observed an increase in the percentage of retransmitted messages. We believe, this is mainly due to the increase in the number of perturbed members with the group size. Because, in these experiments, number of perturbed members equals 25% of the group size $(f = n/4)$.

6 Conclusion

Our study yields some general conclusion about the behavior of basic Pbcast and virtually synchronous multicast protocols. In the first part of the study, we focused on the virtually synchronous Ensemble multicast protocols in the case of soft process failures. We showed that even a single perturbed group member impacts the throughput of unperturbed members negatively. On the other hand, Pbcast achieves the ideal throughput rate even with high percentage of perturbed members. In the second part of the study, we focused on the performance of Pbcast in the case of soft process failures. We showed that the throughput behavior of Pbcast remains stable as we scale the process group size even with high rates of failures. Furthermore, our results confirm that overhead on the correct processes is bounded as the size of process group increases.

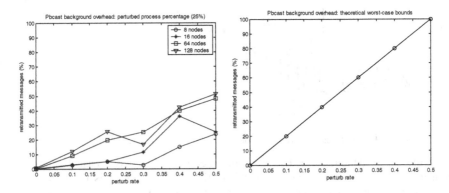

Fig. 7. Percentage of message retransmissions as a function of p. a) Experimental results, b) Theoretical worst-case bounds

References

1. Piantoni, R. and Stancescu, C., 1997, Implementing the Swiss Exchange Trading System, FTCS 27, Seattle, WA, 309–313p.
2. Birman, K.P., Hayden, M., Ozkasap, O., Xiao, Z., Budiu, M. and Minsky, Y., 1999, Bimodal Multicast, ACM Transactions on Computer Systems, 17(2), 41–88p.
3. Birman, K.P. and van Renesse, R., 1994, Reliable Distributed Computing with the Isis Toolkit, New York: IEEE Computer Society Press.
4. Birman, K.P., 1993, The Process Group Approach to Reliable Distributed Computing, Communications of the ACM, 36(12), 37–53p.
5. Van Renesse, R. and Birman, K.P., 1995, Protocol Composition in Horus, Technical Report, TR95-1505, Department of Computer Science, Cornell University.
6. Van Renesse, R., Birman, K.P. and Maffeis, S., 1996, Horus: A Flexible Group Communication System, Communications of the ACM, 39(4), 76–83p.
7. Moser, L.E., Melliar-Smith, P.M., Agarwal, D.A., Budhia, R.K., et.al, 1996, Totem: A Fault-tolerant Multicast Group Communication System, Communications of the ACM, 39(4), 54–63p.
8. Dolev, D. and Malki, D., 1996, The Transis Approach to High Availability Cluster Communication, Communications of the ACM, 39(4), 64–70p.
9. Hayden, M., 1998, The Ensemble System, Ph.D. dissertation, Cornell University Dept. of Computer Science.
10. Lidl, K., Osborne, J. and Malcome, J., 1994, Drinking from the Firehose: Multicast USENET News, USENIX Winter 1994, 33–45p.
11. Floyd, S., Jacobson, V., Liu, C., McCanne, S. and Zhang, L., 1997, A Reliable Multicast Framework for Light-weight Sessions and Application Level Framing, IEEE/ACM Transactions on Networking, 5(6), 784–803p. http://www-nrg.ee.lbl.gov/floyd/srm.html

12. Speakman, T., Farinacci, D., Lin, S. and Tweedly, A., 1998, PGM Reliable Transport Protocol, Internet-Draft.
13. Paul, S., Sabnani, K., Lin, J. C. and Bhattacharyya, S., 1997, Reliable Multicast Transport Protocol (RMTP), IEEE Journal on Selected Areas in Communications, special issue on Network Support for Multipoint Communication, 15(3), http://www.bell-labs.com/user/sanjoy/rmtp2.ps
14. Lin, J.C. and Paul, S., 1996, A Reliable Multicast Transport Protocol, Proceedings of IEEE INFOCOM '96, 1414-1424p. http://www.bell-labs.com/user/sanjoy/rmtp.ps
15. Cristian, F., Aghili, H., Strong, R. and Dolev, D., 1985, Atomic Broadcast: From Simple Message Diffusion to Byzantine Agreement. Proc. 15th International FTCS, 200-206p.
16. Lamport, L., 1978, The Implementation of Reliable Distributed Multiprocess Systems, Computer Networks, 2, 95–114p.
17. Birman, K.P. and Joseph, T.A., 1987, Exploiting Virtual Synchrony in Distributed Systems, Proceedings of the 11th Symposium on Operating System Principles, New York: ACM Press, 123–128p.
18. Birman, K.P., 1997, Building Secure and Reliable Network Applications, Manning Publishing Company and Prentice Hall, Greenwich, CT. http://www.browsebooks.com/Birman/index.html
19. Hayden, M. and Birman, K.P., 1996, Probabilistic Broadcast, Technical Report, TR96-1606, Department of Computer Science, Cornell University.
20. Liu, C., 1997, Error Recovery in Scalable Reliable Multicast, Ph.D. dissertation, University of Southern California.
21. Lucas, M., 1998, Efficient Data Distribution in Large-Scale Multicast Networks, Ph.D. dissertation, Dept. of Computer Science, University of Virginia.
22. Ozkasap, O., Xiao, Z. and Birman, K.P., 1999.a, Scalability of Two Reliable Multicast Protocols, Technical Report, TR99-1748, Department of Computer Science, Cornell University.

On the Rearrangeability of Multistage Networks Employing Uniform Connection Pattern

Rza Bashirov

Department of Mathematics,
Eastern Mediterranean University
Gazi Magusa, TRNC (via Mersin-10, Turkey)

Abstract. In this paper, the rearrangeability of $(2s - 1)$-stage networks is proved. This result is used to prove that $(2\log_n N{-}1)$-stage nonsymmetric networks employing uniform connection pattern, two passes through s-stage networks with the same kth and $(s{-}k{+}1)$st stages, and $2\log_n N{-}1$ circulations through single-stage networks are rearrangeable.

1   Introduction

Multistage networks with the capability of performing all the $N!$ permutations on N elements are known as rearrangeable networks. Much has been written about rearrangeable multistage networks. In particular, the rearrangeability of back-to-back omega [1], back-to-back baseline [2] and back-to-back butterfly [3] networks has been extensively studied.

Shuffle-exchange networks are made of (2×2)-switches and based on connection pattern called perfect shuffle, first introduced in [4]. It is well-known that at least $2\lg N - 1$ shuffle-exchanges[1] are necessary for rearrangeability [5]. Are $2\lg N - 1$ shuffle-exchange stages also sufficient for rearrangeability? This has been conjectured to be true and has actually been proved for $N \leqslant 8$. In [6], by using algebraic techniques it was shown that $(3\lg N)$-stage shuffle-exchange network is rearrangeable. It was observed that $3\lg N - 1$ stages are indeed sufficient for rearrangeability of shuffle-exchange network [7]. It was shown later that $3\lg N - 3$ stages are sufficient for $N = 16$ and 32 [8]. The rearrangeability of $3\lg N - 4$ shuffle-exchange stages was established using a constructive approach [9]. In [10], the authors used the balanced-matrix method to prove that $3\lg N - 4$ shuffle-exchange stages are sufficient for rearrangeability.

During the past decade, many attempts have been made to prove the conjecture on the rearrangeability of $2\lg N - 1$ shuffle-exchange stages. An incorrect proof of the conjecture was given in [11]. The error was in assuming that $(2\lg N - 1)$-stage shuffle-exchange network is topologically equivalent to a serial cascade of two baseline networks. Another claim for a proof of the conjecture, made in [12], also turned out to be incorrect. Attempts to prove the conjecture on rearrangeability of $(2\lg N - 1)$-stage shuffle-exchange network for $N \geqslant 8$

---

[1] $\lg N = \log_2 N$.

T. Yakhno (Ed.): ADVIS 2000, LNCS 1909, pp. 170–179, 2000.

have so far been unsuccessful and the conjecture still remains open — the best known lower-bound is $3 \lg N - 4$ stages [9,10]. The above mentioned results on rearrangeable networks are summarized in [5,13]. In [5] the authors bring up the following related problem: "Is there any $(2 \lg N - 1)$-stage network employing a uniform connection pattern, that can be shown to be rearrangeable? No such network is known at present."

In this paper, by using the results on rearrangeability of nonsymmetric networks obtained in [14] we prove that $2 \log_n N - 1$ stages are sufficient for the rearrangeability of a network made of $(n \times n)$-switches, that employs a uniform connection pattern. This particularly implies the rearrangeability of $(2 \lg N - 1)$-stage shuffle-exchange network. The proof is constructive and establishes the type of connection pattern between stages.

In Section 2, we deal with the terminology to be used in the present paper. In Section 3, we prove the main result. In Section 4, we are concerned with application of this result.

2   Nomenclature

In this section we will discuss the terminology used in the sequel. The set theoretic approach used in this paper comes from [15].

We say that a multistage network is *symmetric*, if first-to-middle stages of the network is an inverse of its middle-to-last stages. In the opposite case, the network is *nonsymmetric*.

Consider a $(2s-1)$-stage network with N inputs and N outputs labeled from 0 to $N - 1$. Let $X = \{0, \ldots, N - 1\}$. Suppose that R is a partition of N into r pairwise disjoint subsets $N_1, \ldots, N_r$. In fact, R defines a stage consisting of r crossbar switches. Now consider two consecutive stages, respectively partitioned by R_k and R_{k+1}. The output x of the stage partitioned by R_k is linked to the input $\varphi(x)$ of the stage partitioned by R_{k+1}, where φ is a permutation from X to X. Denote by $\varphi(B)$ the set $\{\varphi(x) : x \in B \subseteq X\}$. Let $\varphi(R_k)$ be the partition of X such that $\varphi(R_k) = \{\varphi(B) : B \in R_k\}$. We will say that $\varphi(B)$ *intersects* R_k if $A \in R_{k+1}$ implies $\varphi(B) \cap A \neq \emptyset$. If so, the switch determined by B in the stage partitioned by R_k is directly connected to all the switches in the stage partitioned by R_{k+1} over interstage channels in φ. Similarly, $\varphi(R_k)$ *crosses* R_{k+1} if $B \in R_k$ implies that $\varphi(B)$ intersects R_{k+1}. If so, then each switch in the stage partitioned by R_{k+1} can be reached by every switch in the previous stage, partitioned by R_k, over an interstage channel in φ. Let R and S be a partitions of X. We will say that R *strictly refines* S, denoted $R \succ S$, if $R \neq S$ and each element of S is the union of elements of R. Similarly, R *refines* S, denoted $R \succeq S$, if either $R \succ S$ or $R = S$. The symbol $_AR$ will denote the partition $A \subseteq X$ induced by R, such that $_AR = \{C \cap A : C \in R\}$.

3 Rearrangeable Nonsymmetric Networks

In this section we will derive under certain conditions our main result, which concerns the proof of the rearrangeability of $(2s - 1)$-stage nonsymmetric networks.

Consider a $(2s - 1)$-stage network for $s \geqslant 2$. Assume that the ith stage of the network is partitioned by R_i, $i = 1, \ldots, 2s - 1$, and is connected to the $(i + 1)$th stage over interstage channels in φ_i, $i = 1, \ldots, 2s - 2$. Suppose that R^k, $k = 1, \ldots, s - 1, s + 1, \ldots, 2s - 1$, are the partitions of N such that

$$R^{s-1} \succcurlyeq R^{s-2} \succcurlyeq \cdots \succcurlyeq R^1, \tag{1}$$

$$R^{s+1} \succcurlyeq R^{s+2} \succcurlyeq \cdots \succcurlyeq R^{2s-1}, \tag{2}$$

$$R_s = R^{s-1} = R^{s+1}. \tag{3}$$

Suppose that

$$\Psi^k = \begin{cases} \{N\}, & k = 0 \\ \varphi_k^{-1} \cdots \varphi_{s-1}^{-1}(R^k), & k = 1, \ldots, s - 1 \\ \varphi_k \cdots \varphi_s(R^{k+1}), & k = s, \ldots, 2s - 2 \\ \{N\}, & k = 2s - 1 \end{cases}$$

are partitions of X such that

(c1) $_{\varphi_{k-1}(B)}\Psi^k$ crosses $_{\varphi_{k-1}(B)}R_k$ for $B \in \Psi^{k-1}$, $k = 1, \ldots, s - 1$,
(c2) $_{\varphi_{k+1}^{-1}(B)}\Psi^k$ crosses $_{\varphi_{k+1}^{-1}(B)}R_{k+1}$ for $B \in \Psi^{k+1}$,
      $k = s, \ldots, 2s - 2$,
(c3) $R_k = R_{2s-k}$, $k = 1, \ldots, s - 1$

with $\varphi_0 = \varphi_{2s-1} = e$, where e is the identity permutation.

Theorem 1 *A $(2s - 1)$-stage network satisfying (1)-(3) and (c1)-(c3) is rearrangeable.*

The proof is by induction on the number of the stages [14].

Although a multistage network satisfying conditions (1)-(3) and (c1)-(c3) is rearrangeable, it is technically difficult to verify the rearrangeability by using Theorem 1. This is because this theorem does not provide a technique for defining the partitions $R^1, \ldots, R^{2s-1}$. However, we can define these partitions for a particular class of nonsymmetric $(2s-1)$-stage networks.

Let $n_1, \ldots, n_{2s-1}$ be positive integers such that

$$\prod_{k=1}^{s} n_k = \prod_{k=s}^{2s-1} n_k = N \tag{4}$$

where $n_k \geq 2$ for all k. Suppose the partitions R_k, introduced in the previous section, are as follows:

$$R_k = \{A_{kt}\},$$

$$A_{kt} = \{x : tn_k \leq x < (t+1)n_k\}, \tag{5}$$

where $t = 0, \ldots, \frac{N}{n_k} - 1$, $k = 1, \ldots, 2s - 1$. Let us define partitions $R^k = \{D_{ki}\}$, where

$$D_{ki} = \left\{ x : \frac{iN}{\prod_{h=1}^{k} n_h} \leq x < \frac{(i+1)N}{\prod_{h=1}^{k} n_h} \right\} \tag{6}$$

for $i = 0, \ldots, \prod_{h=1}^{k} n_h - 1$, $k = 1, \ldots, s - 1$, and

$$D_{ki} = \left\{ x : x \in A_{st}, t \equiv i \left(\bmod \frac{N}{\prod_{h=s}^{k-1} n_h} \right) \right\} \tag{7}$$

for $i = 0, \ldots, \frac{N}{\prod_{h=s}^{k-1} n_h} - 1$, $k = s+1, \ldots, 2s-1$, $t = 0, \ldots, \frac{N}{n_s} - 1$.

Lemma 1 *The partitions $R^1, \ldots, R^{s-1}$, defined by (6), and the partitions $R^{s+1}, \ldots, R^{2s-1}$, defined by (7), satisfy (1) and (2), respectively.*

Proof. From (6), we have

$$D_{ki} = \left\{ x : \frac{iN}{\prod_{h=1}^{k} n_h} \leq x < \frac{(i+1)N}{\prod_{h=1}^{k} n_h} \right\} = \left\{ x : \frac{in_{k+1}N}{\prod_{h=1}^{k+1} n_h} \leq x < \frac{(i+1)n_{k+1}N}{\prod_{h=1}^{k+1} n_h} \right\}$$

$$= \bigcup_{j=in_{k+1}}^{(i+1)n_{k+1}-1} \left\{ x : \frac{jN}{\prod_{h=1}^{k+1} n_h} \leq x < \frac{(j+1)N}{\prod_{h=1}^{k+1} n_h} \right\} = \bigcup_{j=in_{k+1}}^{(i+1)n_{k+1}-1} D_{k+1,j}.$$

Consequently, every element of R^k is the union of elements of R^{k+1}, which means that R^{k+1} refines R^k for $k = 1, \ldots, s-1$. That is why the partitions, defined by (6), satisfy condition (1).

On the other hand, it follows from (7) that

$$D_{k+1,i} = \{ x : x \in D_{kj}, j \equiv i \,(\bmod\, n_k) \},$$

where $i = 0, \ldots, \frac{N}{\prod_{h=s}^{k} n_h} - 1$. Hence, R^k refines R^{k+1}, $k = s+1, \ldots, 2s-1$.

Combining these results, we obtain that the partitions defined by (6) and (7) satisfy conditions (1) and (2), respectively, as claimed. □

We now consider the partitions $R^k(l) = \{B_{li}^k\}$ where

$$B_{li}^k = \left\{ x : x \in A_{lt}, t \equiv r \left(\bmod \prod_{h=1}^{l-1} n_h \right), i \leq \frac{r}{\prod_{h=k+1}^{l-1} n_h} < i+1 \right\} \tag{8}$$

for $i = 0, \dots, \prod_{h=1}^{k} n_h - 1$, $l = k+1, \dots, s$, $k = 1, \dots, s-1$, and

$$B_{li}^{k} = \left\{ x : x \in A_{lt}, t \equiv r \left(\bmod \prod_{h=k+s-l}^{2s-1} n_h \right), i \leq \frac{r}{\prod_{h=s}^{l-1} n_h} < i+1 \right\} \qquad (9)$$

for $i = 0, \dots, \prod_{h=k}^{2s-1} n_h - 1$, $l = s, \dots, k-1$, $k = s+1, \dots, 2s-1$. The double inequalities in (8) and (9) are not defined when $l = k+1$ and $l = s$, respectively. In these two cases they are substituted by $i \leq r < i+1$.

Lemma 2 *For $k = 1, \dots, s-1, s+1, \dots, 2s-1$ the following equality holds*

$$R^k(s) = R^k.$$

Proof. Substituting l by s in (8), we obtain

$$B_{si}^{k} = \left\{ x : x \in A_{st}, t \equiv r \left(\bmod \prod_{h=1}^{s-1} n_h \right), i \leq \frac{r}{\prod_{h=k+1}^{s-1} n_h} < i+1 \right\}$$

$$= \left\{ x : x \in A_{st}, \ t \equiv r \left(\bmod \frac{N}{n_s} \right), i \leq \frac{r}{\prod_{h=k+1}^{s-1} n_h} < i+1 \right\}$$

$$= \left\{ x : x \in A_{st}, \ i \leq \frac{t}{\prod_{h=k+1}^{s-1} n_h} < i+1 \right\}$$

$$= \left\{ x : tn_s \leq x < (t+1)n_s, i \prod_{h=k+1}^{s-1} n_h \leq t < (i+1) \prod_{h=k+1}^{s-1} n_h \right\}$$

$$= \left\{ x : i \prod_{h=k+1}^{s} n_h \leq x < (i+1) \prod_{h=k+1}^{s} n_h \right\}$$

$$= \left\{ x : \frac{iN}{\prod_{h=1}^{k} n_h} \leq x < \frac{(i+1)N}{\prod_{h=1}^{k} n_h} \right\}.$$

Then $B_{si}^{k} = D_{ki}$. Consequently, $R^k(s) = R^k$ for $k = 1, \dots, s-1$.

The proof of the statement for $k = s+1, \dots, 2s-1$ may be obtained as in the case considered above. Since $l = s$, from (9) we obtain

$$B_{si}^{k} = \left\{ x : x \in A_{st}, t \equiv r \left(\bmod \prod_{h=k}^{2s-1} n_h \right), i \leq r < i+1) \right\}$$

$$= \left\{ x : x \in A_{st}, \ t \equiv i \left(\bmod \prod_{h=k}^{2s-1} n_h \right) \right\} = \left\{ x : x \in A_{st}, \ t \equiv i \left(\bmod \frac{N}{\prod_{h=s}^{k-1} n_h} \right) \right\}.$$

Hence, $R^k(s) = R^k$ for $k = s+1, \dots, 2s-1$. $\qquad \square$

Suppose now that the lth and the $(l+1)$st stages of the network are connected over interstage channels in φ_l. Suppose also that for $x \in B_{li}^{k}$

$$x \equiv t \left(\bmod \frac{N}{n_{l+1}} \right) \text{ if and only if } \varphi_l(x) \in A_{l+1, t}$$

where $i = 0, \ldots, \prod_{h=1}^{k} n_h - 1$ and either $k = 1, \ldots, s-1$, $l = k+1, \ldots, s$ or $k = s+1, \ldots, 2s-1$, $l = s, \ldots, k-1$.

Lemma 3 *Under the above condition, the following equality holds*

$$B_{l+1,i}^k = \varphi_l \left(B_{li}^k \right).$$

Proof. If $x \in B_{li}^k$, from (5) and (8) we obtain

$$i \prod_{h=k+1}^{l} n_h + \alpha \prod_{h=1}^{l} n_h \leq x < (i+1) \prod_{h=k+1}^{l} n_h + \alpha \prod_{h=1}^{l} n_h$$

where α is a nonnegative integer. Suppose $x \equiv t \left(\mathrm{mod} \frac{N}{n_{l+1}} \right)$. From this condition and the above inequality we have

$$i \prod_{h=k+1}^{l} n_h + \beta \prod_{h=1}^{l} n_h \leq t < (i+1) \prod_{h=k+1}^{l} n_h + \beta \prod_{h=1}^{l} n_h$$

where β is some nonnegative integer. Clearly, the index t determined above matches with that for $A_{l+1,\, t} \in B_{l+1,\, i}^k$, which means $\varphi_l(x) \in B_{l+1,\, i}^k$. Consequently, $B_{l+1,\, i}^k = \varphi_l(B_{li}^k)$ for $k = 1, \ldots, s-1$, $l = k+1, \ldots, s$.

One can verify that the statement of the lemma is also correct when $k = s+1, \ldots, 2s-1$, $l = s, \ldots, k-1$. This can be done by simply comparing x from (5) and (9) with the index t of $A_{l+1,\, t}$. $\qquad\square$

Corollary 1 *The following identity holds:*

$$R^k(l+1) = \varphi_l(R^k(l)).$$

where either $k = 1, \ldots, s-1$, $l = k+1, \ldots, s$ or $k = s+1, \ldots, 2s-1$, $l = s, \ldots, k-1$.

Proof. This result follows from Lemma 3 by taking into consideration that permutation φ_l is a one-to-one mapping. $\qquad\square$

A sufficient condition for rearrangeability of $(2s-1)$-stage network is given by the following theorem.

Theorem 2 *Given a $(2s-1)$-stage network, $s \geq 2$, suppose its stages are defined by partitions described by (5). Suppose the kth stage of the network is made of $\frac{N}{n_k}$ $n_k \times n_k$ crossbar switches where $n_1, \ldots, n_{2s-1}$ satisfy (4), and let $n_k = n_{2s-k}$, $k = 1, \ldots, s-1$. Furthermore, the kth and the $(k+1)$st stages are interconnected over channels in permutation φ_k such that $x \equiv t \left(\mathrm{mod} \frac{N}{n_{k+1}} \right)$ if and only if $\varphi_k(x) \in A_{k+1,t}$. Then the network is rearrangeable.*

Proof. The basic idea of the proof is to show that the so defined networks satisfy the conditions of Theorem 1.

Assume that partitions $R^1, \ldots, R^{s-1}$ and $R^{s+1}, \ldots, R^{2s-1}$ are defined by (6) and (7). By Lemma 3, we know that these partitions satisfy the conditions (1) and (2).

Letting $k = s - 1$, from (6) we have

$$D_{i,s-1} = \left\{ x : \frac{iN}{\prod_{h=1}^{s-1} n_h} \leq x < \frac{(i+1)N}{\prod_{h=1}^{s-1} n_h} \right\} = \{x : in_s \leq x < (i+1)n_s\} = A_{st}.$$

Consequently, $D_{i,s-1} = A_{st}$, and, therefore, the partitions R^{s-1} and R_s are made of the same elements, which means that $R^{s-1} = R_s$. On the other hand, if $k = s + 1$, from (7), we get

$$D_{s+1,i} = \left\{ x : x \in A_{st} : t \equiv i \left(\mod \frac{N}{n_s} \right) \right\} = \{x : x \in A_{st}\}$$

which implies $R^{s+1} = R_s$. By combining the above observations, we conclude that (3) holds for the proposed networks.

Now let us show that condition (c1) holds as well. That is, $_{\varphi_{k-1}(B)}\Psi^k$ crosses $_{\varphi_{k-1}(B)}R_k$ for each $B \in \Psi^{k-1}$, and $k = 1, \ldots, s-1$. From (6), it easily follows that $\varphi_{k-1}(B) = \varphi_k^{-1} \cdots \varphi_{s-1}^{-1}(D_{k-1, i})$. On the other hand, Lemma 4 implies that $D_{k-1, i} = B_{si}^{k-1}$, and hence $\varphi_{k-1}(B) = \varphi_k^{-1} \cdots \varphi_{s-1}^{-1}(B_{s, i}^{k-1})$. By repeatedly applying Corollary 1, we obtain

$$\varphi_k^{-1} \cdots \varphi_{s-1}^{-1}(B_{s,i}^{k-1}) = \varphi_k^{-1} \cdots \varphi_{s-2}^{-1}(B_{s-1,i}^{k-1}) = \cdots = \varphi_k^{-1}\left(B_{k+1,i}^{k-1}\right) = B_{ki}^{k-1}.$$

Consequently, $_{\varphi_{k-1}(B)}R_k =_{B_{ki}^{k-1}} R_k = \{A_{kt}\}$ where $t \equiv i \left(\mod \prod_{h=1}^{k-1} n_h\right)$ and $i = 0, \ldots, \prod_{h=1}^{k-1} n_h - 1$. Any x from the above set is of the form $x = in_k + \alpha \prod_{h=1}^{k} n_h + \delta$ where $\alpha = 0, \ldots, \prod_{h=k+1}^{s} n_h$, $\delta = 0, \ldots, n_k - 1$. On the other hand, the condition $x \equiv t \left(\mod \frac{N}{n_{k+1}}\right)$ implies that $x - t = \beta \prod_{h=1}^{k} n_h$ where β is a nonnegative integer. By combining this fact with the property of x obtained above, one can verify that $t = in_k + \delta + \gamma \prod_{h=1}^{k} n_h$ where γ is a nonnegative integer. By applying Lemma 2 and Lemma 3, it is easy to show that $_{\varphi_{k-1}(B)}\Psi^k = \varphi_k^{-1}\left(B_{ki}^{k-1}R^k\right) = \varphi_k^{-1}\{A_{k+1, t}\}$ for $t \equiv j \left(\mod \prod_{h=1}^{k} n_h\right)$, $j = 0, \ldots, \prod_{h=1}^{k} n_h - 1$. Now it is clear that $t = j + \beta \prod_{h=1}^{k} n_h$ where $\beta = 0, \ldots, \prod_{h=k+2}^{s} n_h$. It is obvious that $x - t = (\alpha - \beta)n_1 \cdots n_k$. Hence, $_{\varphi_{k-1}(B)}\Psi^k$ crosses $_{\varphi_{k-1}(B)}R_k$ for $B \in \Psi^{k-1}$, $k = 1, \ldots, s-1$, that is, condition (c1) holds for the proposed networks.

The proof of validity of condition (c2) for the proposed networks is similar to the one for condition (c1). Finally, it is assumed in the statement of the theorem that condition (c3) is correct for considered networks. We have shown that conditions (1)–(3) and (c1)–(c3) are fulfilled, which completes the proof. □

4   Applications

In this section we will be concerned with the applicability of Theorem 2. A network proposed by this theorem can be described by several properties. First, the kth stage of a network is made of $\frac{N}{n_k}$ $(n_k \times n_k)$-switches. Second, the connection pattern used to interconnect the kth and the $(k+1)$st stages of a network links the first output of the first switch in the kth stage to the first input of the first switch in the $(k+1)$st stage, the second output of the first switch in the kth stage to the first input of the second switch in the $(k+1)$st stage, etc. Likewise, the outputs of the second switch in the kth stage are connected to the first inputs of the next the n_k switches in the $(k+1)$st stage. Continuing in like fashion, the outputs of the ith switch in the kth stage are connected to the first idle inputs of the next n_k switches in the $(k+1)$st stage.

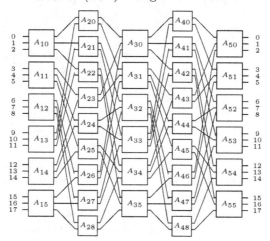

Fig. 1. 5-stage $3 \times 2 \times 3 \times 2 \times 3$ network

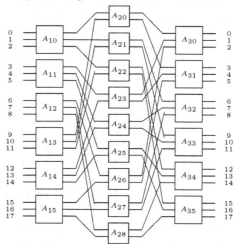

Fig 2. 3-stage $3 \times 2 \times 3$ network

Given N, the number of left (or right) subnetworks of a network satisfying the conditions of Theorem 2 is equal to the sum of the numbers of generalized permutations of factors $n_1, \ldots, n_s$ over all possible factorizations of N. Fig. 1 shows one of these networks with $n_1 = n_3 = n_5 = 3$, $n_2 = n_4 = 2$.

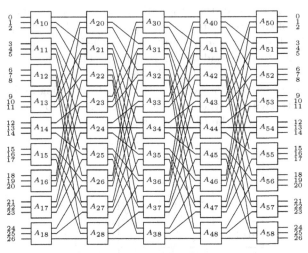

Fig. 3. 5-stage $3 \times 3 \times 3 \times 3 \times 3$ network

By setting $n_k = n_{s+k-1}$, $k = 1, \ldots, s$, in Theorem 2, we get a network whose kth and $(s+k-1)$st stages are made of the same switches. This network uses the same connection pattern to link the kth and the $(k+1)$st stages in the left subnetwork and the $(s+k-1)$st and the $(s+k)$th stages in the right subnetwork. Indeed, such network is a cascade of two copies of the same s-stage network. Now let us interconnect the outputs of the last stage of the s-stage network to the inputs of its first stage via wrap around channels such that the ith output is linked to the ith input for all i's. One pass through the $(2s-1)$-stage network can be processed in two passes through the wrap around s-stage network. The rearrangeability of the $(2s-1)$-stage network implies that of two passes through the wrap around s-stage network. For example, one pass through the 5-stage $3 \times 2 \times 3 \times 2 \times 3$ network in Fig.1 can be processed in two passes through the $3 \times 2 \times 3$ network in Fig.2.

Now let us assume that $n_k = n$, $k = 1, \ldots, 2s - 1$. A nonsymmetric network proposed by Theorem 2, that satisfies this condition is a serial cascade of $(2 \log_n N - 1)$ stages where each stage is made of $\frac{N}{n}$ $(n \times n)$-switches. Such a network uses a uniform connection pattern to link any two consecutive stages. One pass through such network can be processed in $2 \log_n N - 1$ circulations through a single-stage network made of $\frac{N}{n}$ $(n \times n)$-switches that employs exactly same connection pattern between its output and input terminals as does $(2 \log_n N - 1)$-stage one. The rearrangeability of the $(2 \log_n N - 1)$-stage network implies that of a single-stage one. For instance, one pass through the 5-stage network in Fig. 3 can be done by a cycling through its single stage $2 \log_3 27 - 1 = 5$ times.

If $n = 2$ the connection pattern is shuffle-exchange. This means that the $(2 \lg N - 1)$-stage shuffle-exchange network is rearrangeable. This is the solution of the problem conjectured in [7].

5 Summary

New results on the rearrangeability of $(2s - 1)$-stage networks are presented in this paper. The main result is the proof of the rearrangeability of $(2s - 1)$-stage nonsymmetric networks. This result is used to show that $(2 \log_n N - 1)$-stage nonsymmetric network made of $(n \times n)$-switches that employs uniform connection pattern, two passes through networks with the same kth and $(s - k + 1)$st stages, and $2 \log_n N - 1$ circulations through single-stage networks employing relevant connection patterns are rearrangeable.

References

1. D. H. Lawrie. Access and alignment of data in an array processor, *IEEE Transactions on Computers*, **C-24**(12), 1975, pp. 1145–1155.
2. C. Wu, T. Feng. On a class of multistage interconnection networks, *IEEE Transactions on Computers*, **C-29**(8), 1980, pp. 694–702.
3. L. R. Goke, G. J. Lipovski. Banyan networks for partitioning multiprocessor systems, in *Proceedings of 1st annual symposium on computer architecture*, 1973, pp. 21–28.
4. H. S. Stone. *Parallel processing with perfect shuffle*. IEEE Transactions on Computers, **C-20** (2), 1971, pp. 153–161.
5. A. Varma, C. S. Raghavendra. *Interconnection Networks for Multiprocessors and Multicomputers*. IEEE Computer Society Press, 1994.
6. D. S. Parker. *Notes on shuffle/exchange type switching networks*. IEEE Transactions on Computers, **C-29** (3), 1980, pp. 213–222.
7. C. L. Wu, T. H. Feng. *The universality of shuffle/exchange network*. IEEE Transactions on Computers, **C-30** (5), 1981, pp. 324–332.
8. C. K. Kotari, S. Lakshmivarahan. *A note on rearrangeable networks*. Technical report, School of Engineering and Computer Science, University of Oklahoma, 1983.
9. A. Varma, C. S. Rahavendra. *Rearrangeability of multistage shuffle exchange networks*. IEEE Transactions on Communications, **36** (10), 1988, pp. 1138–1147.
10. N. Linital and M. Tarzi. *Interpolation between bases and the shuffle exchange networks*. European Journal of Combinatorics, **10**, 1989, pp. 29–39.
11. F. Soviš. *On rearrangeable networks of shuffle/exchange type*. Computers and Artificial Intelligence, **7** (4), 1988, pp. 359-373.
12. H. Cam, J. A. B. Fortes. Rearrangeability of shuffle/exchange networks, in *Proceedings Frontiers of Massively Parallel Computation*, 1990, pp. 303–314.
13. Frank K. Hwang. *The mathematical theory of nonblocking switching networks networks*. World Scientific, 1999.
14. R. Bashirov. On the rearrangeability of $(2s - 1)$-stage nonsymmetric interconnection networks, in *Proceedings 2000 International Conference on Parallel and Distributed Processing Techniques and Applications, June 2000*, pp. 907–912.
15. V. E. Beneš. *Mathematical theory of connecting networks and telephone traffic*. New York, Academic Press, 1965.

Numeric Constraint Resolution in Logic Programming Based on Subdefinite Models

Evgueni Petrov

Institut de recherche en informatique de Nantes
2 rue de Houssinière
BP 98000, Nantes, cedex 3 France
petrov@irin.univ-nantes.fr

Abstract. Subdefinite computations introduced in [12] are a method of representation and processing of partial specifications. With respect to Constraint Programming, subdefinite computations are a generic constraint resolution techniques. The theoretic part of the paper describes, in turn, subdefinite computations, their properties, and a method for integration of subdefinite computations into logic programming. The practical part describes an implementation of subdefinite computations with subdefinite numbers in the logic programming system ECL^iPS^e (Interval Domain library), empirically compares the Interval Domain library to a state-of-art solver of non-linear constraints and describes an application of the Interval Domain library to a problem from financial mathematics. This research is supported by grant 98–06 from l'Institut d'A. M. Liapounov d'informatique et de mathématiques appliquées.

1 Introduction

Because people usually express their knowledge in an implicit way employing partial information, computers need special means in order to "understand" such partial specifications. Few years ago in the field of Constraint Programming (CP), it has been proposed to simply add a control mechanism to these specifications provided they are sufficiently formal [11].

This approach has merged into object oriented, logic and imperative programming, and that produced such constraint logic and imperative programming systems and object oriented libraries as CLP(BNR), ECL^iPS^e, PROLOG-IV, Kaleidoscope, ILOGSolver, NeMo+ [4,14,5,10,15,2]. Besides that, there exist such systems as UniCalc, Numerica which entirely belong to CP proper [16,1].

Subdefinite computations introduced in [12] are a method of representation and processing of partial specifications. With respect to Constraint Programming, subdefinite computations are a generic constraint resolution techniques. Given a set of constraints, SD computations produce a compact description of some set containing all the solutions to the constraints (Section 2).

ECL^iPS^e is a logic programming system offering some powerful facilities for implementation of constraint satisfaction techniques (Section 3.1).

T. Yakhno (Ed.): ADVIS 2000, LNCS 1909, pp. 180–190, 2000.
© Springer-Verlag Berlin Heidelberg 2000

The theoretic part of the paper describes, in turn, subdefinite computations, their properties, and a method for integration of subdefinite computations into logic programming. The practical part describes an implementation of subdefinite computations with subdefinite numbers in the logic programming system ECL^iPS^e (Interval Domain library), empirically compares the Interval Domain library to a state-of-art solver of non-linear constraints and describes an application of the Interval Domain library to a problem from financial mathematics (Sections 3, 4).

2 Constraints and Subdefiniteness

In the context of this paper, a *constraint* is an expression formed of some relation, constants and variables. A *constraint satisfaction problem* (CSP) is a finite set of constraints. A *solution* to a CSP C is a valuation of the variables under which each constraint in C holds. The expression $\text{sol}(C)$ denotes the set of all solutions to a CSP C. A CSP C is (globally) consistent iff $\text{sol}(C) \neq \emptyset$.

SD computations have been introduced in the early 1980's and are intensely studied by our colleagues from Ershov Institute of Informatics Systems Russian Acad. Sci. and Russian Research Institute of Artificial Intelligence. With respect to constraint programming, subdefinite computations are a method for resolution of the CSP's over partially specified (subdefinite) data. This section describes SD computations and establishes the relationship between CSP's over subdefinite and ordinary data.

Let the symbols $\{Q, \ldots\}$, $\{a, \ldots\}$ denote some relations over some set $\mathcal{D}$ and, respectively, its elements. Let $\{x, y, z, \ldots\}$ be a set of variables. A *subdefinite extension* is an injection $(\cdot)^\star$ from $2^{\mathcal{D}}$ into $2^{\mathcal{D}}$ which meets the conditions:

$$\emptyset^\star = \emptyset, \quad d \subseteq d^\star, \quad (d^\star)^\star = d^\star, \quad d_1 \subseteq d_2 \Longrightarrow d_1^\star \subseteq d_2^\star.$$

Convex closure for finite set domains in [7], hull closure for numeric domains in [3] are examples of subdefinite extensions.

Usually, a subdefinite extension is strictly injective (see the preceding examples). Images of subsets of $\mathcal{D}$ under a subdefinite extension $(\cdot)^\star$ form a set $\mathcal{D}^\star$ of so-called *subdefinite values*. Each relation Q over $\mathcal{D}$ and each element $a \in \mathcal{D}$ define a relation $Q^\star$ over $\mathcal{D}^\star$ and a subdefinite value $a^\star \in \mathcal{D}^\star$ as follows:

$$a^\star = \{a\}^\star, \quad \text{for all} \quad d_1^\star, d_2^\star, \ldots \in \mathcal{D}^\star \quad \left(Q^\star(d_1^\star, d_2^\star, \ldots) \iff Q \cap \prod d_i^\star \neq \emptyset \right).$$

Finally, each CSP C over $\mathcal{D}$ is transformed to the following CSP $C^\star$ over $\mathcal{D}^\star$:

$$\{Q^\star(\ldots x^\star \ldots) | \quad Q(\ldots x \ldots) \in C\}.$$

The latter CSP is called a *subdefinite model* of the CSP C. Figure 1 shows a scheme of transformation of a CSP to a subdefinite model. Theorem 1 relates solutions to C and $C^\star$.

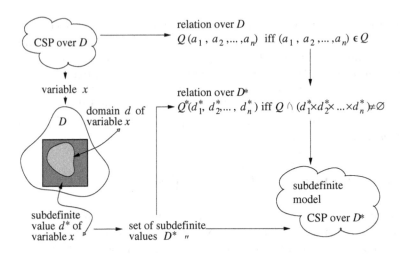

Fig. 1. Transformation of a CSP over $\mathcal{D}$ to a subdefinite model

Th. 1 (Approximation) *For any set $\mathcal{D}$, any SD extension $(\cdot)^\star$, any CSP C over $\mathcal{D}$, each solution γ to C, each solution α to $C^\star$, and any mapping α' from the variables to $\mathcal{D}^\star$, the following statements hold:*

1. $\{x \mapsto \{\gamma(x)\}^\star\}$ *is a solution to* $C^\star$
2. $\{x \mapsto (\alpha(x) \cup \alpha'(x))^\star\}$ *is a solution to* $C^\star$

The theorem has the following corollary. For any set $\mathcal{D}$, any SD extension $(\cdot)^\star$, any consistent CSP C over $\mathcal{D}$, there is such a solution α to the subdefinite model $C^\star$ of C that $\mathrm{sol}(C) \subseteq \{\gamma |\ \gamma(x) \in \alpha(x)$ for each variable $x\}$. Such solutions are called *subdefinite solutions* to the CSP C over $\mathcal{D}$ with respect to $(\cdot)^\star$.

Given a CSP C and a particular SD extension $(\cdot)^\star$, SD computations generate a chain $\alpha = \{x \mapsto \mathcal{D}^\star\} \supseteq \alpha' \supseteq \ldots$ of subdefinite solutions to C (the inclusions apply to each variable). In this chain, each α is transformed to α' by a *calculation function* of some constraint $Q(\ldots y \ldots) \in C$ (y takes k-th place) as follows:

$$\alpha'(y) = \mathrm{Pr}_k(Q \cap \prod d_i^\star)^\star, \quad \alpha'(x) = \alpha(x) \quad \text{for} \quad x \neq y.$$

Above (a) Pr_k projects a subset a Cartesian product onto k-th coordinate, (b) $d_i^\star$ is either $a^\star$ for some constant a or $\alpha(x)$ for some variable x, depending upon the content of the i-th place in the constraint. This calculation function *reads* the variables occurring in the constraint and *writes* y.

Each CSP defines a network of calculation functions which is similar to networks of constraints proposed by other authors. The network contains nodes of two types, variables (constants) and calculation functions. Each arc in the network connects a variable (a constant) to a calculation function. If the calculation function reads the variable (the constant), then the arc is directed toward the calculation function. Otherwise, the arc is directed toward the variable. The network splits into a number of overlapping "stars", the center of each star being a

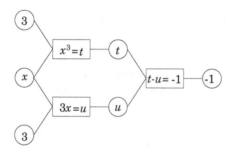

Fig. 2. Equation $x^3 - 3x + 1 = 0$ as a graph

calculation function and the rays reaching the variables this function reads and writes. An alternative decomposition of the network, which puts a variable to the center of each star, is also possible.

SD computations propagate each change of the domain of each variable x to the neighbours of x — the variables written by the calculation functions reading x. This data-driven propagation raises a wave of domain updates in the network of calculation functions. Because the domains never grow, this wave expires and SD computations stop.

Let us apply the technique to $x^3 = t$, $3x = u$, $t - u = -1$ which are equivalent to $x^3 - 3x + 1 = 0$ (see Fig.2). If we use interval domain representation, i.e., $a^\star = [\min a, \max a]$, then the six calculation functions are

$$t := t \cap [\min x^3, \max x^3],$$
$$x := x \cap [\sqrt[3]{\min t}, \sqrt[3]{\max t}],$$
$$x := x \cap [\min u/3, \max u/3],$$
$$u := u \cap [3\min x, 3\max x],$$
$$u := u \cap [1 + \min t, 1 + \max t],$$
$$t := t \cap [\min u - 1, \max u - 1].$$

If we set $t = u = x = [-100, 100]$, then after few chaotic movements, domain updates start looping over x, t, u (see Fig.2) and finally produce $t = [-6.385\ldots, 3.596\ldots]$, $u = [-5.638\ldots, 4.596\ldots]$, $x = [-1.879\ldots, 1.532\ldots]$. The latter is guaranteed to contain all real roots of the equation.

3   SD Computations in ECL^{i}PS^e

ECL^{i}PS^e is an abbreviation for *ECRC Common Logic Programming System*. It is a Prolog-based system whose aim is to serve as a platform for creation of various extensions of logic programming. Implementation of SD computations in ECL^{i}PS^e includes a data structure for networks of calculation functions and a control mechanism for data-driven computations.

3.1 ECL^{i}PS^e and Its Specific Data Structures

ECL^{i}PS^e offers two new data structures: *meta-term* and *delayed goal*. Using meta-terms and delayed goals, applications can organize additional information and control flows independently of Prolog standards. Papers [9,8] give more details of usage of meta-terms and delayed goals in ECL^{i}PS^e.

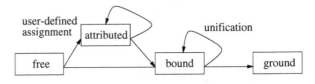

Fig. 3. The set of states of a variable in ECL^{i}PS^e

A meta-term consists of two or more terms, the first term visible to "everyone", called *Prolog value* of the meta-term, and the others, called *meta-attributes*, visible only to few tools which convert meta-terms to standard Prolog data and vice versa. A meta-term is written like T{name1:T1,...} where T is its Prolog value, T1 is its meta-attribute name1, etc. Unlike the standard Prolog, ECL^{i}PS^e allows for a variable a finer set of states than {free, bound, ground}. Figure 3 shows the possible transitions between the basic states of a variable in ECL^{i}PS^e.

A delayed goal represents a Prolog goal that should be invoked in the future. There are three major operations with delayed goals: creation, scheduling for execution, and execution of all scheduled goals. A scheme in Fig. 4 shows these and other operations on (delayed) goals and their effect.

A delayed goal is written like 'GOAL'(G) where G is the goal that has been delayed and 'GOAL' is a label indicating that fact. The unification procedure processes each delayed goal as an atom — a unique, indecomposable object, — though, outside unification, this object has the same components as G, e.g.

```
[eclipse 1]:
        make_suspension(X=Y, 3, S), make_suspension(X=Y, 3, T), S = T.
no (more) solution.
[eclipse 2]: make_suspension(X=Y, 3, S), write(S), Y=1.
GOAL(X = Y)
X = X
S = 'GOAL'(X = 1)
Y = 1

Delayed goals:
        X = 1
yes.
```

3.2 Implementation of SD Computations

Each network of calculation functions can be decomposed into the overlapping stars of two types. Therefore, in order to map the network of calculation functions

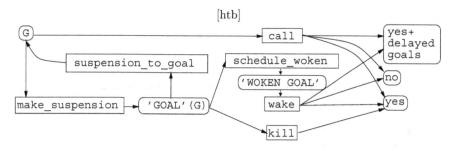

Fig. 4. Operations on goals in ECL^{i}PS^e

into data structures available in ECL^{i}PS^e, it suffices to work out a representation for stars of each type. Figure 5 shows the rules for computing a representation of the network of calculation functions. The upper part of the figure describes the transformation of a calculation function of a constraint $Q(x \ldots y \ldots z)$ (y takes i-th place). The lower part of the figure describes the transformation of a variable which is read by calculation functions F, G, H, etc. Because in the network of calculation functions the stars of different types overlap, the rules in Fig. 5 are mutually recursive and the representation they describe may be a cyclic term. ECL^{i}PS^e is able to process such cyclic terms correctly because they contain delayed goals which, in fact, break the cycles (see Section 3.1).

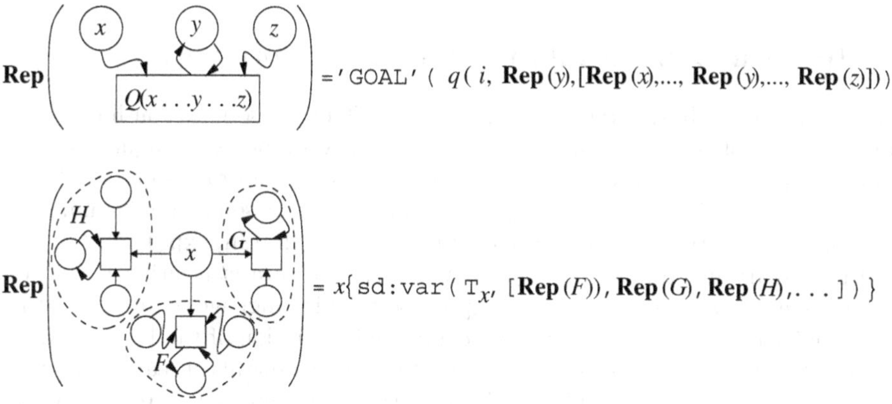

$$\textbf{Rep} \left(\begin{array}{c} x \quad y \quad z \\ Q(x \ldots y \ldots z) \end{array} \right) = \text{'GOAL'} (\; q(\; i, \textbf{Rep}(y), [\textbf{Rep}(x), \ldots, \textbf{Rep}(y), \ldots, \textbf{Rep}(z)]))$$

$$\textbf{Rep} \left(\begin{array}{c} H \\ x \\ G \\ F \end{array} \right) = x\{ \texttt{sd:var}(\; T_{x'}\; [\textbf{Rep}(F)), \textbf{Rep}(G), \textbf{Rep}(H), \ldots]) \}$$

Fig. 5. A representation for the network of calculation functions

Data-driven invocation and execution of calculation functions is modeled with the help of operations on delayed goals. The calculation function of a constraint $Q(\ldots y \ldots)$ (y takes k-th place) reading the variables $\bar{v} = (\ldots y \ldots)$ and writing y is transformed into an ECL^{i}PS^e predicate which updates y as specified in Section 2 and operates on delayed goals. Figure 6 presents the ECL^{i}PS^e code of such a predicate. In order to avoid technical details, the description is given modulo the following predicates:

```
q(K, Y, V):-
  delay(q(K, Y, V), V),
  read_domain(Y, TY),
  maplist(read_domain, V, TV),
%
% compute a new domain TYO for Y --- depends on the predicate
%
  ( TYO == TY -> true ;
    write_domain(Y, TYO),
    read_delayed_goals(Y, FY),
    write_delayed_goals(Y, []),
    schedule_woken(FY),
    wake
  ).
```

Fig. 6. A simplified code of a calculation function

1. `delay(G, T)` — for each variable in the term `T`, add the delayed goal `'GOAL'(G)` to the list of delayed goals stored in the meta-attribute `sd` of this variable;
2. `read_domain(V, `T_v`)` (`write_domain(V, `T_v`)`) — read from (write into) the meta-attribute `sd` of `V` the domain T_v;
3. `read_delayed_goals(V, `F_v`)` (`write_delayed_goals(V, `F_v`)`) — read from (write into) the meta-attribute `sd` of `V` the list of delayed goals F_v;

4   Interval Domain Library and Its Performance

Interval Domain library (ID) is an extension of ECL^{i}PS^e solving non-linear constraints over integer and real numbers. The library supports a language of numeric expressions, equations and inequalities. A *numeric expression* is either a number, or a variable, or a Prolog term constructed of some simpler numeric expressions with the help of the unary functors cos, sin, tan, exp, log, $\sqrt{\cdot}$, $(\cdot)^2$ and the binary functors $+$, $-$, $*$, $/$, $\char`^$. A *numeric constraint* is either $e = e'$, or $e \leq e'$ where e, e' are two numeric expressions. This language is interpreted in the standard way. See [17] for a more complete description of the ID library.

The ID library implements SD computations over intervals of real and integer numbers. The bounds of real intervals are always numbers exactly representable in the double precision format.

The ID library consists of Prolog and C++ parts implementing SD computations and, respectively, interval calculation functions, and an interface between the two. The C++ part of the library has been adopted from the kernel of the UniCalc solver previously developped by our colleagues from Russian Research Institute of Artificial Intelligence.

The table in Fig. 7 shows the performance of the ID library and the Numerica system on the set of problems from [16]. Numerica is a powerful solver of polynomial constraints which combines the interval Newton method and symbolic transformations from interval analysis with consistency techniques from

	i1	i2	i3	i4	b40	b80	b160	b320	m10	m20	m40	m80
n	10	20	20	10	40	80	160	320	10	20	40	80
	[−2,2]	[−1,2]	[−2,2]			[−1,1]				[−4,5]		
Δt	0.11	0.26	0.24	51.91	20.70	45.51	91.29	198.94	2.07	8.69	54.60	170.98
$\dfrac{\Delta t}{\Delta t'}$	·	·	·	·	·	2.19	2.00	2.18	·	4.19	6.28	3.13
Δt	0.06	0.30	0.31	73.94	9.79	22.13	48.30	113.71	4.07	24.49	192.81	1752.64
$\dfrac{\Delta t}{\Delta t'}$	·	·	·	·	·	2.26	2.18	2.35	·	6.02	7.87	9.08

Fig. 7. Comparison of the ID library with the Numerica system

constraint programming. Each column of the table shows (from top to bottom) the identifier (1), dimension (2), initial domains (3), amount of time spent and slow-down wrt the preceding instance for the ID library (4–5) and Numerica (6–7). Time has been measured on a machine with a 100MHz Pentium processor, 32Mb of memory, under Linux (for the ID library) and with a 40MHz Super-SPARC processor under Solaris (for the Numerica system). The typical values of SPECfp95 for these machines are 2.59 and, respectively, 1.38.

Each of the problems i1, i2 i3, i4 is a sparse system of polynomial equations. The number of variables in each system is either 10 or 20, the degree of its highest monome does not exceed 11. Each of the problems b40, b80, b160, b320 is to find the common zero of Broyden functions of 40, 80, 160 and, respectively, 320 variables. The i-th Broyden function of n variables is given by the following expression

$$x_i(2 + 5x_i^2) + 1 - \sum_{j \in [1,n] \cap [i-5,i+1]} x_j(1 + x_j).$$

Each of the problems m10, m20, m40, m80 is to approximate the values $x_j = x(t_j)$ of a function x in the n=10, 20, 40, 80 points $t_j = j/(n + 1)$. The function satisfies the following integral equation

$$x(t) = -\frac{1}{2}\left((1 - t)\int_0^t \tau(x(\tau) + \tau + 1)^3 d\tau + t\int_t^1 (1 - \tau)(x(\tau) + \tau + 1)^3 d\tau\right).$$

Despite a simpler resolution technique, the ID library competes well with Numerica on the problems of fixed dimension and copes even better with scalable problems.

4.1 Asset Allocation

Investment in an asset is associated with some expected return as well as some risk which measures how much our actual return may deviate from that expected value. Portfolio theory says that returns on a combination of assets (portfolio)

```
[eclipse 3]: Y=[Y1,Y2,Y3,Y4,Y5,Y6,Y7,Y8],Y ** 0..infty,
T=L*(Y1/18.5+Y2/5+Y3/16+Y4/23+Y5/30+Y6/26+Y7/20+Y8/24-1)+sqr(Y1)+
2*Y1*(Y2*0.45+Y3*0.7+Y4*0.2+Y5*0.64+Y6*0.3+Y7*0.61+Y8*0.79)+sqr(Y2)+
2*Y2*(Y3*0.27-Y4*0.01+Y5*0.41+Y6*0.01+Y7*0.13+Y8*0.28)+sqr(Y3)+
2*Y3*(Y4*0.14+Y5*0.51+Y6*0.29+Y7*0.48+Y8*0.59)+sqr(Y4)+
2*Y4*(Y5*0.25+Y6*0.73+Y7*0.56+Y8*0.13)+sqr(Y5)+
2*Y5*(Y6*0.28+Y7*0.61+Y8*0.75)+sqr(Y6)+
2*Y6*(Y7*0.54+Y8*0.16)+sqr(Y7)+2*Y7*Y8*0.44+sqr(Y8),
L*==-2*T,Y1/18.5+Y2/5+Y3/16+Y4/23+Y5/30+Y6/26+Y7/20+Y8/24-1*==0,
maplist(diff(T,[L|Y]),Y,DT),checklist(*=<(0),DT),
multiply(Y,DT,YDT),checklist(*==(0),YDT),locate([],Y).
```

Fig. 8. Query for optimum asset allocation

vary less then returns on a single asset. That fact leads us to the following question: what portfolio should we decide to invest in, in order to risk as little as possible?

Constructing the minimum risk portfolio is a quadratic programming problem, though in certain cases a simpler technique suffices [6]. Here ID library is employed to construct minimum risk portfolio assuming no short sales and riskless lending-borrowing rate.

Let assets and invested amounts be denoted by integers 1, 2, 3, etc. and x_1, x_2, x_3, etc. Each asset i is described by standard deviation of its actual returns σ_i. Besides that, each pair of assets $(i\ j)$ is described by a correlation coefficient ρ_{ij} indicating how returns on these assets move together. We have $-1 \leq \rho_{ij} \leq 1$ and $\rho_{ii} = 1$ for all i and j.

Let y_i be $x_i\sigma_i$. Then square of standard deviation of actual returns σ^2 is $\Sigma_{ij} y_i y_j \rho_{ij}$. The amounts x_i's must be non-negative and sum to 1. So, we are to minimize σ^2 subject to $y_i \geq 0$ and $\Sigma_i y_i/\sigma_i = 1$.

Let us make some observations. Each minimum of σ^2 subject to $\Sigma_i y_i/\sigma_i = 1$ is as well a minimum of $\theta = \sigma^2 + \lambda(\Sigma_i y_i/\sigma_i - 1)$ with free variables λ and y_i's. Each minimum of θ subject to $y_i \geq 0$ meets Kuhn-Tucker conditions:

$$\frac{\partial\theta}{\partial\lambda} = 0, \quad y_i\frac{\partial\theta}{\partial y_i} = 0, \quad \frac{\partial\theta}{\partial y_i} \geq 0.$$

So, we have $0 = \Sigma_i y_i \frac{\partial\theta}{\partial y_i} = 2\theta + \lambda$, or $\lambda = -2\theta$.

These equations and inequalities are directly passed to ID library for resolution. Figure 8 shows a query which constructs the optimum portfolio of large stocks (S&P's index including dividends, asset 1), bonds (Lehman Brothers aggregate index, asset 2), small stocks (CRSP, asset 8), emerging market (asset 5), mutual stock funds in Canada, Japan, Pacific region, and Europe (assets 3, 4, 6, 7). The example comes from [6], pp.105–109. For this particular data, solution takes 165 seconds on a 100MHz Pentium under Linux. In general, time consumed depends mainly on correlation coefficients and is within thirty minutes for problems with up to 30 assets.

5   Conclusion

This paper contributes to Constraint Logic Programming a generic implementation of a constraint satisfaction method parameterized by domain representation (so-called subdefinite computations). In order to be more precise, we describe the implementation in terms of ECL^{i}PS^e data types "delayed goal" and "metaterm" which extend the standard set of Prolog data types. Thus we refer only to industrial logic programming systems (SICStus, ECL^{i}PS^e, etc.) which offer a similar functionality.

Subdefinite computations are intensely studied by our colleagues from Russian Institut of Artificial Intelligence and A. P. Ershov Institut of Informatics Systems of Russian Acad. Sci. Our generic implementation of subdefinite computations opens a direct way to Constraint Logic Prorgamming for the software developped by these teams. For example, we have successfully applied our implementation to plug into ECL^{i}PS^e an advanced set constraint solver developped in Russian Institute of Artificial Intelligence [13].

Acknowledgements. We are cordially grateful to our colleagues Alexandre Vassilievitch Zamulin, Tamara Petrovna Kashevarova, Yuri Alexeevitch Zagorul'ko from A. P. Ershov Institute of Informatics Systems SB RAS, Alexandre Semenovitch Narin'yani from Russian Research Institute of Artificial Intelligence for their kind attention, support, comments during numerous seminars in Novosibirsk (Russia). We gratefully acknowledge the truly valuable support of professor Frédéric Benhamou, Laurent Granvilliers and the Constraints team from l'Institut de recherche en informatique de Nantes (France). We sincerely thank l'Institut d'A. M. Liapunov d'informatique et de mathématiques appliquées for the financial support provided for the Project 98–06.

References

1. A. Babichev, O. Kadyrova, T. Kashevarova, and A. Semenov. Unicalc — as a tool for solving problems with inaccurate and sub-definite data. *Interval Computations*, 3(5):13–16, 1992.
2. F. Benhamou, F. Bouvier, A. Colmerauer, H. Garreta, B. Gilleta, J.-L. Massat, and G. Narboni. *Le manuel de Prolog IV*. PrologIA, 1996.
3. F. Benhamou, F. Goualard, L. Granvilliers, and J.-F. Puget. Revising hull and box consistency. In *Proc. Int. Conf. on Logic Programming*. The MIT Press, 1999.
4. F. Benhamou and W. J. Older. Programming in CLP(BNR). In *Proc. Int. Conf. Principle and Practice of Constraint Programming*, Newport, USA, 1994.
5. ECRC. *ECL^{i}PS^e 3.5: ECRC Common Logic Programming System. User's Guide.*, 1995.
6. E. J. Elton and M. J. Gruber. *Modern Portfolio Theory and Investment Analysis*. John Willey & Sons, Inc., New York, 1995.
7. C. Gervet. Interval propagation to reason about sets: definition and implementation of a practical language. In *Constraints*, volume 3, pages 191–244. Berlin: Kluwer Academic Publishers, 1997.

8. C. Holzbauer. Metastructures versus attributed variables in the context of extensible unification. In *Proc. 3rd Int. Work. Programming Languages Implementaion and Logic Programming*, pages 260–268, 1992.

9. S. Le Huitouze. A new data structure for implementing extensions to prolog. In *Proc. 2nd Int. Work. Programming Languages Implementaion and Logic Programming*, volume 456 of *Lect. Notes Comp. Sci.*, pages 136–150. Springer-Verlag, 1990.

10. G. Lopez, B. Freeman-Benson, and A. Borning. Kaleidoscope: A constraint imperative programming language. In B. Mayoh, E. Tyugu, and J. Penjaam, editors, *Constraint Programming*, volume 131 of *NATO Advanced Science Institute Series, Series F:Computer and System Sciences*, pages 313–329. Springer-Verlag, 1994.

11. B. Mayoh, E. Tyugu, and J. Penjaam, editors. *Constraint Programming*, NATO Advanced Science Institute Series, Series F: Computer and System Sciences. Springer-Verlag, 1994.

12. A. S. Narinyani. Subdefiniteness and basic means of knowledge representation. *Computers and Artificial Intelligence*, 2(5):443–452, 1983.

13. E. Petrov and T. Yakhno. Extensional set library for ECL^{i}PS^e. In *Proc. Int. Conf. Practical Application of Constraint Logic Programming*, pages 253–270. The Practical Application Company Ltd, 2000.

14. J.-F. Puget. A C++ implementation of CLP. In *Proc.SPICIS*, Singapore, 1994.

15. I. E. Shvetsov, V. V. Telerman, and D. M. Ushakov. NeMo+: Object-oriented constraint programming environment based on subdefinite models. In *Principles and Practice of Constraint Programming, Proc. 3rd Int. Conf.*, volume 1330 of *Lecture Notes in Computer Sciences*, pages 534–548. Springer-Verlag, 1997.

16. P. Van Hentenryck, L. Michel, and Y. Deville. *Numerica: a Modelling Language for Global Optimization*. The MIT Press, Cambridge, MA, 1997.

17. T. M. Yakhno, V. Z. Zilberfaine, and E. S. Petrov. Applications of ECL^{i}PS^e: Interval Domain library. *The ICL Systems Journal*, pages 35–50, Nov. 1997.

Combining Constraints and Consistency Techniques in Knowledge-Based Expert Systems

Ilié Popescu

Département d'informatique
Université du Québec à Hull
Hull, Québec
CANADA, J8X 3X7

Abstract. Knowledge-Based Expert Systems (KBES) have long been widely used to perform tasks that normally require human knowledge and intelligence. One important issue that has not been addressed satisfactorily in the existing KBESs is that they try to make posing queries simple by letting the users specify what they want to compute rather than how to compute it. In this paper, we show that the solutions computation process can be modeled with Constraint Satisfaction Problem (CSP) techniques, employing their simple representation schemes and consistency techniques. The motivation behind this is the desire to build up a computation model for reducing the vast amount of deductions required by a KBES when executed on a logic program system. A key idea is to represent the relations among the rules as constraints and to integrate the rule chaining with constraint solving. In this integration, the constraints are regarded as special facts at each node of the solutions graph, and the constraints propagation may cause firing of rules. In this way the model allows the solutions graph to grow progressively by enumerating the solutions of the system of constraints and validating the rules associated to these constraints. The approach to accomplish this is to spend more time in each node of the solutions graph by reducing the sets of possible values for not-yet-assigned variables. The model is introduced as a general control mechanism and realizes an a priori pruning in the solutions graph. This is done by assuming that the only rules to be considered are those arising from the propagation of their constraints and by computing only the rules that acquired some domain-dependent information about the significance of various domain interactions.

1 Introduction

The Knowledge-Based Expert Systems (KBES) provide a framework in which it is possible to express many combinatorial problems encountered in artificial intelligence. Many problems to which artificial intelligence techniques are applied can be described as constraint satisfaction problem (CSP). The constraint satisfaction problem is known to be a NP-complete problem. If we want to design KBESs capable of behaving intelligently in some environment, we need an unambiguous language for expressing knowledge about this environment together with some well-understood way of manipulating this knowledge. Logic programming languages are the most powerful and general formal description languages known. They are widely used to

T. Yakhno (Ed.): ADVIS 2000, LNCS 1909, pp. 191-200, 2000.
© Springer-Verlag Berlin Heidelberg 2000

build KBESs to perform tasks that normally require human knowledge and intelligence.

In the CSP research, many efficient constraint algorithms have been produced [9, 14]. A number of languages and systems based on the model have been developed and used for solving real-life problems (PROLOG III, PROLOG IV, ILOG SOLVER, CLP(R), CHIP). These languages are primarily used for modeling and solving real-life problems, such as planning and resource allocations.

In this paper we present a method to compute the solutions of a KBES using domains of concepts and consistencies techniques in a CSP. The motivation behind this is the desire to build up a new model for reducing the vast amount of computation required by a KBES when executed on a logic programming system. In their logical formulation, the representation of complex objects supports powerful data abstraction but are generally slow in processing a goal because they handle a huge research space to manipulate knowledge between the initial state of the goal and its final state. However, only relatively few programming languages [1, 2, 10, 19] embody concept domains and constraints for keeping the complexity of the solution space under control. It is well-known that the logic programming paradigm well supports this programming style. Therefore introducing domain concepts and constraints into KBES looks quite promising.

The KBES methodologies try to model the human cognition by computer programs involving primarily symbolic computation. Acquisition and efficient use of a large variety of such knowledge in a KBES have dominated the research in this field [4, 17, 20, 21, 23, 24]. Much of this work is motivated by the desire to conceive KBESs as logical combinations of objects under control (states of the system) and all aspects of the future behaviour when combined with incremental knowledge of the future inputs. The computation of a state through a KBES consists in acquiring sets of inputs from constraints gathering it and logically weighting all together with respect to the system of constraints associated with this state. A logically weighted sum of these computations constitutes the attributes to a rule activation. What the system is doing is the calculation of the current state of the system, the determination of a strategy to pass from the current state to the next state (i.e. the control problem), and the generation of control parameters to the control policy. The process of acquiring knowledge and translating it into a self-consistent rule base provides an insight into the reasoning process.

When processing a goal in a KBES, current optimization techniques mainly rely upon the syntactic knowledge and storage details. They do not utilize meaningfully the properties of an application domain, which are likely to play a key role during the optimization process. Current KBES techniques do not take advantages of semantic information during the goal optimization and processing.

In this paper, we are concerned with the use of domain-specific semantic knowledge, and with the use of systems of constraints on the variables implied in each rule into the system. The computation of a goal or a query is a sequence of derivations that combines logic programming and constraint system solving.

2 Knowledge-Based Expert Systems and Constraint Satisfaction Problem

In this section, first we present the theoretical background of KBES and CSP, and highlight some problems in efficiently handling constraints in a solution subgraph of a KBES.

2.1 Notations and Definitions

Consider a KBES where the rules are expressed in first order logic as a set of Horn rules. The paradigm of rule-based constraint programming defines a class of logic programs where each instance of a logic program is obtained by the specification of its structure of computation. In this paper, by computation structure, we mean a tuple of the form $S = \{V, C, D^n, F, R\}$ formed by a set of variables V, a set of constants C, the domain of the problem D^n, a set of function symbols F and a set of relation symbols R.

Definition 1: *A term* is defined inductively as follows:
- A variable or a constant is a term
- If $t_1, ..., t_n$ are terms and f is an n-ary function symbol, then $f(t_1, ..., t_n)$ is a term.

Definition 2 : *An atom* is defined as follows: if $t_1, ..., t_n$ are terms and r is n-ary relation symbol then $r(t_1, ..., t_n)$ is an atom.

Definition 3 : *A literal* is either an atom or a negated atom.

Definition 4 : *A Horn rule* or simply *a rule* is a sentence of the form

$$H(\bar{X}) :- L_1(\bar{X}_1), ..., L_n(\bar{X}_n) \tag{1}$$

where $H(\bar{X})$ is an atom - the head of the rule, $L_1(\bar{X}_1), ..., L_n(\bar{X}_n)$ are literals and $\bar{X}, \bar{X}_1, ..., \bar{X}_n$ are vectors of terms. All variables implied in the rule are universally quantified and the commas between literals denote conjunction. We refer to $L_1(\bar{X}_1), ..., L_n(\bar{X}_n)$ as the body of the rule. If the body of a rule is empty, the rule is called fact.

Definition 5 : A *substitution* is a finite set (possibly empty) of pairs of the form $x_i = t_i$, where x_i is a variable and t_i is a term, $x_i \circ x_j$ for any $i \circ j$, and x_i does not occur in t_j for any i and j.

The result of applying a substitution $\theta : x = t$ to a term A, denoted by $A\theta$, is the term obtained by replacing every occurrence of x by t in A. For any substitution $\theta : \{x_1 = t_1, x_2 = t_2, ..., x_n = t_n\}$ and term B, the term $B\theta$ denotes the result of simultaneously replacing in B each occurrence of the variable by x_i by t_i, $1 \le i \le n$; the term Bq is called an *instance* of B.

Recall that a term t is a common instance of two terms t_1 *and* t_2 if there exist substitutions θ_1 *and* θ_2 such that t is a common instance of $t_1\theta_1$ *and* $t_2\theta_2$. A most general unifier (mgu) of two terms is a unifier such that the associated common instance is most general.

2.2   Knowledge-Based Expert Systems

Consider a KBES as a set of rules of the form (1). We require that the rules are safe, i.e., a variable that appears in $\overline{X}$ must also appear in $\overline{X}_1 \cup \overline{X}_2 \cup ... \cup \overline{X}_n$.

A KBES has three main components [7, 11, 24]:

- A working memory representing the current state of the problem at hand.
- A rule memory representing a set of rules which test and/or alter the working memory.
- An inference engine which applies the rules to the working memory.

Searching for a solution in a KBES can be represented as a AND/OR-graph. We concentrate ourselves on the problem how to find a connected subgraph of an AND/OR-graph. We concentrate ourselves on the problem how to find a connected subgraph of an AND/OR-graph which is consistent, i.e. such that all the nodes satisfy the consistency requirements (constraints).

One of the major shortcomings of the KBES methodologies is that it is very difficult to make use of disjunctive information and/or to reason about choices.

Let us review some useful notions about graphs. Let $G = (N, A)$ be an oriented graph. The set of nodes N is $N = N_{-and} \cup N_{-or}$ where N_{-and} N_{-or} are disjoint sets, i.e. $N_{-and} \cap N_{-or} = \varnothing$. The set of arcs A is made up of binary relations on the cartesian product $N \times N$, $A \subseteq N \times N$.

Nodes are called N_{-and} or N_{-or} according to the way sets of their successors are interpreted, logical AND or logical OR respectively.

We define a solution graph on the graph G as a subgraph $S_P = (N_P, A_P)$ with the root $P \in N_P$ $(N_P \subseteq N, \ A_P \subseteq A)$ such that the following properties hold.

Solution Graph

i. $\forall a$, if $(a \in N_P \wedge a \in N_{-or})$ then
$$\exists b, (b \in N_P \wedge (a,b) \in A_P)$$

ii. $\forall a, \forall b$, if $(a \in N_P \wedge a \in N_{-and} \wedge b \in N_P)$ then
$$\exists (a,b) \in A_P$$

iii. $\forall a$, if $\left(a \in N_P \wedge a \text{ is a leaf node} \right)$ then

$$\exists b, (b,a) \in A_P$$

iv. N_P contains at least one leaf node.

In other words, a solution graph is such that if a node $n \in N_P$ is a N_{-or} node then at least one of its successors belongs to S_P. In this way, the solution graph can be seen as a collection of routes from the seat P to a leaf node.

2.3   An Algorithm for Chronological Backtrack Searching

Let us present the known algorithm for chronological backtracking of an AND/OR graph as a logical program. We include this algorithm presentation because later we attempt to enhance the algorithm by incorporating constraint satisfaction techniques to search procedure to prune off substantial portions of the search tree by identifying solution sets that cannot satisfy input constraints.

In our formulation of problems, a solution is obtained by applying rules to state descriptions until a set of knowledge describing the goal is obtained. All of the AND/OR graph states can be modeled by the following search process:

i.   The start node is associated with the initial state description of the problem to be solved (the goal).

ii.   Expanding a N_{-or}
        If n is a N_{-or} node, then exactly one of its successors is developed. Pointers are set up from each successor back to its parent node. These pointers indicate a path back to the start node when the goal node is found.

iii.   Expanding a N_{-and}
        If n is a N_{-and}, then all its immediate successors are developed. The successor nodes are checked to see if they meet the goal's description state.

If a goal node has not yet been found, the process of expanding nodes and setting up pointers continues.

It is important to realize that all the nodes of the searched AND/OR graph can in any moment be in one of the following cases:

- The node is a goal node in which case the pointers are traced back to the start node to produce a solution path.
- The node is consistent with a potential solution path but has some of its successors unexplored.
- The node violates some constraint; in this case, the node is not part of a solution.

2.4   Depth First Search Algorithm

The basic depth first search algorithm as described in [18] is defined by two lists OPEN and CLOSED to keep record of the nodes which have their predecessors explored but not their successors and of the nodes which have all their successors explored. In any moment during searching, nodes on OPEN and CLOSED are those forming a potential solution graph.

Depending of the state-space representation of a problem, there are many approaches to implement the depth first search algorithm. Our representation is based on idea of neighborhood and the state-space is a collection of edges between nodes : edge (X,Y). To find a solution path, from a given node A, to some goal node:

- if A is a goal node then Path = [A]
- if there is a successor node, Z of A such that there is a path, Path1, from Z to a goal node, then $Path = [A|Path]$.

The depth-first search is simple and easy to program. It is widely used to build knowledge-based expert systems as long as the combinatorial complexity of the problem at hand does not require the use of problem-specific constraints.

3 Using Constraints in Knowledge-Based Expert Systems

The Constraint Satisfaction Problem provides a framework in which it is possible to express many combinatorial problems encountered in artificial intelligence. CSP is known to be a NP-complete problem. However, by imposing restrictions on the constraint interconnections [5, 6, 8, 15, 16], or on the form of the constraints [3, 12, 13, 22], it is possible to obtain restricted versions of the problem that are tractable. The fundamental mathematical structure required to express constraints is the "relation".

Definition 6 : A Constraint Satisfaction Problem consists of :
- A finite set $V = \{v_1, v_2 ... v_n\}$ of n variables v_i which are also referred to as constraint variables.
- A finite domain $D^n = D_1 \times D_2 \times ... \times D_n$ of values, where each domain D_i contains all the possible values for the variable v_i.
- A finite set $C = \{C_1, C_2, ..., C_k\}$ of k constraints, each constraint $C_i = R_i(v_1, ..., v_m)$ is an m-ary relation over D^n.

For each constraint C_i, the tuples in R_i indicated the allowed combinations of simultaneous values for its variables.

A solution to CSP is a function from the variables V to the domain D^n such that the image of each constraint scope is an element of the corresponding constraint relation. The main approach used for solving CSPs is to embed constraint propagation techniques in a backtrack search environment, where backtrack search performs the search for a solution and consistency check techniques prune the search space.

Consistency techniques are characterized by using constraints to remove inconsistent values from the domains of variables.

Let us consider a current state of a KBES characterized by the tuple (R_i, C_i), where R_i is a set of rules of the form (1) and C_i is a set of constraints.

We introduce a consistency technique as following :

S1: Node Consistency

> *Node consistency* is simply a state where the domains of each variable occurring in C_i have been reduced to the set of all possible values that satisfy C_i .

The following algorithm enforces node consistency.

Algorithm S1

for each variable

> **if** $\left(C_i = \varnothing\right)$
>> **then** do nothing
>> **else**
>>> **for** each value $a_i \in D_i$
>>>> **if** $\left(a_i \text{ satisfies } C_i\right)$
>>>>> **then** do nothing
>>>>> **else** $D_i = D_i - \{a_i\}$

The time-complexity of this algorithm is where n is the number of variables occurring in C_i and m is the maximum size of the domain.

S2: Transition Consistency

> *Transition consistency* ensures that the transition from the current state to a new state is made only if the constraints connecting the new state to the current state are satisfied.

> Let (R_i, C_i) et (R_j, C_j) be the two states et C_{ij} the set of transition constraints.

> The transition from the current state (R_i, C_i) to a new state (R_j, C_j) occurs if and only if for each value of a variable in (R_i, C_i) there exists a value for a variable in (R_j, C_j) which satisfies the set of transition constraints C_{ij} . The following algorithm implements this idea.

Algorithm S2

repeat
   removed = false
    for each variable v_i in C_i
     for each variable v_i in C_i
      if $\left(C_{i\,j} = 0 \right)$
        then do nothing
        else
         for each value $a_i \in D_i$
          for each value $a_j \in D_j$
           if $\left(a_i, a_j \text{ satisfy } C_i \right)$
            then do nothing
            else

$$\left\{ D_i = D_i - \langle a_i \rangle; \text{removed} = \text{true} \right\}$$

until not removed.

The idea behind this algorithm is to use constraints and domains for keeping the complexity of the solutions space under control. This is done by allowing the transition consistency algorithm to progressively restrain the variable domains by eliminating the values which do not satisfy the transition constraints. The approach to accomplish this is to spend more time in each current state of graph to reduce the sets of possible values for not-yet-assigned variables.

The algorithm cycles through each potential transition to determine the values of the current state variables which satisfy the constraints connecting the two states. If a value is found for which this is not true, it is removed. If any values are removed, the algorithm restarts from the beginning, since the removal might cause the violation of some constraints already tested.

S3: Path Consistency

Path Consistency requires that for any path $S_0 \to S_1 \to \dots S_{n-1} \to S_n$, then for any value assignments for the variables occurring in S_0 and S_n, there must exist value assignments for each variable occurring in $S_2, \dots, S_{n-1}$, such that all constraints on adjacent states are satisfied.

Path consistency algorithms are not necessary. A dynamic application of state and transition consistency presented here performs the same function in a controlled manner.

The computational level in a KBES can be viewed as an implementation of the depth first search algorithm using CSP and consistency techniques. As we underlined above, the depth first search algorithm when executed as a logic program leads to a generate and test or a standard backtracking approach, which exhibits pathological behaviours and its performances drastically decrease as the problem size grows. Its rigidity, especially its ability to recognize objects, to plan its actions, to manipulate and adapt in an unstructured environment makes it useless in real world problems.

4 Conclusions

One motivation behind the spurt of activity in the area of CSP is the desire to develop models that cope with the complexity of a NP-complete problem. The consistency techniques approach substantially improves the KBES's solutions computation by introducing the domain concept inside the logic programming. From a practical point of view it allows the addition of domain-specific knowledge and the introduction of constraints transition that embody the idea of a priori pruning.

At each stage of the search, a goal features correspond to the validation of a set of constraints that supports the input specification, a set of rules to change the current state into a new one and a set of transition constraints to eliminate the rules which are not part of any solution. Thus, the inference rules are oriented toward the prevention of failures and allow both, an early detection of failures and a reduction in backtracking checks. In most cases, consistency techniques improve the efficiency of a KBES and must be considered as a fundamental approach for solving real problems. The solution computation may still be too expensive time wise when the KBES is very large. In such cases a useful technique is that the variable domains schemes are supplied/encoded by the user in such a way that the more important domains validation are established at higher level, while the establishment of less important validations are delegated to lower levels. Judging the importance of consistency techniques, integrating node, transition and path consistency into a KBES enhance its efficiency. The failures are detected precisely by applying node and transition consistency algorithm avoiding expensive backtracking and avoiding much of redundant work. The search space will be smaller and consequently the memory space will be less overdrawing.

References

[1] F. Benhamou, Touraïvane, "Prolog IV : langage et algorithmes", JFPL 95.
[2] A. Colmerauer, "An introduction to Prolog III", Communications of the ACM, 30(7), 1990, pp. 2-68.
[3] M.C. Cooper, D.A. Cohen and P.G. Jeavons, "Characterizing tractable constraints, Artificial Intelligence 65 (1994), pp. 347-361.
[4] R. Dechter, "Enhancement schemes for constraint processing : Backjumping, learning, and curset decomposition", Artificial Intelligence 41, 1990, pp. 273-312.
[5] R. Dechter and J. Pearl, "Network-based heuristics for constraint satisfaction problems, Artificial Intelligence 34(1), 1988, pp. 1-38.

[6]   E.C. Freuder, "A sufficient condition for backtrack-bounded search", J. ACM 32, 1985, pp. 755-761.

[7]   J. Giarratano and G. Riley, "Expert Systems: principles and programming", PWS Publishing, 1994.

[8]   M. Gyssens, P.G. Jeavons and D.A. Cohen, "Decomposing constraint satisfaction problems using database techniques, Artificial Intelligence 66(1), 1994, pp. 57-89.

[9]   P.V. Hentenryck, Y. Deville and C.-M. Teng, "A generic arc consistency algorithm and its specializations", Artificial Intellligence 27, 1992, pp. 291-322.

[10] J. Jaffar, S. Michaijlov, P.J. Stuckey and R.H.C. Yap, "The CLP(R) Language and System", ACM Transactions on Programming Languages and Systems, Vol. 14, no 3, 1992, pp. 339-395.

[11] J. Jaffar and J. Lassez, "Constraint logic programming", POPL-87, 1987.

[12] P.G. Jeavons and M.C. Cooper, "Tractable constraints on ordered domains", Artificial Intelligence 79 (2), 1995, pp. 327-339.

[13] L. Kirousis, "Fast parallel constraint satisfaction", Artificial Intelligence 64, 1993, pp. 147-160.

[14] A.K. Mackworth, "Consistency in networks of relations", Artificial Intelligence 8, 1977, pp. 99-118.

[15] U. Montanari, "Networks of constraints: fundamental properties and applications to picture processing", Inform. Sci. 7, 1974, pp. 95-132.

[16] U. Montanari and F. Rossi, "Constraint relaxation may be perfect", Artificial Intelligence 48, 1991, pp. 143-170.

[17] U. Montanari, F. Rossi, "Perfect Relaxation in Constraint Logic Programming", Proc. of the Eight International Conference on Logic Programming, 1991, pp. 223-237.

[18] N.J. Nilson, "Problem-Solving Methods in Artificial Intelligence", McGraw-Hill, 1971.

[19] W. Older, F. Benhamou, "Programming in CLP(BNR)", PPCP '94, Newport, RI (USA), 1994.

[20] I. Popescu, "Hierarchical Neural Network for Rules Control in Knowledge-Based Expert Systems", International Journal of Neural, Parallel and Scientific Computations, Dynamic Publishers, 1995, pp. 379-391.

[21] I. Popescu and M. Zaremba, "Efficient Method for Solving Systems of Constraints in Intelligent Systems", International Journal of Computers and Their Applications, 1995, pp. 96-103.

[22] P. van Beck and R. Dechter, "On the minimality and decomposability of row-convex constraint networks", J. ACM 42, 1995, pp. 543-561.

[23] F. Rossi, "Constraint satisfaction problems in logic programming", SIGART Newsletter, 106, 1988.

[24] P. Van Hentenryck, "Constraint Satisfaction in Logic Programming, MIT Press, 1989.

Declarative Modelling of Constraint Propagation Strategies

Laurent Granvilliers[1] and Eric Monfroy[2]*

[1] IRIN
B.P. 92208 – F-44322 Nantes Cedex 3 – France
Laurent.Granvilliers@irin.univ-nantes.fr
[2] CWI
Kruislaan 413 – 1098 SJ Amsterdam – the Netherlands
Eric.Monfroy@cwi.nl

Abstract. Constraint propagation is a generic method for combining monotone operators over lattices. In this paper, a family of constraint propagation strategies are designed as instances of a single algorithm shown to be correct, finite, and strategy-independent. The main idea is to separate complex reasoning processes (that exploit some knowledge of constraints and solvers in order to accelerate the convergence) from basic fixed-point algorithms. Some sequential interval narrowing strategies are expressed in this framework and parallel implementations in different execution models are discussed.

1 Introduction

Constraint propagation [22,17,12] is a cornerstone algorithm of constraint programming, that combines a set of elementary solvers associated with possibly heterogeneous constraints. Given a dependence relation between constraints and solvers, new information (modification of a domain, entailment of a constraint) is delivered to the set of constraints, until a stable state is reached, *i.e.*, no further information can be inferred. One may single out three frameworks dealing with constraint propagation: subdefinite models from Narin'yani [18], narrowing from Benhamou [2], and chaotic iteration (CI) from Apt [1]. In the following, we consider the narrowing formalism since its level of generality is in adequation with our approach.

The essential property of constraint propagation is confluence or strategy-independence. In other words, the ordering of solver applications does not influence the output which is characterized in terms of a common fixed-point of the solvers. Although the basic narrowing algorithm is general enough, the design of a new strategy often requires the implementation of a novel algorithm. Furthermore, this new algorithm must be fitted in the formalism afterwards. We therefore believe there is a need for a component-based algorithm that better

---

* This research was carried out while Eric Monfroy was visiting IRIN.

T. Yakhno (Ed.): ADVIS 2000, LNCS 1909, pp. 201–215, 2000.
© Springer-Verlag Berlin Heidelberg 2000

embodies the notion of strategy. Each component programs must also verify a given set of minimal properties in order to guarantee termination and confluence.

In this paper, we design SCP, a two level-algorithm that specifies a family of constraint propagation strategies. Its main feature is the dissociation of what we call the strategy (for expressiveness reasons) from the computation of common fixed-points of elementary solvers (for re-use of programs and efficiency reasons). A strategy is a complex reasoning process based on some knowledge related to solver behaviours, constraint forms, dynamic information, ... A strategy aims at selecting the "best" solvers to consider at some point. Fixed-points are computed by specialized algorithms implemented in different execution models, and may be re-used by any strategy.

The rest of this paper is devoted to the description of some already implemented as well as new strategies to demonstrate the capabilities of our framework. On the one hand, some interval narrowing algorithms are shown to be instances of SCP: the most famous is certainly the core engine of Numerica [21] that implements box consistency for different interval forms of constraints; two related algorithms from the authors [13,7] describe variant strategies that accelerate the computation of box consistency. Furthermore, two new strategies are introduced: hull and box consistency are really interleaved, as is suggested in the seminal paper discussing their combination [3]; and an efficient implementation of box_φ consistency [8] is proposed: the strategy iteratively decreases the value of φ to 0, and a fixed-point is computed for each of these values.

On the other hand, the generic IFP algorithm that computes common fixed-points of solvers is designed. It uses the selection-intensification heuristics proposed in [9], that is an adaptation for parallelism of the one in [11]. Four phases are iteratively processed: selection by application, detection of fixed-point, intensification, and finally propagation. The main idea is to dynamically select a subset of the solvers that must be applied, to intensify the use of the best ones, and to propagate only after a long step of computation in order to limit the updating costs. Two implementations of IFP are proposed: a parallel one in the Bulk-Synchronous-Parallel model [9] and a parallel and distributed one [15].

We believe that the main contribution of this paper is the design of a constraint propagation algorithm that better connects the narrowing model with *ad hoc* strategies. Furthermore, the component algorithms are specified by declarative properties that are independent from any particular implementation. Finally, specialized implementations of fixed-point algorithms are proposed.

The outline of this paper is the following: Section 2 introduces some notions from constraints and constraint propagation; the SCP algorithm is presented in Section 3, followed by a selection-intensification IFP algorithm. Some interval narrowing strategies are described in Section 4. Parallel implementations of IFP are discussed in Section 5. Finally, we conclude in Section 6.

2 Preliminaries

In the following, let $\mathcal{D}$ be a set called the universe, $\mathcal{F}$ a set of operation symbols, $\mathcal{R}$ a set of relation symbols, $\Sigma = \langle \mathcal{D}, \mathcal{F}, \mathcal{R} \rangle$ a structure, and $\mathcal{X} = \{x_1, x_2, \ldots, x_n\}$ a set of variables.

In order to model the approximations performed by the algorithms, Benhamou *et al.* [5,2] have introduced the notion of *approximation domain*.

Definition 1 (Approximation domain). *An* approximation domain D *is a subset of the power-set of* $\mathcal{D}$, *closed under intersection, containing* $\mathcal{D}$, *and such that set inclusion on D is a well-founded relation.*

A *constraint* is a conjunction of atomic formulas made on Σ and $\mathcal{X}$ in the usual way. Given a constraint c, let ρ_c denote the underlying relation and v_c the set of variables occurring in c. Given an approximation domain D, a *constraint satisfaction problem* (CSP) is given by a vector of domains $d = (d_1, \ldots, d_n)$ from D^n, where $x_i \in d_i$, and a set of constraints $C = \{c_1, c_2, \ldots, c_m\}$. In order to intersect constraint relations, the relation ρ_c associated with any constraint $c(x_{i_1}, \ldots, x_{i_k})$ is extended to the set $\{(a_1, \ldots, a_n) \in \mathcal{D}^n \mid (a_{i_1}, \ldots, a_{i_k}) \in \rho_c\}$. Thus, a *solution of the CSP* is an element that belongs to the intersection of all the constraint relations extended in this way.

The pruning of CSP's domains is performed by narrowing operators/functions. Given $d \in D^k$, let $d_\times$ denote the product $d_1 \times \cdots \times d_k$.

Definition 2 (Narrowing operator). *Let D be an approximation domain, c a constraint and k the arity of c. A* narrowing operator *for c is a mapping $f : D^k \to D^k$ that verifies the following properties for every $d, d' \in D^k$:*

$$
\begin{array}{rl}
\textit{contractance:} & f(d) \subseteq d \\
\textit{correctness:} & \rho_c \cap d_\times \subseteq f(d)_\times \\
\textit{monotonicity:} & d \subseteq d' \Rightarrow f(d) \subseteq f(d')
\end{array}
$$

f is said to depend on every variable occurring in c.

Definition 3 (Extension of narrowing operator). *Let D be an approximation domain and c a constraint over* $x_{i_1}, \ldots, x_{i_k}$. *Let us consider a narrowing operator f for c. Then, f is extended to $f^+ : D^n \to D^n$ as follows. For all $d \in D^n$, we have:*

$$
\begin{array}{ll}
f^+(d)_j = f(d_{i_1}, \ldots, d_{i_k})_j & \textit{if } j \in \{i_1, \ldots, i_k\} \\
f^+(d)_j = d_j & \textit{otherwise}
\end{array}
$$

In the rest of this paper, we will often write f instead of f^+ when no precision is required.

Constraint propagation is a pruning algorithm that can be described in terms of fixed-point computation of a set of narrowing operators associated with the constraints of a CSP. More formally, let F be a finite set of narrowing operators and d the vector of variables' domains. We have the following results [2]:

- $\cap_{f \in F} f : D^n \to D^n$ is a narrowing operator, and then the set of fixed-points of $\cap_{f \in F} f$ is a lattice; thus, given $d \in D^n$, the greatest fixed-point $gfp(F, d)$ of $\cap_{f \in F} f$ included in d does exist;
- given $d \in D^n$, the limit of any fair computation that applies the functions from F on the domains is $gfp(F, d)$; furthermore, the computation is finite since set inclusion over the approximation domain is well-founded.

The generic constraint propagation algorithm (see Table 1) is implemented by a queue of narrowing operators. An operator is added in the queue each time a domain it depends on is modified by the application of another operator.

Table 1. The generic constraint propagation algorithm.

Propagation $(F : \text{set of narrowing operators}, \ d : D^n) : D^n$
begin
    $Q \leftarrow F$ % *initialisation of propagation list*
    repeat
        choose f in Q % *selection of an operator*
        $d' \leftarrow f(d)$ % *pruning of domains*
        if $d \neq d'$ **then**   % *propagation*
            $Q \leftarrow Q \cup \{f \in F \mid f \text{ depends on some } x_i \text{ s.t. } d_i \neq d_i'\}$
            $d \leftarrow d'$
        else $Q \leftarrow Q \setminus \{f\}$ % *fixed-point of f is detected*
        endif
        until $d_\times = \varnothing$ **or** $Q = \varnothing$ % *the global fixed-point is detected*
        return d
end

3 Narrowing Strategies

In this section, we propose a more structured and strategy-oriented version of the generic constraint propagation algorithm. Essentially, the choice of an operator is replaced with a selection of a set of operators with respect to some knowledge and history of computation (strategy). The application of these operators is then scheduled by a specialized fixed-point algorithm. The component algorithms are specified by properties to guarantee termination and confluence. Thus, they permit very different implementations that can exploit features of different hardware configurations, such as nets of machines and parallel machines with shared memory.

We believe that this framework clarifies several constraint propagation systems in a unique formalism and may help designing more complex strategies (or combination of strategies) in an easy way.

3.1 The SCP Algorithm

The well known dilemma of constraint narrowing is to maximize the tradeoff between contraction ratio and computation time. In the same spirit, we identify a similar motivation between selection/reasoning and basic application of narrowing operators. Thus, we propose to clearly distinguish both processes: the SCP algorithm alternates selection of a set of operators with computation of their fixed-point (by the IFP algorithm).

Furthermore, a set of narrowing operators associated with a CSP is generally structured in terms of running costs, reduction capabilities, and the kind of constraints the operators manipulate. We then consider a static partition of such a set of operators into *classes* to which some priority/importance criterion are assigned. These classes, some knowledge of operators, and an history of computation are the static/dynamic information used by the strategy of SCP. Let us formalize now these notions.

Definition 4 (Class of operators). *Consider a set F of operators. A* class *of operators from F is an element of $\mathcal{P}(F)$.*

The notion of class is voluntarily very wide and may tackle various characteristics of operators:

- solvers modelled by different types of narrowing operators, *e.g.*, hull consistency and box consistency-based techniques (see Section 4.2);
- operators processing heterogeneous constraints, *e.g.*, constraints over intervals of real numbers, Boolean constraints, or integer constraints;
- operators with different running costs/reduction capabilities; in this case, classes can determine estimated running times or strongness of functions (see Section 4.1).

An history simply stores the classes of narrowing operators that have already been considered in SCP.

Definition 5 (History of computation). *Consider a set Cl of operators classes. An* history of computation *is a possibly empty sequence $cl_1, \ldots, cl_n$ of elements of Cl. Let $\mathcal{H}$ denote the set of histories on Cl and $\varnothing_H$ the empty history. We also consider the usual concatenation operation of two sequences $\cdot : \mathcal{H}^2 \to \mathcal{H}$.*

The knowledge represents both static information related to the narrowing operators, and a static strategy for selecting some classes with respect to previous computation.

Definition 6 (Knowledge of operators). *A knowledge K about a set F of narrowing operators with respect to a set Cl of classes is defined by:*

- *a mapping $\alpha_K : F \to Cl$, called the* partitioner *of K;*
- *a mapping $\gamma_K : \mathcal{H} \times \mathcal{P}(Cl) \to \mathcal{P}(Cl)$, called the* class selector *of K, which associates to an history H and a set of classes Cl a non empty set of classes Cl' such that $Cl' \subseteq Cl$.*

Cl is said to be known by *K.*

Table 2. Declarative algorithm modelling narrowing strategies.

SCP $(F$: set of operators, $d : D^n$, K : knowledge about a superset of $F)$: D^n
begin
 $Q \leftarrow F$ % *propagation list of operators to be applied*
 $Hist \leftarrow \varnothing_H$ % *history of computation*
 repeat
  % *Selection of a subset of operators and modification of the history*
  % *where K represents the knowledge related to the operators*
  $(F', Hist) \leftarrow$ Selection $(Q, K, Hist)$

  % *Contraction of domains by the fixed-point algorithm* IFP, *where Arg_{fp}*
  % *is a list of arguments depending on specific implementations*
  $d' \leftarrow$ IFP $(F', d, Arg_{\mathrm{fp}})$

  % *Update of domains and propagation list*
  if $d \neq d'$ **then**
   $Q \leftarrow$ Update (Q, F, F', d, d')
   $d \leftarrow d'$
  else
   $Q \leftarrow Q \setminus F'$
  endif
 until $d_\times = \varnothing$ **or** StopCriterion (F, Q, d)
  % *end of the algorithm when the fixed-point gfp(F, d) is computed*
 return d
end

The SCP algorithm (see Table 2) represents a static (fixed-point) strategy: it iteratively selects a set of operators (from the propagation list) with respect to a knowledge and an history of computation (Selection function), and calls the IFP algorithm to compute their fixed-point. Let us remark that SCP may implement the generic constraint propagation algorithm. The minimal properties to be verified by the component algorithms of SCP are the following:

1. Given Q a set of narrowing operators, K a knowledge about operators, and H an history of selections, Selection(Q, K, H) returns (F', H') such that $F' \subseteq Q$, $F' \neq \varnothing$, and $H' = Cl \cdot H$ where Cl is a set of classes known by K.
2. Given F and Q two sets of narrowing operators such that $Q \subseteq F$, and $d, d' \in D^n$ such that $d' = gfp(F, d)$, we request:

$$\{f \in F \mid f(d') \neq d'\} \subseteq \text{Update}(Q, F, F', d, d') \subseteq F \setminus F'$$

The leftmost set inclusion corresponds to the correctness property, *i.e.*, each function that may modify d' must be applied again. The rightmost set inclusion means that each operator from F can be considered, except the ones from F' for which a fixed-point has just been computed (efficiency reason).
3. Given $d \in D^n$, and F, Q two sets of operators, StopCriterion(F, Q, d) is verified if and only if we have $f(d) = d$ for all $f \in Q$.

4. Given F' a set of narrowing operators, d from D^n, and a list of arguments Arg_{fp}, IFP $(F', d, Arg_{fp}) = gfp(F', d)$.

Algorithm SCP verifies the fundamental termination and confluence properties.

Proposition 1. *Consider a set of narrowing operators F, and a vector of domains $d \in D^n$. Assume that the Selection, IFP, Update, and StopCriterion functions terminates. Then, the SCP algorithm terminates and computes in d the greatest fixed-point of the narrowing operators from F.*

Proof. The SCP algorithm terminates in finite time since at each iteration either the cardinal of Q is strictly decreased (if d is not modified), or d is strictly contracted. The proof then follows since set inclusion over an approximation domain is a well-founded relation.

Though declarativity is our final goal, some less formal properties than the aforementioned ones may clarify this framework. We make the link with standard implementations of selection and update algorithms, and the stopping criterion. In this section, we only consider generic algorithms, whereas some more domain and application specific techniques will be tackled in Section 4.

The Update function. It selects all the operators that can possibly modify the current vector of domains. A "nearly-perfect"[1] one is the following:

$$\text{Update}(Q, F, F', d, d') = \{f \in F \setminus F' \mid f(d') \neq d'\}$$

In practice, the perfect update function is beyond reach since the condition $f(d') \neq d'$ can not be estimated without application of f, and a good approximation is generally implemented by:

$$(Q \setminus F') \cup \{f \in (F \setminus F') \mid f \text{ depends on some } x_i \text{ s.t. } d_i \neq d'_i\}$$

The StopCriterion function. It detects that the desired common fixed-point of the input operators has been reached. The minimal property specified above is not realistic since it requires applying all the functions without considering the new computed domains. In practice, the stop criterion often requires the propagation list to be empty:

$$\text{StopCriterion}\,(F, Q, d) = \text{true} \quad \text{iff} \quad Q = \varnothing$$

The Selection function. It selects a non-empty[2] subset of the narrowing operators still to be applied, and relies on three notions: operator classes, knowledge of the operators, and history of the performed computation. More formally, let us

---

[1] This function is not totally perfect, since two functions f and g such that $f(d) \subseteq g(d)$ can be added to Q.

[2] Termination of SCP is no more guaranteed if an empty set of operators is selected.

consider a set of narrowing operators Q, a given knowledge K of Q for some classes Cl, and an history of computation H. Then, we have:

$$\mathsf{Selection}(Q, K, H) = (\{f \in Q \mid \exists cl \in Cl', \, \alpha_K(f) = cl\}, \, Cl' \cdot H)$$

where $Cl' = \gamma_K(\{cl \in Cl \mid \exists f \in Q, \alpha_K(f) = cl\}, H)$. Let us explain the meaning of the previous formula. The set $\{cl \in Cl \mid \exists f \in Q, \alpha_K(f) = cl\}$ represents all the operator classes that are present in Q and known by K. We denote it Cl''. Applying class selector γ_K on Cl'' and H determines Cl', the set of classes that will be next considered. The set $\{f \in Q \mid \exists cl \in Cl', \alpha_K(f) = cl\}$ represents the narrowing operators (from Q) of the Cl''s classes, i.e., the input of the next call of IFP. Finally, the history is updated by concatenating Cl' to the previous history.

The IFP function. It represents the reduction phase of the SCP algorithm. The only requirement is that it computes a fixed-point of all the narrowing operators given as input. Thus, several algorithms are possible candidates, such as the AC3-like [12] constraint propagation algorithm of [2], the chaotic iteration algorithm [1], or the subdefinite models algorithm from [18]. Some other algorithms, either dedicated to specific hardware configuration, based on specific computation models, or specialized for specific domains have also been proposed, such as:

- the intensification-based parallel constraint propagation for interval constraints in the BSP model [9];
- the distributed [15] and parallel constraint propagation algorithms based on asynchronous iterations [16], that is a generalization of chaotic iteration for parallel computation;
- the coordination-based chaotic iteration algorithm of [14] which presents constraint propagation as coordination of cooperative agents.

In the SCP algorithm, we thus consider a generic IFP algorithm that can be instantiated by the aforementioned methods. A call to IFP is made as follows: IFP $(F', d, Arg_{\mathrm{fp}})$ where F' is a set of narrowing operators, d a vector of domains, and Arg_{fp} is a list of implementation specific arguments.

In the next section, we will propose a particular IFP algorithm based on pre-selection of narrowing operators with respect to their dynamic behaviour, and intensive use of the selected ones.

3.2   The IFP Algorithm

Computation of fixed-points of sets of operators by the IFP algorithm is based on the selection-intensification heuristics [11,9]. From a sequential point of view [11], the idea is to intensify the use of the most contracting operators (dynamic criterion). More generally, for instance from a parallel point of view [9], the essential property is that the fixed-point is detected between both phases of selection where all operators are independently applied (suitable property for

Table 3. Intensification-based fixed-point algorithm.

IFP $(F :$ set of functions, $d : D^n$, $Arg_{\mathrm{fp}} :$ list of arguments$) : D^n$
begin
    $Q \leftarrow F$ % *initialisation of propagation list*
    fixed-point $\leftarrow$ *false*
    repeat
        % *All operators from Q are applied in order to select a subset F' of Q*
        % *The new domains are stored in d'*
        $(F', d') \leftarrow$ SelectionByApplication $(Q, d, Arg_{\mathrm{fp}})$

        if $d' = d$ **then**
            % *The fixed-point is detected after the selection phase*
            %   *since all operators from Q have been applied*
            fixed-point $\leftarrow$ *true*
        else
            % *The previously selected operators (in F') are iteratively applied on d'*
            % *The end of this phase needs no further specification*
            %   *since it does not influence the detection of the fixed-point*
            $d'' \leftarrow$ Intensification $(F', d', Arg_{\mathrm{fp}})$

            % *Standard propagation phase*
            $Q \leftarrow \{f \in F \mid f$ depends on some x_i s.t. $d_i \neq d''_i\}$
            $d \leftarrow d''$
        endif
    until fixed-point
    return d
end

---

parallelism) and intensification that may not compute a fixed-point (weak property that permits the design of a large variety of algorithms). The IFP algorithm is described in Table 3. The generic functions that are part of IFP must verify the following properties:

1. Given Q, d and $(F', d') \leftarrow$ SelectionByApplication $(Q, d, Arg_{\mathrm{fp}})$, we demand that $F' \subseteq Q$ and $d' = \cap_{f \in Q} f(d)$. In practice, the last condition will be implemented by the application of all operators from Q.
2. Given F, d, and $d' =$ Intensification $(F, d, Arg_{\mathrm{fp}})$, we have:

$$gfp(F, d) \subseteq d' \subseteq \cap_{f \in F} f(d)$$

The leftmost inequality is a correctness property and the rightmost one specifies that all operators from F must be applied at least once (for efficiency).

Algorithm IFP verifies the fundamental termination and confluence properties.

Proposition 2. *Consider a set of narrowing operators F, and a vector of domains $d \in D^n$. Assume that the* SelectionByApplication *and* Intensification *functions terminates. Then, the* IFP *algorithm terminates and computes in d the greatest common fixed-point of the narrowing operators from F.*

Proof. Termination of the IFP algorithm in finite time is immediate since either the fixed-point is detected after the selection, or d is strictly contracted. The proof then follows since set inclusion over an approximation domain is a well-founded relation.

Remark 1. Both the SCP and the IFP algorithms are confluent since they implement particular constraint propagation strategies.

4   Using Classes for Interval Constraints

This section illustrates the framework presented in Section 3, describing narrowing/propagation algorithms for interval constraints. The approximation domain is the set of closed intervals bounded by floating-point numbers.

Let us informally introduce the main notions used in the following. Box consistency [4,20,3] is a local consistency notion that specifies the consistency of the bounds of a variable's domain d with respect to a constraint c and an interval extension of c. An interval extension of c is an interval relation such that every interval vector that contains a solution to c belongs to the relation. In other words, no solution of c is lost when one of its interval extension is considered. The classical narrowing algorithm for box consistency combines a binary splitting of d with a Newton-like iterative method (extended over intervals) for searching for the leftmost and rightmost consistent values a and b; as a consequence all values in $d \setminus [a, b]$ are removed from the domain (see [10] for a presentation of bound consistencies over continuous domains).

Hull consistency [6] is another local consistency notion, stronger than box consistency, but generally implemented for primitive constraints, *i.e.*, constraints resulting from the decomposition of the user's constraints. The performed approximations are then weaker but obtained at a lower cost. The combination of both techniques, discussed in [3], is then an interesting feature to accelerate constraint propagation while keeping the same final precision.

We now illustrate our notions of classes and Selection function on existing and new strategies for interval constraints.

4.1   Weak and Strong Operators

In this section, a static ordering on operators is established with respect to their estimated reduction power. This implies an underlying set of classes to be used by the Selection function of SCP.

Definition 7. *Let D be an approximation domain and f, c, k (resp. f', c', k') a narrowing operator f for a k-ary constraint c. We say that f is stronger than f', written $f \gg f'$, iff $\forall d \in D^n, f^+(d) \subseteq f'^+(d)$.*

The StopCriterion function can then be refined as follows:

$$\text{StopCriterion}(F, Q, d) = \text{true} \quad \text{iff} \quad Q = \varnothing \text{ or } \forall f \in Q, \exists g \in F \setminus Q, g \gg f$$

The second aforementioned condition means that there exists a function g stronger than f that needs no application (since it does not belong to Q). Thus, it is not worth applying f.

Weak and strong functions. We now describe in our framework the box consistency strategies developed in [13]. The following operator classes are considered with respect to the algorithm implemented by their narrowing operators: N corresponding to one iteration of interval Newton, N^* to a fixed-point of interval Newton, IBC to box consistency, and $LIBC_r$ to a weak box consistency where the computation of a narrowing operator stops if the domain reduction ratio falls below a given parameter $r \in [0,1]$ during an iteration of interval Newton. The ordering on classes is defined as follows:

$$Cl \preccurlyeq Cl' \quad \text{iff} \quad \forall f \in Cl', \exists g \in Cl, g \gg f$$

Thus, we have:

$$IBC \preccurlyeq LIBC_{r_1} \preccurlyeq \cdots \preccurlyeq LIBC_{r_n} \preccurlyeq N^* \preccurlyeq N \quad \text{with} \quad r_1 < \cdots < r_n.$$

Several strategies are proposed in [13], and can be divided in two categories: hierarchical strategies that consider a class of a given strongness once and then do not use it anymore, and strategies where only the weakest functions are enforced at a time. Let us illustrate both kinds of strategies.

The first one successively computes three fixed-points, for operators respectively from N^*, $LIBC_{10^{-4}}$, and IBC. The knowledge K is given by a partitioner α that returns the class associated with an operator, and a class selector γ_1 that considers ordering $\preccurlyeq$ as follows:

$$\gamma_1(Cl, \{cl\} \cdot H) = \{\max_{\preccurlyeq}(\{cl' \in Cl \mid cl' \preccurlyeq cl\})\}$$
$$\gamma_1(Cl, \varnothing_H) = \{\max_{\preccurlyeq}(Cl)\}$$

Strategies from the second category implement γ_2 which always selects the weakest class represented in Q:

$$\gamma_2(Cl, H) = \{\max_{\preccurlyeq}(Cl)\}$$

Note that in [13], the update function of the CI algorithm is modified in order to realize some strategies from the first category. To be fully formal, a new proof of termination and fixed-point of the new CI algorithm should be given. On the contrary, these strategies naturally fit in our framework, and thus properties are ensured.

Box_φ consistency. This strategy consists in gradually refining the interval domains by decreasing parameter φ of box_φ consistency. We thus consider n classes of operators $Box_{\varphi_1}, \dots, Box_{\varphi_n}$ with $\varphi_1 < \cdots < \varphi_n$, such that for each constraint c of a CSP, there is an operator able to contract the domains of all the variables of c in each class Box_{φ_i}. Using the ordering $\preccurlyeq$ defined in the previous paragraph, the classes are ordered as follows:

$$Box_{\varphi_1} \preccurlyeq Box_{\varphi_2} \preccurlyeq \cdots \preccurlyeq Box_{\varphi_n}$$

Using the class selector γ_1 defined above, the fixed-point of operators from Box_{φ_n} will be first computed, then the one from $\text{Box}_{\varphi_{n-1}}$, and so on till Box_{φ_1}. Note that classes IBC and Box_0 (given $\varphi = 0$) are equivalent. Box consistency may then be computed by ending the computation with the operators from Box_0.

4.2   Box Consistency Operators

Contrary to the previous section, we now consider a set of classes such that some operators from different classes are not compelled to be comparable. As a consequence, we assume a StopCriterion function as defined in Section 3.

Numerica. In Numerica [21], two classes of operators are considered: Box_n that represents operators enforcing box consistency on natural extensions of constraints, and Box_t that groups operators enforcing box consistency on Taylor extensions of constraints. While the operators of these classes are not comparable, the experiments suggest that operators from Box_t are more tightening than operators from Box_n when intervals are narrower. The strategy of Numerica then alternates the computation of fixed-points for both classes Box_n and Box_t, until a fixed point is reached. It can be realized using the following class selector:

$$\gamma(Cl, \varnothing_H) = \{\text{Box}_n\}$$
$$\gamma(Cl, \{\text{Box}_t\} \cdot H) = \{\text{Box}_n\}$$
$$\gamma(Cl, \{\text{Box}_n\} \cdot H) = \{\text{Box}_t\}$$

Interleaving hull and box consistency. In an analogous way, Algorithm BC4 [3] interleaves computation of hull and box consistency on natural extensions until a fixed point is reached. Let Hull_n denote the class of operators that enforce hull consistency on natural extensions. This strategy can be formalized with the following γ selector:

$$\gamma(Cl, \varnothing_H) = \{\text{Hull}_n\}$$
$$\gamma(Cl, \{\text{Box}_n\} \cdot H) = \{\text{Hull}_n\}$$
$$\gamma(Cl, \{\text{Hull}_n\} \cdot H) = \{\text{Box}_n\}$$

Combining interval forms. Here we propose to merge the box consistency strategies described in [21,3,8,7]. A new class is defined, namely HullBox_n containing operators performing box and hull consistency on natural extensions. This strategy can be formalized with the following γ selector:

$$\gamma(Cl, \varnothing_H) = \{\text{HullBox}_n\}$$
$$\gamma(Cl, \{\text{HullBox}_n\} \cdot H) = \{\text{Box}_t\}$$
$$\gamma(Cl, \{\text{Box}_t\} \cdot H) = \{\text{HullBox}_n\}$$

As a consequence, box and hull consistency are really interleaved as proposed in [7], and the Taylor extensions are used after the natural extensions.

5 Parallel Constraint Propagation

The design of a parallel algorithm generally strongly depends on the execution model (synchronous/asynchronous, local/shared memory) as well as characteristics of parallel machines (number of processors, bandwidth) and may demand a non trivial adaptation of a sequential algorithm to fit in the model. Thus, for re-use of parallel programs, we aim at extracting from constraint propagation algorithms the component parts that remain unchanged whatever is the strategy, namely the computation of a fixed-point of a set of operators.

5.1 Intensification-Based Parallel Interval Narrowing

The bulk-synchronous-parallel (BSP [19]) algorithm of [9] computes a fixed-point of a set of operators performing box consistency for constraint projections (couples constraint/variable). Thus, an operator is able to contract only one domain.

This algorithm alternates an asynchronous selection phase, a global synchronization to localize the selection results on one processor, and a sequential intensification phase that is exactly the generic constraint propagation algorithm. The SelectionByApplication function first performs a balanced distribution of the queue Q of operators among all processors. All operators are then locally applied on the initial domain d. Each contracted domain leads to the selection of the most contracting (with respect to the domain width) operator for this domain. All contractions are intersected in order to keep the tightest domains (for efficiency). They are then located on one processor together with the selected operators in order to initialize the intensification phase. More formally, we have:

$$(F', d') = \text{SelectionByApplication} \, (Q, d, Arg_{f_{fp}})$$

such that $d' = \cap_{f \in Q} f(d)$, and $F' = \{f_{i_1}, \ldots, f_{i_k}\}$ such that $d'_{i_j} \neq d_{i_j}$, f_{i_j} is the operator contracting d_{i_j} the most, for $j \in \{1, \ldots, k\}$, and $d'_h = d_h$ for $h \in \{1, \ldots, n\} \setminus \{i_1, \ldots, i_k\}$.

5.2 Combining Strategy and Parallel Execution Models

The next step in parallel constraint propagation is the design of algorithms that use all characteristics of our framework, namely a non trivial strategy in SCP, and parallel selection and intensification algorithms in IFP. We then propose a new algorithm implementing box consistency where the strategy is the one for box_φ consistency (see Section 4.1), the selection of IFP is the one associated with the BSP model (see Section 5.1), and the intensification is performed by the asynchronous and distributed chaotic iteration algorithm of [15]. This algorithm is necessarily efficient since all techniques have been shown to be powerful heuristics, and needs no further description since all the component algorithms verify the required specifications.

6 Conclusion

A high-level modelling of constraint propagation strategies is allowed through the design of a parameterized algorithm. It implements a fix strategy that applies some component algorithms specified by properties such as termination and confluence. It then enables to describe already existing strategies in a more comprehensive way, to compare complex strategies in terms of declarative properties, and to define/combine/re-use open new strategies. This is exemplified on some algorithms dealing with interval constraint narrowing and parallel constraint propagation.

We identify two directions for further research. On the one hand, this framework could serve as a basis for development of more powerful interval narrowing strategies using other interval extensions, propagation algorithms, and heuristics. On the other hand, a similar work could be done for branch and prune (bisection) algorithms, alternating bisection of domains based on different heuristics with propagation.

Acknowledgments. We thank Gaétan Hains for previous work with Laurent Granvilliers on similar ideas for parallel constraint propagation, and Frédéric Goualard who has carefully read a draft version of this paper. The research exposed here was supported in part by the project 98/06 of the French/Russian A.M. Liapunov Institute.

References

1. Krzysztof Apt. The Essence of Constraint Propagation. *Theoretical Computer Science*, 221(1-2):179–210, 1999.
2. Frédéric Benhamou. Heterogeneous Constraint Solving. In *Proceedings of International Conference on Algebraic and Logic Programming*, volume 1139 of *LNCS*, pages 62–76, Aachen, Germany, 1996. Springer.
3. Frédéric Benhamou, Frédéric Goualard, Laurent Granvilliers, and Jean-François Puget. Revising Hull and Box Consistency. In *Proceedings of International Conference on Logic Programming*, pages 230–244, Las Cruces, USA, 1999. The MIT Press.
4. Frédéric Benhamou, David McAllester, and Pascal Van Hentenryck. CLP(Intervals) Revisited. In *Proc. of International Logic Programming Symposium*, 1994.
5. Frédéric Benhamou and William Older. Applying Interval Arithmetic to Real, Integer and Boolean Constraints. *Journal of Logic Programming*, 32(1):1–24, 1997.
6. John Cleary. Logical Arithmetic. *Future Computing Systems*, 2(2):125–149, 1987.
7. Laurent Granvilliers. Towards Cooperative Interval Narrowing. In *Proceedings of International Workshop on Frontiers of Combining Systems*, volume 1794 of *LNAI*, pages 18–31, Nancy, France, 2000. Springer.
8. Laurent Granvilliers, Frédéric Goualard, and Frédéric Benhamou. Box Consistency through Weak Box Consistency. In *Proceedings of IEEE International Conference on Tools with Artificial Intelligence*, pages 373–380, Chicago, USA, 1999. IEEE Computer Society.

9. Laurent Granvilliers and Gaétan Hains. A Conservative Scheme for Parallel Interval Narrowing. *Information Processing Letters*, 2000. Forthcoming.

10. Olivier Lhomme. Consistency Techniques for Numeric CSPs. In *Proc. of International Joint Conference on Artificial Intelligence*, 1993.

11. Olivier Lhomme, Arnaud Gotlieb, and Michel Rueher. Dynamic Optimization of Interval Narrowing Algorithms. *Journal of Logic Programming*, 37(1–2):165–183, 1998.

12. Alan Mackworth. Consistency in Networks of Relations. *Artificial Intelligence*, 8(1):99–118, 1977.

13. Éric Monfroy. Using Weaker Functions for Constraint Propagation over Real Numbers. In *Proceedings of ACM Symposium of Applied Computing*, pages 553–559, San Antonio, USA, 1999.

14. Éric Monfroy. A Coordination-based Chaotic Iteration Algorithm for Constraint Propagation. In J. Carroll, E. Damiani, H. Haddad, and D. Oppenheim, editors, *Proceedings of the 2000 ACM Symposium on Applied Computing (SAC'2000)*, pages 262–269, Villa Olmo, Como, Italy, March 2000. ACM Press.

15. Éric Monfroy and Jean-Hugues Réty. Chaotic Iteration for Distributed Constraint Propagation. In *Proceedings of ACM Symposium of Applied Computing*, pages 19–24, San Antonio, USA, 1999.

16. Éric Monfroy and Jean-Hugues Réty. Itérations Asynchrones: un Cadre Uniforme pour la Propagation de Contraintes Parallèle et Répartie. In F. Fages, editor, *Proceedings of Journées Francophones de Programmation Logique et Contrainte*, pages 123–137. Hermès, 1999. In French.

17. Ugo Montanari. Networks of Constraints: Fundamental Properties and Applications to Picture Processing. *Information Science*, 7(2):95–132, 1974.

18. Aleksandr Narin'yani. Sub-definiteness and Basic Means of Knowledge Representation. *Computers and Artificial Intelligence*, 5, 1983.

19. Leslie Valiant. A Bridging Model for Parallel Computation. *Communications of the ACM*, pages 103–111, 1990.

20. Pascal Van Hentenryck, David McAllester, and Deepak Kapur. Solving Polynomial Systems Using a Branch and Prune Approach. *SIAM Journal on Numerical Analysis*, 34(2):797–827, 1997.

21. Pascal Van Hentenryck, Laurent Michel, and Yves Deville. *Numerica: a Modeling Language for Global Optimization*. MIT Press, 1997.

22. David Waltz. Generating Semantic Descriptions from Drawings of Scenes with Shadows. In P. H. Winston, editor, *The Psychology of Computer Vision*. McGraw Hill, 1975.

Clustering of Texture Features for Content-Based Image Retrieval

Erbug Celebi and Adil Alpkocak

Dokuz Eylul University
Department of Computer Engineering
35100 Bornova, Izmir, TURKEY
{celebi,alpkocak}@cs.deu.edu.tr

Abstract. Content-based image retrieval has received significant attention in recent years and many image retrieval systems have been developed based on image contents. In such systems, the well-known features to describe an image content are color, shape and texture.

In this paper, we have studied an approach based on clustering of the texture features, aiming both to improve the retrieval performance and to allow users to express their queries easily. To do this, the texture features extracted from images are grouped according to their similarities and then one of them is chosen as a representative for each group. These representatives are then given to users to express their query. Besides the detailed descriptions of clustering process and a summary of results obtained from the experiments, a comparison about statistical texture extraction methods and effects of clustering to them are also presented.

1 Introduction

Image databases are becoming increasingly popular due to large amount of images that are generated by various applications and the advances in computation power, storage devices, scanning, networking, image compression, and desktop publishing. The typical application areas of such systems are medical image databases, photo clip archives, art images, textile pattern archive, photojournalism, WWW, and etc. All these fields need better techniques and mechanism to store and retrieve such huge amount of images.

The early implementation of image databases were based on simply giving descriptive keywords to each image, and allowing users to make query on these keywords for accessing the images. However, this method has some deficiencies such as subjectivity and labor-intensive nature of the keywords assigning process. Moreover, it is sometimes almost impossible to describe content of an image by words, especially for textures found in an image. As a solution to these problems, content-based image retrieval method is proposed based on image features extracted automatically from images. These well-known features are local color, global color, structures in the image (i.e. shape.) and textures found in an image. In this study we focused on textural features of images and aimed both to improve retrieval performance and help users to express their queries easily.

T. Yakhno (Ed.): ADVIS 2000, LNCS 1909, pp. 216–225, 2000.

An image database may contain thousands of textured images. The main problem a user faced is locating the images having similar texture given in the query. More clearly, this problem has two main parts: (1) finding the images having the similar texture given in query and (2) specifying a texture in query. A good image retrieval system dealing with textures must provide solutions to both problems.

Specifying the requirement is not a trivial task for querying images since image are hard to describe in nature. Moreover, the information is meaningful only when it can be retrieved through an expressive query. In forming an expressive query for texture, it is quite unrealistic to expect the user to draw the texture (s)he wants. In our approach, all the textures extracted from the database are classified into clusters and one of the textures in each cluster is chosen as a representative. These representative textures can be presented to the user and asked to choose the closest one that of the requested image. In this way, textures can be used for expressive querying.

Although several researchers have been working on texture classification, none of them is aiming to use texture classification for expressive querying. Smith et. al. [9] proposed a method for classification and discrimination of textures based on the energies of image subbands. They also proposed that texture classification may be used for indexing large databases.

In this study, we have used gray level single-textured images to extract their features and construct a feature vector by using co-occurrence matrixes for each textured image. Then, feature vectors are clustered into groups by using hierarchical clustering techniques and one of the textures is selected as a representative for each group. These representatives are then used in user interfaces for query forming process to narrow down the entire search space. In this way, a much smaller subset of the whole database is given to user to query. Cluster based retrieval also has some advantages on retrieval performance. Our experimentation showed that the results are promising.

In the remainder of this paper, first, texture features and their extraction methods are discussed. In section three clustering process of texture features are presented. The results obtained from our experimentation are presented in section four. Section five gives a look to future works and concludes the paper.

2 Representation of Textures Features

Query submitted by the user will be executed on feature vectors rather than on images themselves. Therefore, the representation of texture is directly related to system performance. However, the features must be extracted first via texture analysis that provides an algorithm for extracting texture features from images.

Haralick proposed the usage of co-occurrence matrixes for texture feature representation [4]. This approach explored the gray level spatial dependencies of texture. It is first constructed a co-occurrence matrix based on the orientation and distance between image pixels and then extracted meaningful statistics from the matrix as texture representation. The feature vector for a texture is constructed from many matrices for different orientation and distance. In the literature there are many other approaches for extracting texture features like wavelet transformations [9], spatial/spatial frequency (s/s-f) subbands [8], Markov Random Fields [2], Gabor wavelet [5].

A feature is an attribute that characterizes a specific property of an object. An n-dimensional feature vector represents an object, where n is the selected numbers of attributes. An object may be image, video, sound and etc. More formally, an image I, can be represented as following feature vector:

$I : < e_0, e_1, e_2, e_3, e_4, e_5, e_6, e_7 ...>$, where each entry ($e_i$) of this vector represents a feature of image I.

Spatial placement of pixels are used to make texture analysis. The information about the spatial placement of pixels can be summarized in two dimensional co-occurrence matrices computed for different distances and orientations. In the following subsection, we give a short description on Gray Level Co-occurrence matrices, which is a well-known statistical texture feature extraction techniques and provide a comparison among them.

2.1 Gray Level Co-occurrence Matrix

The Gray Level Co-occurrence matrix (GLCM) contains the information about gray levels (intensities) of pixels and their neighbors, at fixed distance and orientation. The idea is to scan the image and keep track of gray levels of each of two pixels separated within a fixed distance d and direction θ. But only one distance and one direction generally are not enough to describe textural features. So, we have used more than one direction and distance. It is common to use four directions horizontally and vertically and, two for diagonals. Most of the researchers use four directions and five distances [1, 4]. In our study, we have used four matrices for every value of distance d and four matrices for direction θ. We have represented an image by totally 16 matrices.

Each matrix is 256×256 in size assuming that images are in 256 gray levels. But each of matrix, which is 256×256 in size, requires a huge memory to store and it is a time consuming task to produce the matrix. Therefore, we first convert the images in to 16 gray-levels and then produce a co-occurrence matrix for this new image instead of the original one. This reduction allows us to work with GLCM matrices 16×16 in size. Our experiments show that converting the image from 256 in to 16 gray level does not affect the texture query results. However, these matrices are still containing much data (each matrix has got 16×16=256 entries) and needs to be reduced. What is usually done is, to analyze these matrices and compute a few simple numerical values that encapsulates the information about matrixes. Some statistics computed from GLCM can be used instead of the whole matrix. The gray level co-occurrence matrix is determined as follows:

Let $D_x = \{0, 1, ..., N_x-1\}$ and $D_y = \{0, 1, ..., N_y-1\}$ be the spatial domains of row and column dimensions respectively, where N_x and N_y are the number of pixels in axis X and Y respectively. And, $G = \{0, 1, ..., N_g-1\}$ be the domain of gray levels where N_g is the number of gray levels. The Image I can be represented as a 2D function; $I:D_x \times D_y \rightarrow G$. For abbreviation, a new domain can be defined, as $D \subset N^2$ (where N is the set of Natural numbers) instead of $D_x \times D_y$. Positions and orientations are shown in Figure 1 and Figure 2.

2	2	2	2	2
2	1	1	1	2
2	1	p	1	2
2	1	1	1	2
2	2	2	2	2

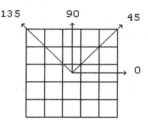

Fig.1. Distances of pixel *p* for co-occurrence matrix   **Fig. 2.** Directions for co-occurrence matrix

In our derivation, we used the following definition for the co-occurrence matrix in distance *d* and direction θ, P(i,j;d,θ) as in Equation 1.

$$p(i,j;d,\theta) = \#\{((x,y),(x',y') \in DxD | d = \|(x,y),(x',y')\|, \theta = \angle((x,y),(x',y')), I(x,y) = i, I \quad (1)$$

where, $P(i,j;d,\theta)$ is the co-occurrence matrix, # stands for the function "number of", *(x,y)* and *(x',y')* are valid image pixel coordinates, *D* is discrete gray scale image domain, *d* is the distance between two pixels, θ is the direction of two pixels.

In Equation 1 we obtained the features of textured images. However, two images with the same texture, but different in size may have different feature vectors. To accomplish this, matrices need to be normalized by the size of images as in Equation 2.

$$P(i,j;d,\theta) = \frac{\#\{((x,y),(x',y') \in D \times D | d = \|(x,y),(x',y')\|, \theta = \angle((x,y),(x',y')), I(x,y) = i, I(x',y') = j\}}{\#\{((x,y),(x',y') \in D \times D | d = \|(x,y),(x',y')\|, \theta = \angle((x,y),(x',y'))\}} \quad (2)$$

In order to use the information contained in the gray level co-occurrence matrices, Haralick defined 14 statistical measures, which is based on textural characteristics like homogeneity, contrast, organized structure, complexity and nature of gray level transitions [4]. However, many authors used only one of the characteristics; variance [1]. We have used only four of GLCM features shown in Table 1. Other features can be found from Haralick's study [4].

Table 1. Four features computed from co-occurrence matrixes

Features	Equations	Features	Equations
Entropy	$-\sum\limits_{i,j} P(i,j) \log P(i,j)$	Homogeneity	$\sum\limits_{i,j}\dfrac{p(i,j)}{1+|i-j|}$
Contrast	$\sum\limits_{i,j}|i-j|^k\, p^\ell(i,j)$	Variance	$\sum\limits_{i,j} P(i,j).(i-j)^2$

Once the texture features are extracted and stored in a database, a similarity function is required to compare texture for similarities. In our study, we have used Euclidean distance metric as similarity function.

3 Clustering Texture Features

The Cluster analysis is a partitioning of data into meaningful subgroups (clusters), when the number of subgroups and other information about their composition or representatives are unknown. A general information for clustering can be found in [10] and [3].

Cluster analysis does not use category labels that tag objects with prior identifiers. In other words, we don't have prior information about cluster seeds or representatives. The absence of category labels distinguishes cluster analysis from discriminant analysis (and classification and decision analysis). The objective of cluster analysis is simply to find a convenient and valid organization (i.e. group) of the data.

There are many application areas of cluster analysis such as image segmentation, image indexing or, in general, object indexing and to allow users to navigate over the images. The main purpose of clustering is to reduce the size and complexity of the data set. Data reduction is accomplished by replacing the coordinates of each point in a cluster with the coordinates of that cluster's reference point (cluster's seed or representative). Clustered data require considerably less storage space and can be manipulated more quickly than the original data. The value of a particular clustering method depends on how closely the reference points represent the data as well as how fast the program runs.

As mentioned before, a clustering schema may represent simply a convenient method for organizing a large set of data so that the retrieval of information may be made more efficiently. Cluster representatives may provide a very convenient summary of the database. In another say, it forms a narrowing down phase of the whole search space.

3.1 Data Types and Data Scales

Clustering algorithms group objects, or data items based on indices of proximity (similarity) between pairs of objects.

Pattern Matrix: If each object in a set of n objects is represented by a set of d measurements each object is represented by a pattern, or d-dimensional vector. The set itself is viewed as $n \times d$ pattern matrix. Each row of this matrix defines a pattern and each column denotes a feature or measurement.

Proximity Matrix: A proximity matrix $[d(i,j)]$ accumulates the pair-wise indices of proximity in a matrix in which each row and column represents a pattern. In proximity matrix, d_{ij} denotes the similarity/dissimilarity between object i and j. Note the matrix is always symmetric.

Similarity and dissimilarity can be summarized as follows:

(*a*) For a dissimilarity: $d(i,i) = 0$, for all i,
(*b*) For a similarity: $d(i,i) = max_k d(i,k)$, for all k.

3.2 Group Similarities

In cluster analysis it is some times convenient to use the distance measurement between groups instead of distance of objects. One obvious method for constructing distance measure between groups is to substitute group mean for the d variables in the formula for inter individual measures such as Euclidean distance or other distance metrics. If, for example, group A has a mean vector $\overline{X}_A = [\overline{x}_{A1}, \overline{x}_{A2}, ..., \overline{x}_{Ad}]$ and group B has a mean vector, $\overline{X}_B = [\overline{x}_{B1}, \overline{x}_{B2}, ..., \overline{x}_{Bd}]$, then one measure of the distance between the two group would be as in Equation 3.

$$d_{AB} = \sqrt{\sum_{i=1}^{d} \left(\overline{x}_{Ai} - \overline{x}_{Bi} \right)^2} \qquad (3)$$

3.3   Hierarchical Clustering

A hierarchical classification is a nested sequence of partitions. In this study we have used exclusive (each object belongs to exactly one group) and agglomerative classification. Agglomerative classification places each object in its own cluster and merges these atomic clusters in to larger and larger clusters. The algorithms start with a set of object and merge them to form the clusters and, ends when there are no object to merge with any cluster. In this study, we modified the algorithms to stop when the desired number of clusters reached.

Single linkage is one of the most popular methods, which is used for agglomerative clustering, and it is also known as *nearest neighbor* technique. The characteristic of the method is that distance between groups is defined as the closest pair of objects. For example distance between a cluster with two objects and an object can be defined as follows:

$d_{(ij)k} = min\,[d_{ik},\,d_{jk}]$

The algorithm starts with searching the proximity matrix and finds the smallest entry. Smallest entry means the most similar textures they can form a cluster. Once they are merged they are considered as a single object. The algorithm works until all the objects are in the same cluster or, the desired number of clusters is reached.

There are different clustering methods, which can be defined according to measurement of distances between clusters, such as centroid clustering, complete-link, group average clustering and single-link clustering as described by [10]. We used group average method in this study. In addition to clustering, we have selected a representative for each cluster to help user to form texture queries. Representatives are selected by the closest object to the average feature vector of each cluster.

4 Experimentation Results

In this section cluster based texture query experiments are presented. We used the well-known methods mentioned in the previous sections. We have implemented an application to observe the results of our study. The purposes of implementing this application is both to provide an example for content-based texture query systems and monitor the effectiveness of a cluster based texture query systems.

4.1 Image Test Bed

There are many texture sets for the purpose of evaluating the textured image retrieval methods. We have used three of them; Brodatz Texture Set, Ohan & Dubes Texture Set and VisTex Texture Set. In our experiments, we have used 14 samples from Brodatz album as a test bed of homogeneously textured images as in some other researches did [7,8]. Each of the 14 sample images has been divided into 25 partly overlapping sub-images of size 170 × 170. We have selected 5 sub-images randomly from each image and prepared totally 70 images. Our expectation was to obtain 14 clusters containing 5 textures, when desired number of clusters is 14. We are also expecting each sub-texture, which was extracted from the same texture, to be in the same cluster, since they are similar.

4.2 Experimentation Results

We have made a series of experiments to compare their performance on our test bed. In our experiments, first we have tested homogeneity, variance, entropy and contrast features of textures. The results obtained from experiments are summarized in Figure-3.

It is clear that the retrieval performance of entropy is the worst among others. However it is hard to say which one is the best. It depends on the texture domain. For instance, precision value of homogeneity for query 7 is better than variance's value. But variance gives better result than homogeneity for query 4.

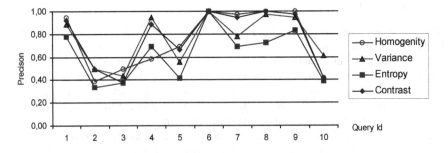

Fig. 3. Precison graph for all features on Texture Query Experiments

Recall and *precision* are popular measures for retrieval evaluation. Recall signifies the proportion of relevant images in the entire database that are *retrieved* in the query. In other words Recall is the ability to retrieve *relevant* textures. Precision is the ability to reject *nonrelevant* textures. A good system should have high precision and recall values as in [6].

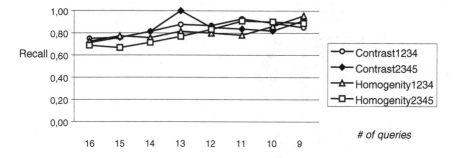

Fig. 4. Recall graph of two feature types and their models

The *X*-axis of the Figure 4 and Figure 5 shows the number of clusters. We used two models; 1234 which means distances, *d*, 1,2,3 and 4 are used to extract features and 2345 means distance 2,3,4 and 5 are used for feature extraction. We did 7 experiments for each feature type to observe the effect of number of clusters in query. We were expecting the performance to be the best, when there are 14 clusters (as explained in previous sections). We could yield two results from these experiments, one is which feature type is better on clustering and other is how many clusters gives the best performance. From Figure 5 we can conclude that the more cluster means the higher value in precision. In another word, ability to reject non-relevant texture increases when we set more clusters. As can be seen from the Figure 5, contrast with model 1234 gives the best performance on precision. The ability to retrieve relevant textures can be measured by the values of recall. As can be seen from the Figure-6, it is best with contrast in model 1234.

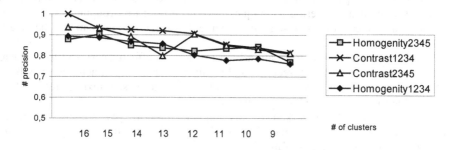

Fig. 5. Precision graph of two feature types and their models

In this study, an application has been developed to see the effects of cluster-based texture retrieval. A typical user interface screen of the application can be seen in Figure-6. In the left side of the screen-shot the representatives of each cluster are shown and, one of the image has been clicked as a query. The query results returned for the query can be seen in right side of Figure-6.

Fig 6 : An example cluster query on gravel155.2.bmp

5 Conclusion and Future Works

In this study, we have developed a system that allows user to input a textured image and retrieves textures from a database similar to the query. We proposed to cluster textures into groups according to their similarities to make the query expressions easier. Hierarchical clustering techniques have been used for grouping textures in to clusters and selected a representative texture for each group to make the query expressions easier.

We have developed an application both to show its effectiveness and to evaluate the system performance by precision and recall measures. Experimentation has been shown that the results are promising and the cluster-based texture retrieval is acceptable for content based image retrieval.

In this study, we have worked on only single textured images. We are planning to extend our study to the domain independent multi-textured images and test its scalability for more images in future.

References

1. Aksoy, S. & Haralick, R. M. (1998). Content Based Image Database Retrieval Using Variance of Gray Level Spatial Dependencies. IAPR International Workshop on Multimedia Information Analysis & Retrieval (MINAR'98), Hong Kong.
2. Andrey P. & Tarroux P. (August 1995). Unsupervised segmentation of Markov random field modeled textured images using selection relaxation. Technical Report: BioInfo-95-03.
3. Everitt, B. S. (1980). Cluster Analysis. John Wiley & Sons.
4. Haralick, R.M. (May 1979). Statistical and structural approaches to texture. Proceedings of the IEEE, 67(5): 786-804.
5. Manjunath B. S. & Ma W. Y. (August 1996). Texture features for browsing and retrieval of image data. IEEE Transactions on Pattern Analysis and Machine Intelligence. Vol. 18, No:8, pp. 837-842
6. Ozkarahan, E. (1986). Database Machines and Database Management. Prentice Hall, New Jersey.
7. Puzicha, J. & Hofmann, T. & Buhman, J.(1998). Histogram Clustering for Unsupervised Segmentation and Image Retrieval.
8. Smith J. R., & Chang S.F. (May 1996). Automated binary texture feature sets for image retrieval. Proceedings of the IEEE. ICASSP-96. Atlanta, GA.
9. Smith, J. R. & Chang, S. F. (1994). Transform features for texture classification and discrimination in large image databases. Proceedings of IEEE International Conference on Image Processing (ICIP-94), Austin, Texas.
10. Jain, A. K. & Dubes, R. C. (1988). Algorithms for Clustering Data, Prentice Hall.

Fuzzy Content-Based Retrieval in Image Databases[1]

Ibrahim Gokcen, Adnan Yazici, and Bill P. Buckles

Dept. of EECS, Tulane University
New Orleans, LA, 70118
{gokcen,yazici,buckles}@eecs.tulane.edu

Abstract Image data is a very commonly used multimedia data type and usually have visual characteristics that have imprecise descriptions. Fuzzy retrieval of images that are stored in an image database is a natural and effective way to access image data. Recently, some work has been done on fuzzy content-based retrieval systems but to the authors knowledge none of them rely on a defined model for fuzzy query processing part. In this paper, an approach for fuzzy content-based retrieval using the Fuzzy Object-Oriented Data (FOOD) model will be described. A novel way of determining the fuzzy values from extracted color features will also be presented.

1 Introduction

Very large collections of images are growing rapidly due to the advent of cheaper storage devices and the Internet. For example, satellites send tens of images of earth each second and these images are stored in huge databases for future retrieval. Retrieving images from such large collections efficiently and effectively based on their content has become an important issue recently.

Image data is fuzzy in nature and in content-based retrieval this property creates some problems such as: (1) Descriptions of image contents usually involve inexact and subjective concepts: After seeing an image, different people would have different descriptions. Because of the diversity of image contents, it's not wise to rely on the descriptions of people (manual description). Computers can produce relatively objective measures of visual characteristics of images, [Wu95]. (2) Providing user's needs to the databases for retrieval, effectively may be naturally fuzzy: The user can provide preferences using numerical values (ex: 0.8 to color, 0.2 to shape) but it's not easy for the users to provide the preferences in this way. A fuzzy linguistic approach may be used to support users' preferences. Thus, the user may specify linguistic qualifiers such as more, less, few, many, some etc. for his/her preference rather than numerical values, [Nepal95]. (3) Finite set of recognizable feature values by human, is a restricted subset of what the image actually may have: User can only recognize a finite set of feature values. For example when querying by color, user always refers to a set of colors, which fall within the human perceptual range. Feature values that fall in the human perception range should be used to overcome this problem, [Nepal95]. (4) Usually imprecision and vagueness exist in descriptions of the images and in some

---

[1] This work was supported in part by a grant from NASA/Goddard Space Flight Center. #NAG5-8570 and in part by DoD EPSCoR and the State of Louisiana under grant F49620-98-1-0351.

T. Yakhno (Ed.): ADVIS 2000, LNCS 1909, pp. 226-237, 2000.

of the visual features. At archiving (a concrete description is associated with each stored image, so as to retain the important visual characteristics), the descriptions are generally imprecise and quantization of visual features (object features) can also be vague. In querying (queries are specified as abstract specifications of visual properties, distinguishing the classes of objects that must be retrieved) again deliberate imprecision and vagueness is needed to cope with the incomplete knowledge of the users. In the final stage, model checking (comparing abstract query specifications against the concrete image descriptions) involves combination of all kinds of approximations.

Imprecision and vagueness of both image descriptions and query specifications usually impair a definite decision about the satisfaction of a query. Relaxation of model-checking rules is needed to overcome this limit with the introduction of a score, quantifying the degree of truth, by which the available description permits a decision about a given query. [Corridoni98] This implies that we are not looking for an exact match but we are looking for the nearest matches.

Some previous studies have been done on this subject. Some researchers tried to define an efficient way of fuzzy query processing and similarity computation between the query and images in the database. Some other researchers worked on the limited values of visual properties and defined representation systems for that task. However, none of them based their fuzzy query processing part on an object-oriented database model. Two of such works are explained briefly in section 2.

This work uses some of the results produced by previous studies [Nepal95, Nepal99, Corridoni98]. The difference of it is the underlying fuzzy object-oriented database (FOOD) model to compute the similarity between the images in the database and the queries. A feature extraction step is necessary to describe the image objects, by using some fuzzy linguistics descriptors. An approach to extract color feature using Neural Networks is also described.

In the coming section, some existing fuzzy content-based retrieval systems and query languages, which can handle fuzziness is presented. Next, the system that is used in this study is introduced briefly. In section 4, we describe the architecture and the other details of the FOOD model. In the last section, we give the conclusions and possible future extensions of this study.

2   Previous Studies

There are three general approaches to content-based retrieval. One can provide symbolic captions that describe the contents as does CHABOT [Ogle94], QBIC [Faloutoso94,Flickner95], and WebSeer [Swain96]. One can provide an exemplar image and perhaps specify which aspects of the exemplar for which to find a match [Ardizzone96,Faloutoso94,Flickner95,Smith94]. The aspects to match might be the color content or texture and perhaps the image layout (i.e., where in the image certain colors or textures occur). Finally, either at the time the image is ingested into the database or dynamically as required, a subsymbolic content description of the image can be computed. That is to say, features such as the color histogram, the fractal dimension of the texture at strategic points, or the frequency domain coefficients can be extracted. Then, at query time, a match for user input features can be searched for

[Ogle95,Faloutoso94,Flickner95,Smith94,Kelly94,Koutsougeras98]. This requires a measure of similarity or distance as we shall see.

We are interested in the case for which subsymbolic content descriptors are matched against user input feature vectors using fuzzy set concepts [Wu98,Nepal95,Nepal99]. In the most general case, we can describe an object within the image as both an ordinary feature vector (e.g., <intensity, coarseness, hue, ... >) using crisp values. This is called feature space. We could then specify a mapping between feature space and fuzzy space – the same vector but for which the domains are fuzzy linguistic values (e.g., "high" or "fine") or fuzzy sets (e.g., "near 50"). An image, I, is then described as a set of objects and an object is a triple $O = (F, A, I)$ where F is expressed in feature space and A is expressed in fuzzy space. A fuzzy query, $O^*=(-,A^*,-)$, is then matched against the set O as is done in FQP [Wu98] using a similarity metric.   Typical metrics include the Euclidean distance, rectilinear distance, "chessboard" distance, or (if the vectors are uncorrelated real values) the cosine of the angle between A and A^*. The specific measure used in FQP is based on the fuzzy correlation measure.

Lastly, metrics can be difficult to define depending on the feature vector and its semantics. Real, integer, and categorical component values present different issues. There is some theory about metric spaces in mathematics but it is of little help unless one has deep semantic knowledge of the variables one is employing to define features. One manner of doing this is to limit queries to one sort of feature, say, color. If A consists only of color, let it be represented as a set for image i of the form $A = \{(red,\mu_{red}(i)), (blue,\mu_{blue}(i)), ...\}^2$ [Nepal95,Nepal99]. One might even go one step further and assign linguistic values for ranges of $\mu_a(i)$ such as *mostly* for [0.7,1.0]. Measuring the distance between A and A^* is done with a compatibility function as described by Nepal [Nepal95]. A user may specify more than one A^* joined by *and, or,* and *not.*

3 Test-Bed System: NASA Prototype

Adapting the available technology of text-based databases to images has led to the technique of attaching descriptive, often statistical, information (metadata) to images to facilitate the search. Image is represented by metadata and the metadata can be rapidly extracted from an image, during its ingest. Metadata deals with the content of the image and includes features such as the mean irradiance, cloud cover percentage and texture measures. There is also geodata, which deals with the context of the image.

Issues with this schema are: selection of metadata to store online that can be used for content-based querying, determination of an extensible database architecture, that is capable of supporting experimental query techniques, and management of uncertainty, that is inherent in image data.

---

[2] The fuzzy membership value, $\mu_{red}(i)$, might be computed by converting the image to HSV representation then quantizing the H-band into the ranges for *red, blue, yellow,* etc. The membership for *red* would then be the percent of pixels that fall within the range for *red.*

There are two classes that are defined in the prototype – image and tile. An object of class Image contains the image itself and the geodata associated with it. An object of class Tile contains the content-oriented data together with the Ids of image/sub-image, which the object describes. There is a one-to-many relationship between Tile and Image objects. Each tile object contains the metadata for a portion of the image and each tile object is stored separately from the image object to which it refers. [Koutsougeras98].

The images are divided into tiles to make more use of the local properties in the images. The histogram result, for example, can be nearly the same for very different two images. Dividing the images into tiles and calculating the histogram values of each tile is a more efficient way of querying based on color content. The local features of each tile, such as color feature and cloud pixels, are kept in the tile object information, while the global features of the image such as day, year, and location are kept in the image object. Some possible queries on the attributes given in class definitions are as follows: (1) Query based on Histogram (requires feature extraction), i.e., *find the images, which are mostly green.* (2) Query based on cloudpixels (requires feature extraction), i.e., *find the images, which are more-or-less cloudy.* (3) Query based on day or year (fuzzy query), i.e., *find the images that are dated in the middle of any month between 1998-2000.*

The features can also be combined in the queries and some relevance values may be applied to them to represent the query better. For example, if we are looking for a hurricane in the last two years, we have to look for images that have many white pixels in them and they should have been in the last two years. We can also specify the last two years by using the linguistic variable *near*. The content and definitions of classes for the example test-bed system are as follows:

```
CLASS tile
INHERITS OBJECT
PROPERTIES
        Tileid      // tile identity
        Imageid     // image identity
        Color       // dominant color in the tile
        Cloudpixels          // percentage of cloudy pixels
        Wavelet     // wavelet coefficients
        ...
RANGES
Color = < Red, Blue, Green, Pink ..>
CloudPixels = [VeryCloudy, MoreOrLessCloudy, FewCloudy]
...
RELEVANCE
        Color = 2, Cloudpixels = 1
        ...
METHODS
        IMAGEID(), PIXMEAN(), COLOR()
        ...
END


CLASS image
INHERITS OBJECT
PROPERTIES
        DAY       // day at which the image is obtained
        YEAR      // year in which the image is obtained
        THEIMAGE          // image itself
        HISTOGRAM         // fuzzy attribute to denote color feature
        ...
```

RANGES
        HISTOGRAM = {ManyRed, ManyGreen, FewBlue, SomeRed, SomeGreen, ... }
        DAY = [Beginning, Middle, End]
        YEAR = [recently, near, far]
        ...
RELEVANCE
        HISTOGRAM = 2,   DAY = 1, YEAR = 1
        ...
METHODS
        DAY(), YEAR(), THEIMAGE()
        ...
END

4 Architecture of the System and Necessary Algorithms

In this section before we describe the architecture of the system, we should first summarize the fuzzy object-oriented database model that is utilized here. The details of this model are given in [Yazici99].

4.1 The Fuzzy Object-Oriented Data (FOOD) Model

The fuzzy object-oriented database model (FOOD) that is briefly described here is similarity-based. For each fuzzy attribute, a fuzzy domain and a similarity matrix are defined. Similarity matrices are used to represent the relation within the fuzzy attributes. The domain, *dom*, is the set of values the attribute may take, irrespective of the class it falls into. The range of an attribute, *rng*, is the set of allowed values that a member of a class, i.e. an object, may take for an attribute. In general $rng \subseteq dom$. A range for each attribute of the class is defined as a subset of a fuzzy domain. The range definition for attribute a_i of class C is represented by the notation, $rng_C(a_i)$, where $a_i \in Attr(C) = \{a_1, a_2, ..., a_n\}$. $Attr(C)$ refers to the attributes of class C. Similar objects are grouped together to form a class and fuzziness at object/class and class/superclass levels are represented this way. The idea of fuzziness extends in the relation of an object with the class of which is created as an instance. An object belongs to a class with a degree of membership.

Fuzziness may occur at three different levels in this fuzzy object-oriented database model; the attribute level, the object/class level and the class/superclass level. We are specifically interested in the fuzziness at attribute and object/class level regarding this paper. Therefore we only describe fuzziness at both of these levels.

Attribute Level
At the attribute level, there are different types of uncertainty of the attribute values. Fuzzy attributes may take a set of fuzzy values having one of the AND, OR and XOR semantics, which are denoted by the following logical operators: AND:<.......>, OR:{.....} and XOR:[......]

Every class has a range definition for each of the fuzzy attributes, with the corresponding relevance values, indicating the importance of that attribute in the definition of that class. In this way an "approximate" description of the class is given.

An attribute of a class is allowed to take any value from the domain, without considering the range values. In this model, semantics is associated with the range definitions to permit a more precise definition of a class. Relevance weights are assigned for each attribute, and they show the significance of the range definition of that attribute on the class definition.

Object/Class Relations

The object/class level denotes the membership degree of an object to a class. The main feature that distinguishes the fuzzy classes from crisp classes is that the boundaries of fuzzy classes are imprecise. The imprecision of the attribute values causes imprecision in the class boundaries. Some objects are full members of a fuzzy class with a membership degree 1, but some objects may be related to this class with a degree between 0 and 1. In this case they may still be considered as instances of this class with the specified degree in the range [0,1]. In the model, a formal range definition, indicating the ideal values for a fuzzy attribute is given in the class definition. However, an attribute of an object can take any value from the related domain. So, the membership degree of an object to the class is calculated using the similarities between the attribute values and the class range values, and the relevance of fuzzy attributes. The relevance denotes the weight of the fuzzy attribute in determination of the boundary of a fuzzy class. If an object has the ideal values for each fuzzy attribute, then this object is an instance of that class with a membership degree of 1. Otherwise, it is either an instance with a membership degree less than 1, or it is not an instance at all (when the membership degree is smaller than the threshold value) depending on the similarities between attribute values and formal range values. The closer the attribute value to the range, the higher the membership degree of the object.

To calculate the membership degree of an object to a class, we must calculate the inclusion degrees of attribute values in the range of attributes. Since the attribute values may be connected through *AND, OR,* or *XOR* semantics, the inclusion value depends on the attribute semantics.

The more similar an object's attribute value to the range definitions, the higher the class/object membership degree. But how is this distance determined? The membership degree of the object o_j to class C is determined by the following formula:

$$\mu_C(o_j) = \frac{\sum INC(rng_C(a_i) / o_j(a_i)) * RLV(a_i, C)}{\sum RLV(a_i, C)}$$

where $INC(rng_C(a_i)/o_j(a_i))$ is the value of the inclusion taking into account the semantics of multivalued attribute values (as will be described below), $RLV(a_i, C)$ is the relevance of attribute a_i to the class C and is given in the class definition. The weighted-average is used to calculate the membership degree of objects. All attributes, therefore, affect the membership degree proportional to their relevance.

Computation of Inclusion Values

The formulas used to calculate inclusion degrees are briefly given below for the AND, OR, and XOR connection semantics. If $o_j(a_i) = \varnothing$, then INC=0 for all semantics, where o_j is an object and a_i is an attribute of object o_j. Otherwise:

1. AND semantics: AND semantics requires that all of the instances exist simultaneously. If an object has all of its values in the range, the inclusion degree is one. Otherwise, it is less than one depending on the similarities. The formula for AND semantics is:

$$INC(rng(a_i)/o_j(a_i)) = Min[Min[Max(\mu_s(x,y))], Min[Max(\mu_s(z,w)]],$$

$$\forall x \in rng(a_i),\ \forall y \in o_j(a_i),\ \forall z \in o_j(a_i),\ \forall w \in rng(a_i).$$

2. OR semantics: OR semantics uses a subset of the range definition. If similarities among the attribute values decrease, the inclusion degree also decreases. This is because, when similarity among the attribute values increases, the uncertainty decreases. This property forces objects to have close and therefore meaningful attribute values. The formula for OR semantics:

$$INC(rng(ai)/oj(aI)) = Min[Max(\mu S(x,z)), Threshold(oj(ai))]$$

$$\forall x \in oj(ai),\ \forall z \in rng(ai)$$

The threshold value indicates the minimum level of similarity between the elements of object attribute and it can be formulated as follows:

$$Threshold(oj(ai)) = Min[\mu S(x,z)]$$

$$\forall x,\ \forall z \in oj(aI)$$

3. XOR semantics: XOR semantics forces only one of the entries in the range to be true. Assuming equal probabilities for the elements of the attribute value, the inclusion degree is formulated as follows:

$$INC(rng(ai)/oj(aI)) = Avg[Max(\mu S(x, y))]$$

$$\forall x \in oj(ai),\ \forall y \in rng(ai).$$

4.2 The Architecture of the System

Query Processing

The user specifies the query using the web interface and this module produces the necessary values for the fuzzy processing module. To be able to modify this part of the system, without considering the fuzzy processing module, it is specified separately.

Feature Extraction

The feature extraction step is used to extract features F^i from the image where i is the number of the feature taken into consideration. These features may be histograms, texture, wavelet signatures etc. Which feature is used is not important for the purpose of this paper and we chose the histogram feature. The concepts presented in this paper can be extended very easily to include other features as well. The method presented here is a novel one and is quite specific for the application. A general tool, based on a different approach is available in Berkeley Digital Library Project [Berkeley].

In computing the histogram of a particular tile several tools can be used such as Khoros, MATLAB or any other image processing tool. Because of its text output generation feature, we chose to use MATLAB. The MATLAB function *imhist(I,y)*, which is present in the Image Processing Toolbox, creates the histogram of a tile denoted by the image variable I and divide the number of pixels to y number of bins in the interval 0-255 of grayscale values. 0 corresponds to black and 255 corresponds to white. The values in between are the values of varying gray colors. MATLAB function *imread* can be used for reading an image of type jpeg to an image variable.

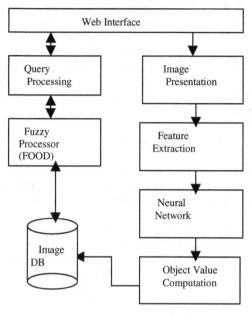

Fig. 1. The architecture of the system

I = imread('c:\ibrahim\110061.jpeg')

[counts, x] = imhist(I,8)

This piece of code assigns the values of the 8 bins to the variable *counts*. An example value for counts for a mostly green tile is: counts = 3513 12229 30452 15490 9341 2008 462 143

Neural Network (Athena)

After getting the number of pixels of each bin (value vector), Athena neural network model is used to classify each value vector. The value vectors from sample tiles will be produced and they will be used to train the network. A sample input vector of the training set is:

3513 12229 30452 15490 9341 2008 462 143 G

In this vector the first 8 elements denote the input to the neural network and G (Green) denotes the class of that input. Here the output is the most dominant color in a specific tile. Athena works based on partitioning the input space into regions so it never misclassifies the training set, whenever a dense enough input set is provided [Koutsougeras88].

Object Value Computation

When the training phase is complete, we end up with a classification of the input space for all the tiles in the database. Now we have to convey this information to the global image. The images should be queried by multiple colors (like ManyGreen and SomeBlue) to make the queries more realistic. The way to do this is to use the crude ratio of number of tiles whose dominant color is C_i (i={1, 2, ... , 9}) to the total number of tiles. Since there are a known number of tiles in an image, we can determine the overall fuzzy space representation of the image by using a threshold method.

As an example, let's think of an image with 16 tiles. Let's assume that the histogram classes of the 16 tiles are 10 Green, 1 Yellow, 4 Blue and 1 Pink. In this case, the ratios of the variables are {G= 0.625, Y= 0.06, B= 0.25, P= 0.06}. Now we can use a threshold value T of 0.1 and assign the fuzzy linguistic variables to the image in the following fashion:

Many = [0.7-1.0], Some = [0.4-0.7], Few = [0.1-0.4]

In this way object value of our image in the database (as described in FOOD) becomes:

o_1(histogram) = { SomeGreen, FewBlue }

Let's assume that we have done a similar kind of computation for 2 other images and found their object values as below. Note that the objects are all described with OR semantics.

o_2(histogram) = {SomeRed}, o_3(histogram) = {SomeBlue, SomeGreen}

Computation of the object values for the histogram attribute of each image, based on the tile histogram information, is done only once at the beginning. The only thing to compute at query time is the inclusion value. How to compute the inclusion is also described in the previous section. For that computation we need the similarity matrix. The similarity matrix consists of the relations of fuzzy variables to each other. We assume that there are possible 9 colors to choose from (i.e.: red (R), green (G), blue (B), black (BLA), purple (P), pink (PI), orange (O), white (W), yellow (Y)) and 3 quantifiers for each color (Many (M), Some (S), Few (F)) so the size of the matrix is 27x27. A part of an example similarity matrix is given below.

Table 1. Similarity matrix for color

Hist	MR	SR	FR	MB	SB	...
MR	1	0.7	0.4	0	0	
SR	0.7	1	0.6	0	0	
FR	0.4	0.6	1	0	0	
MB	0	0	0	1	0.7	
SB	0	0	0	0.7	1	
...						

Image Presentation and Fuzzy Processing

Among all of the possible attributes, only the fuzzy retrieval based on color is discussed here. The other attributes are handled in a similar fashion. The first step is to compute the histogram values of all the tiles in the database and store them in fuzzy representation. Images can be added to the database using the web interface and for each such image the image presentation algorithm is applied. After all the images are processed in this fashion and the query is processed using the query processing module, the fuzzy processing algorithm is applied to each image object and the query object.

Algorithm for image presentation
```
For each image in the database
        For each tile in the image (specified number of tiles in
                each image)
                Calculate the 8-bin histogram value
                Determine the dominant color for the tile using
                        the trained Athena Neural Network
        Calculate ratios of each color in the image using the
                dominant tile color information
Define    the    range    of    the    image    using    the    ratios    and
correspondences
```

Algorithm for fuzzy processing
```
For each image in the database
        Calculate the inclusion value of the query
                INC(range_image(histogram)/query(histogram))
        Sort all the matched objects representing the images
                based on their inclusion values in descending
                order
        Output the ones that have value >= threshold
```

4.3   Example Query

Let's assume that the user posed the query as {FewGreen, FewBlue}. In the description of the FOOD system, the inclusion value is calculated for object/class relations to determine the membership degree of an object to a class. We have the objects in the database, which are corresponding to the objects in FOOD scheme. Computation of the membership degree of an object to a class is used to find the similarity between the query and the objects in the database. For instance: rng_{Image}(histogram) = {FewGreen, FewBlue}. Using the example three objects and the query by the user, the membership degrees (similarities) of each image object to the query object are calculated step by step as shown below.

$INC(rng_{Image}$(histogram)$/o_1$(histogram))
$= Min [Max (\mu_s(FB, FB), \mu_s(FB, SG), \mu_s(FG, FB), \mu_s(FG, SG)), Threshold (o_1$(histogram))]
$= Min [Max (1, 0, 0, 0.6), Min (\mu_s(FB, SG))] = Min [1, 0.3] = 0.3$
$INC(rng_{Image}$(histogram)$/o_2$(histogram))
$= Min [Max (\mu_s(FG, SR), \mu_s(FB, SR)), Threshold (o_2$(histogram))]
$= Min [Max (0, 0), Min (\mu_s(FB, SG))] = 0$
$INC(rng_{Image}$(histogram)$/o_3$(histogram))
$= Min [Max (\mu_s(FG, SB), \mu_s(FG, SG), \mu_s(FB, SB), \mu_s(FB, SG)), Threshold (o_1$(histogram))]
$= Min [Max (0, 0.6, 0.6, 0), Min (\mu_s(SB, SG))] = Min [0.6, 0.5] = 0.5$

In our example, since we are only considering the histogram feature, the membership degrees of the objects are the same as the inclusion values. However, with the inclusion of other features such as CloudPixels, Days, these features will also effect the membership degree in accordance with their relevance values.

The final step for the query processing is to sort the inclusion values for matching image objects and output the ones that are equal or greater than the threshold value. If we choose the threshold as 0.3, the objects will be retrieved in the following way.

1. Object denoted by o_3 representing image 3
2. Object denoted by o_1 representing image 1

5   Conclusion and Future Work

In this paper we presented an approach for fuzzy content-based retrieval in image databases using the FOOD model. A novel way of calculating the histogram value for each tile is given for that purpose. There may also be different approaches such as using a back-propagation network to make use of more than one output values and calculate the range values of each tile directly.

The only feature that is included here, is the histogram value. For all other features such as cloud pixels, texture, day, year etc. different kinds of algorithms can be devised and range values of each of the image in the database for those features can be calculated. In this way more complex queries can be formulated to retrieve images, not just only based on color but also on the other features.

Despite these possible feature extensions, the proposed methods and concepts work well on the existing application. The extensions to improve the efficiency and correctness of fuzzy similarity matching is our ongoing research topic.

References

[Ogle95] V. E. Ogle and M. Stonebraker, "Retrieval from relational database of images," *IEEE Computer* Vol. 28, No. 9, Sept. 1995, pp. 40-56.

[Flickner95] M. Flickner, H. Sawhney, W. Niblack, J. Ashley, Q. Huang, B. Dom, M Gorkani, J. Hafner, D. Lee, D. Petkovic, D. Steele and P. Yanker, "Query by image and video content: The (QBIC) system," *IEEE Computer* Vol. 28, No. 9, Sept. 1995, pp. 23-32.

[Kelly94] P. M. Kelly and T. M. Cannon, "CANDID: Comparison algorithm for navigating digital image databases," *Proc. 7th Working Conf. on Scientific and Statistical Database Management*, Charlottesville VA, Sept. 1994, pp. 252-258.

[Faloutoso94] C. Faloutoso, R. Barber, M. Flickner, J. Hafner, W. Niblack, D. Petkovic and W. Equitz, "Efficient and effective querying by image content," *Intelligent Information Systems* Vol. 3, 1994, pp. 231-262.

[Swain96] M. J. Swain, C. Frankel and V. Athitsos, "WebSeer: An image search engine for the world wide web," Technical Report TR-96-14, Univ. of Chicago, July 1996.

[Ardizzone96] E. Ardizzone, M. L. Cascia and D. Molinelli, "Motion and color based video indexing and retrieval," Proc. Intern. Conf. On Pattern Recognition, Austria, Aug. 1996.

[Smith94] J. R. Smith and S.-F. Chang, "Quad-tree segmentation for texture-based image query," *Proc. Annual ACM Multimedia Conf.*, San Francisco, 1996.

[Wu98] J. K. Wu and D. Nerasimhalu, "Fuzzy Content-based retrieval in image databases". *Information Processing and Management* Vol. 34 No. 5 pp. 513-534. 1998.

[Nepal95] S. Nepal, M. V. Ramakrishna and J. A. Thom, "A fuzzy system for content-based retrieval".

[Nepal99] S. Nepal, M. V. Ramakrishna and J. A. Thom, "A fuzzy object query language (FOQL) for image databases," *Proc. 6th Intern. Conf. On Database Systems for Advanced Applications,"* Hsinchu Taiwan, April 1999.

[Corridoni98] J. M. Corridoni, A. Del Bimbo and E. Vicario, "Image retrieval by color semantics with incomplete knowledge". 1998

[Kulkarni97] S. Kulkarni, B. Verma, P. Sharma and H. Selvaraj, "Content-based image retrieval using a neuro-fuzzy technique".1997.

[Yazici99] A. Yazici and R. George, *Fuzzy Database Modeling.* Physica-Verlag, 1999

[Petry96] F. E. Petry, *Fuzzy Databases Principles and Applications.* Kluwer Academic Publishers, 1996.

[Koutsougeras98] C. Koutsougeras, B. P. Buckles, S. Amer and R. Alba-Flores, "Content-based Search Prototype for Image Databases". *Data Mining and Knowledge Discovery.* Sept, 1998.

[Berkeley] Berkeley Digital Library Project http://elib.cs.berkeley.edu/src/cypress/meets.c

[Koutsougeras88] Koutsougeras, C. and C.A. Papachristou, "Training of A Neural Network Model for Pattern Classification Based on an Entropy Measure*", Proceedings of the IEEE International Conference on Neural Networks (ICNN '88), IEEE,* July 1988.

CLIMS — A System for Image Retrieval by Using Colour and Wavelet Features

O. Kao and I. la Tendresse

Department of Computer Science, Technical University of Clausthal
Julius-Albert-Strasse 4, D-38678 Clausthal-Zellerfeld, Germany

Abstract. In this paper a system called CLIMS (*CLausthal Image Management System*) for content based image retrieval as an important subsystem of a general multimedia database is presented. It offers querying by sketch and image example and uses colour and wavelet based features for the comparison of images. Each image in the database is represented by a set of wavelet coefficients and colour attributes, which form the fundament for the retrieval.

In order to enable efficient similarity search two index structures, VP-Trees and L^q metric, are introduced and discussed. With the extension of the original VP-tree algorithm a ranking of the n most similar images is possible. The efficiency of the proposed retrieval methods is evaluated on a sample, general image catalogue.

1 Introduction

The development of the information technology in the 1990s is often described as a multimedia revolution. Multimedia is the synchronised association of time dependent (dynamic) and time independent (static) media. Dynamic media are audio and video streams as well as animation. Text, graphics and images are part of the static media. Although the hardware (I/O devices, memory, networks etc.) and software infrastructure have improved significantly 90% of the information is still on paper. One of the reasons is the lack of reliable methods for content analysis of the different media types, for example we still do not have suitable image segmentation and analysis algorithms. Furthermore basic mechanisms and technologies for the realisation of multimedia database management systems, e.g. information retrieval are not available. On the other hand subsystems like CD-ROMs with multimedia content, thumbnail systems, video on demand etc. can be seen as a part of a multimedia database [1]. Many existing database systems handle multimedia objects as BLOBs (*Binary Large Objects*). Each object is characterised by a manually compiled set of key words, thus content retrieval bases on full text search in this set. Problems occur due to the disadvantageous reduction of the complex media content on a small number of key words.

Therefore methods for content based information retrieval are necessary. In this paper a system called CLIMS (*CLausthal Image Management System*) for content based image retrieval as an important subsystem of a multimedia database is discussed.

T. Yakhno (Ed.): ADVIS 2000, LNCS 1909, pp. 238–247, 2000.

2   Image Databases

An image database is a system for the archival and retrieval of images and can be found in many areas, for example medical applications, remote sensing, news agencies, authorities, museums etc. It differs from systems for pattern recognition: in such systems the number of possible patterns is small, the accuracy high and the recording conditions are not significantly changing. The results are usually transferred to a control device and automatically evaluated.

In opposition thereof an image database contains a large set of images from different classes, which are taken under various conditions. The goal is to find a number of images, which are similar to a given sample image. High recognition rates are desired but not always possible. Furthermore the system response time is much longer than in the case of the pattern recognition. For the extraction and organisation of the image information suitable analysis and knowledge discovery methods are necessary. A work around of these unsolved problems is realised with the idea of content based image retrieval (CBIR). It is a class of search algorithms based on the extraction and comparison of low level image features, which are combined in logical features in order to represent the image content on a higher abstraction level.

The features can be extracted manually or automatically. In the first case the user marks the interesting areas with sketch tools. Subsequently all images in the database are processed in order to find similar regions. A reasonable system response time is possible if only a small number of images is considered. This is the reason, why many existing systems use a-priori extracted features. These are calculated, when the image is inserted into the database and compared with the corresponding features of the query image.

An important difference between image and conventional databases is given by the querying mechanisms. Languages such as SQL are not very efficient, because the image features are usually abstract numbers and arrays and thus not understandable to users without image processing background. Therefore visual querying methods like query-by-pictorial-example (QBPE), query-by-painting or sketch retrieval [2,3] are preferred. The user creates an example sketch or loads an example image, which are subsequently processed in the same manner as the images in the database.

3   CLIMS

The image database CLIMS is an experimental system for content based retrieval in general image catalogues. The system consists of clear defined modules, which can be easily extended with new methods for feature extraction and retrieval. The CLIMS basic components are the image processing system CLIPS, the relational database system PostgreSQL [4] and a web-based user interface.

CLIPS contains a selection of standard image processing methods such as colour transformations, segmentation etc. Each of this operators is realised as an independent UNIX shell program: it recieves the image data via the standard input and writes the results on the standard output. For a suitable integration of the available filter in the image database the user interface was re-designed and consists now of two web-based modules, which are shown in Figure 1. The draw module offers a selection of basic functions for creation and modification of graphic primitives like lines, ellipses, rectangles etc. In addition to the query by sketch the user can load images in GIF or JPEG format as a starting point for the similarity based search. The browser is used for the visualisation of the query results.

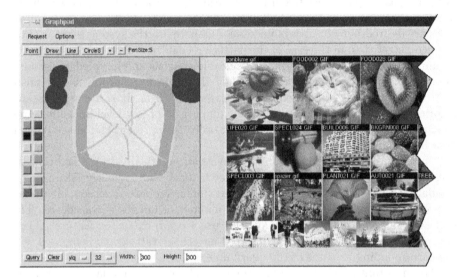

Fig. 1. Graphical user interface: sketching tools and browser for the retrieval results

Once the sample image/sketch is created, the query is started on one of the implemented colour models. Each model offers different advantages and disadvantages for the retrieval of certain image classes, thus further investigations are necessary in order to determine most suitable combinations. We used the PostgreSQL C programming library for the implementation of the functions for the communication with the database server. The basic operations are creation of a new image catalogue, insertion of features, index structures and images and content based search for the n most similar images by using one of the available features.

For each image entered in the database unique identifier and a table column for the values of all available features are created. Subsequently the algorithm for the feature extraction is implemented as a CLIPS filter, tested and finally applied on all images in the database. The calculated results are inserted into the corresponding tables and, if possible, into the index structures, for example

VP-trees (see section 5). These support the similarity search and reduce the computational complexity for the creation of the image ranking.

In the current version of CLIMS two algorithms for feature extraction—based on colour values and wavelet coefficients—are available. The application of the algorithms on the sample image or sketch produces the search parameter. With these values the corresponding index structures, which contain the already calculated values for all images in the database, are searched resulting into a ranking of the n most similar images, where n is a constant value given by the user. In the last step the image raw data is determined, loaded and sent to the user interface for visualisation. Following section presents a detailed discussion of the extraction algorithms.

4   CLIMS Image Features

Colour based features are often used for the comparison of images. A histogram defines e.g. the colour distribution over the image, thus images having similar histograms are considered similar. This method delivers good results, if all examined images belong to the same class, for example landscape images. Otherwise this information is not sufficient for general retrieval.

Another disadvantage is given by the large time and computing effort for the calculation and comparison of histograms in the case of true colour images, which have up to 16,7 Million colours. Therefore such images are pre-processed in order to reduce the number of colours. The most simple algorithm is the fix conversion: the amount of bit per colour channel is decreased. More suitable conversion can be achieved with adaptive methods like the Octree-Downfilter or Median Cut algorithm. The resulting histograms can not be compared directly, because each component of the histogram may represent a different colour. Therefore an assignment of the determined to a given set of reference colours is necessary.

Another colour based approach for image comparison uses statistical colour attributes introduced by STRICKER and ORENGO in [5]. The complex functions for the approximation of the image histogram are replaced by single values like the average E_i, the variance σ_i and the asymmetry s_i. The definition of these elements for the i-th colour channel and the j-th pixel of an image containing N pixel is given by the following Equations:

$$E_i = \frac{1}{N} \sum_{j=1}^{N} p_{ij} \tag{1}$$

$$\sigma_i = \left(\frac{1}{N} \sum_{j=1}^{N} (p_{ij} - E_i)^2\right)^{\frac{1}{2}} \tag{2}$$

$$s_i = \left(\frac{1}{N} \sum_{j=1}^{N} (p_{ij} - E_i)^3\right)^{\frac{1}{3}}. \tag{3}$$

For an image in the RGB colour model the size of these features amounts to three floating point values per channel.

Wavelet coefficients can also be used as features for measuring similarity of two pictures. Every one or two dimensional signal can be represented by wavelet coefficients whose number depends on the subdivision process. This process starts with the highest resolution in the first stage and concludes with the lowest resolution in the last stage. Large wavelet coefficients and/or large differences in the high resolution stage can be compared with high frequencies of the original signal. Small differences in a distinct stage can analogously be a sign for a small contribution of a corresponding frequency in the original signal. Two signals can be matched in the final analysis by consideration of the coefficients on every resolution stage. Specific frequency ranges may be ignored or included in the matching procedure with different weights.

A *Haar-Wavelet-Transformation* is defined as following. Let s_0 be a signal with $2^n, n \in \mathbb{N}$ discrete values $s_{0,k}$:

$$s_0 = \{s_{0,k} \mid 0 \leq k < 2^n\}. \tag{4}$$

If the "mean value/difference" transformation is applied on every pair $a = s_{2k}$ and $b = s_{2k+1}$ a total of 2^{n-1} $(k = 0 \ldots 2^{n-1})$ pairs have to be stored:

$$\begin{aligned} s_{1,k} &= (s_{0,2k} + s_{0,2k+1})/2 \\ d_{1,k} &= s_{0,2k+1} - s_{0,2k} \end{aligned} \tag{5}$$

The original signal s_0 with 2^n values is cut up into two signals: s_1 of the mean values $s_{1,k}$ with 2^{n-1} values and d_1 of the differences $d_{1,k}$ with 2^{n-1} values.

The signal s_1 represents the signal s_0 in a coarser resolution. The difference signal d_1 contains the information for the restoration of the original signal s_0 from s_1. The same transformation can be applied to the signal s_1 resulting into s_2 as a new mean signal and d_2 as a new difference signal. This procedure can be performed n times. At the end of the transformation n detailed signals d_j, $0 \leq j \leq n-1$ with 2^{n-j} coefficients each and one signal s_n on the coarsest resolution stage are available. s_n consists of only one value $s_{n,0}$ which corresponds to the mean value of the entire original signal s_0. This value is called the *DC-component* or the *zero frequency* of the signal.

The wavelet subdivision uses the *haar* functions $h_{j,k}$, $j, k \in \mathbb{Z}$, which are defined as

$$h_{j,k}(x) = \begin{cases} 2^{-j/2} & 2^j(k-1) < x. < 2^j(k-1/2) \\ -2^{-j/2} & 2^j(k-1/2) \leq x < 2^j k \\ 0 & \text{otherwise.} \end{cases} \tag{6}$$

The areas of interest of the *haar* functions $h_{j,k}(x)$ are in range $I_{j,k}$:

$$I_{j,k} = [2^j(k-1), 2^j k]. \tag{7}$$

The characteristic function $\chi_{I_{j,k}}(x)$ with $|I_{j,k}|^{-1/2}$ is used as a weighting function

$$\chi_{I_{j,k}}(x) = \begin{cases} |I_{j,k}|^{-1/2} & x \in I_{j,k} \\ 0 & \text{otherwise.} \end{cases} \tag{8}$$

where $|I_{j,k}|$ is the length of the range $I_{j,k}$. For a given function $f \in L^2(\mathbb{R})$ and a range $I \in \mathbb{R}$, the *haar* coefficient d_I

$$d_I = \int_{-\infty}^{+\infty} f(x)h_I(x)\mathrm{d}x, \tag{9}$$

and the average s_I

$$s_I = \int_{-\infty}^{+\infty} f(x)\chi_I(x)\mathrm{d}x, \tag{10}$$

of f are defined. The numeric computation of the *haar* coefficients necessities $N = 2^n$ discrete values from the function f:

$$s_{0,k} = \sqrt{N} \int_{2^{-n}(k-1)}^{2^{-n}k} f(x)\mathrm{d}x \tag{11}$$

$$d_{1,k} = \frac{1}{\sqrt{2}}(s_{0,2k-1} - s_{0,2k})$$

$$s_{1,k} = \frac{1}{\sqrt{2}}(s_{0,2k-1} + s_{0,2k})$$

We apply this procedure n times according to a pyramid diagram. The computation of the coefficients

$$d_{j+1,k} = \frac{1}{\sqrt{2}}(s_{j,2k-1} - s_{j,2k}) \tag{12}$$

and mean values

$$s_{j+1,k} = \frac{1}{\sqrt{2}}(s_{j,2k-1} + s_{j,2k}) \tag{13}$$

for $j = 0, \ldots, n-1$ and $k = 1, \ldots, 2^{n-j-1}$ requires $2(N-1)$ additions and $2N$ multiplications. Full descriptions and derivations of the specified Equations can be found in [3,9,10,11].

Before the wavelet transformation is applied each image is scaled to a size of 128×128. Fast wavelet algorithms exist for $2^m \times 2^m$, $m \in \mathbb{N}$ images thus a significant speedup can be achieved. The *haar* basis has a further advantage because it is suitable for the representation of large single coloured segments, which are often contained in the sample sketches. The L^q metric, described in the next section, is used for the determination of the nearest neighbours.

5 Index Structures

CLIMS supports VP trees (*Vantage point*) as an index structure for content based retrieval and L^q metrics for the comparison of vectors with wavelet coefficients. VP trees are introduced by CHIUEH in [6] and enable an efficient next neighbour search. For the application in our image database we extended the algorithms, basically a few special cases have to be analysed (see [7]), thus a ranking of the $k > 1$ next neighbours can be generated.

Let M denote a set of n-dimensional vectors with image features as components. In the inside nodes of a VP tree the vantage points are stored. The leaf nodes contain the values of the extracted features. The distribution of these values to the vantage point can be easily demonstrated, if a binary relation is given: all values are sorted according to the distance to the—randomly chosen—vantage point. Let μ be the median of all distances, thus each sub tree, $S_>$ and $S_\leq$ contains half of all points. Both sub trees have to be searched if the distance of the search parameter p to the vantage point v is in the interval $[\mu - \sigma, \mu + \sigma]$.

The next neighbour search starts in the root node by calculating the distance of the search parameter q to the vantage point v and σ is the—estimated—maximum distance. The next neighbour is element of a subset, if following holds:

$$\mu_i - \sigma < d(q, v) \leq \mu_{i+1} + \sigma. \tag{14}$$

All subsets fulfilling this criterion are examined until the next neighbour is found. An unsuccessful search occurs if the value σ is too small, thus the whole process has to be repeated with a modified σ. This adaptation can be performed by addition or multiplication of constant value.

For the evaluation of the image similarity based on the wavelet feature only the m largest coefficients are considered. All images in the database are expanded or reduced to a standard size of 128×128, thus 16384 coefficients are possible. In the current implementation a value of $m = 64$ was found to give good results. In the next step a quantification and standardisation of these coefficients is executed in order to speed up the calculation and minimise the memory requirements. Subsequently the wavelet coefficients are sorted according to the colour model, image channel, value and coordinates and are inserted into the leaf nodes of the index structure.

The determination of the n next neighbours of a query image is performed with the so called L^q metric [8] (modification of the Euclidian distance L^2), which considers deviations resulting from colour shifts, distortions and other errors due to the inaccurate query sketch. Let $Q[0, 0]$ and $T[0, 0]$ denote the coefficients of the scaling function, which depend on the average intensity of the colour channel. Let $\tilde{Q}[i, j]$ and $\tilde{T}[i, j]$ with $(0 \leq i, j \leq 127) \wedge \neg(i = j = 0)$ be the $[i, j]$-th k-largest and normalised wavelet coefficients. We simplify the calculation and set $\tilde{Q}[0, 0] = \tilde{T}[0, 0] = 0$, because these two values do not belong to any wavelet coefficient. The querying metric L^q is defined as:

$$L^q = ||Q, T||_q = w_{0,0}|Q[0, 0] - T[0, 0]| + \sum_{i,j} w_{i,j}|\tilde{Q}[i, j] - \tilde{T}[i, j]| \tag{15}$$

where Q is the query and T the target image. The values $w_{0,0}$ and $w_{i,j}$ are weights, which depend on the importance of the coefficients. They are empirically determined over a number of tests with manually selected images and grouped in classes marked with $W_1, \ldots W_4$. For a detailed description of these weights and of the metrics the reader is referred to [3,8].

After the wavelet decomposition of the query image Q the average intensity I_Q, the indexes and signs of the m largest coefficients for each colour channel c are determined. Subsequently we calculate the difference between the I_Q and I_T for all images T in the database. Finally for each of the wavelet coefficients $\tilde{Q}^c[i,j]$ we search all images, which have one m-largest coefficient with the same sign. The n next neighbours are the first n elements on the list sorted in descending order of the metric values $||Q,T||$.

5.1   Experimental Results

The CLIMS methods for content based retrieval are tested on a database containing 1132 photographs of different categories. For the result evaluation we used the well-known measures Precision p and Recall r, defined as $p = |\mathcal{A} \cap \mathcal{B}|/|\mathcal{A}|$ and $r = |\mathcal{A} \cap \mathcal{B}|/|\mathcal{B}|$ respectively, where $\mathcal{A}$ is the desired and $\mathcal{B}$ the resulting part of the database. The interpretation of these values with respect to image retrieval is complicated, because the definition of $\mathcal{A}$ depends on the subjective user perception.

The retrieval output considers the 32 nearest pictures. The starting points for the test retrievals are 300×300 sketches of real images in the database, which are drawn by 10 different persons. There were no restrictions regarding the colours and draft style. But to prevent that too precise pictures were drafted, each test person had only 3 minutes to prepare the example sketch. Each query starts on the default combination of weights and colour models. If one combination failed in retrieving the correct picture, a next combination is selected. The weights $W_1, \ldots W_4$ are empirically compiled and tested. A detailed description can be found e.g. in [3,8]. Our future work includes a development of automated methods for the adaptation of these values to the image classes in the database.

Table 1. Detailed results of the test scheme for the sketch retrieval

Weights and Hits in the first n pictures				
Colour / Weights	1-8	9-16	17-32	%
1. YUV/W4	32	5	6	43
2. YIQ/W4	14	9	2	25
3. HSL/W4	6	1	2	9
4. YIQ/W1	1	2	2	5
%	53	17	12	**82**

Table 1 summarises the results: approximately 82% of the sketches led to the desired target picture and only 18 remained without a direct hit. However, pictures similar to the target picture are returned in 14 of the 18 cases. If the sketches are considered as hits, a hit rate of 96% is achieved. If only static colour attributes are used for the retrieval, there are resulting values of approximately lower 30%. This method is acceptable only if pictures of a specific class, for example landscapes, are considered. In this case the calculation of precision and recall values is reasonable again and hit rates up to 68% can be achieved. Detailed representations of the performed measurements can be found in [7,3]. Examples of results determined during a similarity search with the wavelet-based features are shown in the Figures 2 and 3.

Fig. 2. City-Query: An example image and the first 12 images of the retrieval

6   Conclusions

CLIMS is a prototype for content based image retrieval in general image catalogues. It offers querying by example sketch and image and uses colour and wavelet based features for the comparison of images. All extracted features are stored in a relational database. A query starts with the application of the extraction algorithms on the given sample image or sketch. Thereby the search parameters are determined, which are subsequently used for the search of the corresponding index structures. These contain the a-priori calculated values for all images in the database. The result is a ranking of the n most similar images. The evaluation of the methods in a general image catalogue result in reasonable recognition rates.

Our future work includes the development and evaluation of other retrieval methods with a-priori and dynamic extracted features as well as the creation of suitable index structures. Furthermore we investigate different soft computing approaches for analysis and comparison of image content. Finally we develop a parallel cluster architecture for the efficient realisation of large image databases.

Fig. 3. Car-Query: The query sketch and the first 12 images of the retrieval

References

1. S. Khoshafian, A. B. Baker. *MultiMedia and Imaging Databases*. Morgan Kaufmann Publishers, 1996.
2. J. Ashley et. al. Automatic and semi-automatic methods for image annotation and retrieval in QBIC. In *Proceedings of Storage and Retrieval for Image and Video Databases III*, pp 24–35, 1995.
3. I. la Tendresse. Wavelet features for content based image retrieval. Master's Thesis, TU Clausthal, 1999.
4. B. Momjian. *PostgreSQL: Introduction and Concepts*. Addison-Wesley, 2000.
5. M. Stricker, M. Orengo. Similarity of color images. In *Storage and Retrieval for Image and Video Databases III*, pp 381–392, 1995.
6. T. Chiueh. Content-based image indexing. In *Proceedings of the 20th VLDB Conference*, pages 582–593, Santiago, Chile, 1994.
7. C. Ruch. Content based search in image databases. Master's Thesis, TU Clausthal, 1997.
8. X. Wen, T.D. Huffmire, H.H. Hu, A. Finkelstein. Wavelet-based video indexing and querying. *Journal of Multimedia Systems*, 7:350-358, Springer, 1999
9. Y. Meyer Wavelets and Operators. In N.T.Peck, E.Berkson and J.Uhl, *Analysis at Urbana*, LNS 137, London Math. Society, 1989
10. I. Daubechies Orthonormal Bases of Compactly Supported Wavelets. *Communications on Pure and Applied Mathematics*, XL1, 1988
11. R. Coifman, G. Beylkin, V. Rokhlin Fast wavelet transforms and numerical algorithms I In *Communications on Pure and Applied Mathematics*, pp 44:141–183, 1991

An Algorithm for Progressive Raytracing

Okan Arıkan[1] and Uğur Güdükbay[2]

[1] Computer Science Division
Department of Electrical Engineering and Computer Science
University of California
387 Soda Hall 1776
Berkeley, CA 94720-1776, USA
[2] Department of Computer Engineering
Bilkent University
Bilkent, 06533 Ankara, Turkey

okan@cs.berkeley.edu, gudukbay@cs.bilkent.edu.tr

Abstract. Progressive generation of images is one of the important research areas of computer graphics. Especially, when the image generation takes too much time the users want to see the progress in the rendering process. The user may decide either to continue the rendering process or stop the rendering according to the current view of the image. This may be due to the fact that either the image is produced to enough detail for the user or the user does not want to continue the rendering process for some reason. This is especially important for progressive transmission of images over the Internet. The images may be progressively transmitted and when the detail level of the image is enough for the user, the transmission process may stop. In this paper, we survey the progressive image generation techniques and present an algorithm for progressive generation of raytraced images. The algorithm utilizes a refinement technique that is similar to the one used in generating interlaced images in a progressive manner.

Keywords: progressive image generation, progressive transmission, raytracing, interlacing.

1 Introduction

Raytracing is a popular image synthesis technique which is used to generate high quality images incorporating shadows, reflections, refractions, texture mapping, etc. [5,8,20]. To address deficiencies of the basic raytracing method (viz. computational complexity and realism problems), some improvements on the basic raytracing algorithm are proposed. Some of these improvements are incorporating shadows, distributed raytracing [3], adaptive depth control [7], spatial coherence [4], first hit speed-up [19], and coupling ray tracing with other global illumination models like radiosity [17]. However, the implementation of a general raytracing algorithm is quite difficult for two reasons. The ray/surface intersection calculations are detailed and the execution time is long. Execution time can

T. Yakhno (Ed.): ADVIS 2000, LNCS 1909, pp. 248–256, 2000.

be reduced for test runs by generating images at a lower resolution than the one finally required, but this would mask errors whose effect is slight and only visible in the final resolution image.

This paper gives a short survey of the methods for generating images in a progressive manner and presents a simple yet effective algorithm for this purpose. Since raytracing calculates the color of pixels independently, the screen can be rendered in any order without compromising efficiency. Thus, the image can be calculated at an initial resolution and then can be detailed further by increasing the resolution. Since the number of rays shot is directly proportional to the total calculation time, user can see the image gaining detail in time and can know the correctness of rendering. This method gives user a chance to see the rendering in progress and gain insight into the quality of final image before waiting the rendering to stop, which may take quite long time.

2   Previous Work in Progressive Image Generation

Generating images in a progressive manner is a research area for computationally intensive rendering methods, such as ray tracing, radiosity, and direct volume visualization methods. Besides, it becomes very important for the Internet since progressive transmission of images over the network saves the users' time a lot and increases network bandwidth utilization. The users may stop the transmission process when the image is transmitted to enough detail for them. Related with this, it is also important for digital content creation for the areas of long-distance education and web-based learning where the progressive generation of educational material on the computers, which is mostly images, is critical due to time restrictions.

In this section we will summarize the progressive image generation techniques in different areas of computer graphics.

2.1   Radiosity

In radiosity there is a great amount of work: here progressive refinement radiosity is the de facto standard radiosity algorithm for generating radiosity images in a progressive manner [1]. There are also some ray tracing based radiosity solutions that utilize progressive refinement [16,17]

2.2   Volume Visualization

There is a great amount of research done especially in the field of volume visualization by ray tracing to generate direct volume visualizations in a progressive manner. Levoy reformulates the front-to-back image-order volume rendering algorithm to use adaptive termination of ray tracing [10]. Another example of rendering volumetric data using progressive refinement is by Laur and Hanrahan [9].

2.3   Raytracing

There are many algorithms for generating ray traced images in a progressive manner. Examples of these are [6,11,12,13,14]. Besides, some public-domain raytracing-based renderers, like RADIANCE [18], offer simple previewing capabilities that allow the generation of a rough approximation to the image very rapidly. The images are refined by tracing more rays into the scene.

Painter and Sloan [12] proposes a technique to incorporate progressive refinement and anti-aliasing to an ordinary raytracer. The algorithm aims to create a high-quality anti-aliased image quickly, whose quality is improved in a progressive manner after the initial image is shown. Their algorithm achieves progressive refinement using statistics and some data structures that allow rapid detection of badly approximated areas that need to be refined. So it is an adaptive refinement algorithm where detailed parts of the image are selected first for refinement. Their method generates the samples of the image stochastically and adaptively. During this process, the samples are evaluated to determine whether further refinement is necessary or not. The evaluation criteria eliminates the plain areas from further consideration. Then, the image is reconstructed by interpolating the samples and the image is filtered and resampled for display.

Raidl and Barth [14] modifies a raytracer for fast previewing during scene composition. Their algorithm controls a strong undersampling so that a very rough approximation can be shown in a very short time. The algorithm exploits similarities to the preceding image to render small changes im the scene. The algorithm is based on a recursive image subdivision technique that uses a heuristic priority calculation to detect changed regions of the image and to prefer them in the adaptive refinement. In this way, changed regions in the image are treated early and other parts are copied from the preceding image until the priority schedule opens them for more accurate calculation.

Haines [6] proposes a technique that first renders everything at a very low resolution and then toss some Gouraud shaded polygons on the screen. Then the algorithm looks for large differences and resolves them by starting from the middle of the picture since generally the central part of the image is the interesting part. In doing this, his algorithm checks different criteria for refinement process, like object corners, colors, shadows, etc.

Pighin et al. [13] proposes a technique for progressive previewing of raytraced images while they are computed using discontinuity meshing. Their algorithm constructs and incrementally updates a constrained Delaunay triangulation of the image plane, which is suggested in [12]. The points in the triangulation correspond to all the image samples that have been computed by the ray tracer, and the constraint edges correspond to various important discontinuity edges in the image. The triangulation is displayed using hardware Gouraud shading, producing a piecewise-linear approximation to the final image. They handle texture mapped surfaces and other regions in the image that are not well approximated by linear interpolation with the aid of hardware texture mapping.

3   The Proposed Algorithm

In this section we propose an algorithm for progressive generation of raytraced images. The algorithm takes samples for low resolution and progressively increase the detail level (and the number of rays shot). For this purpose, we used a very similar refinement that is used in the interlaced GIF images [2]. Thus, the image is calculated line by line and leaving a fixed amount of space in between. These gaps are filled with the image of next rendered line. Then, the non-rendered gaps between lines are halved and newly rendered lines are inserted at each pass. The algorithm is not an adaptive algorithm. In other words, the image is refined uniformly at all parts. The proposed progressive raytracing algorithm is summarized in Fig. 1. Figure 2 shows a simple illustration of the progressive

```
rendered[1..number_of_scanlines] stores whether the
scanline is "raytraced" (1) or "replicated" (0)

initialize rendered array to all 0's

while (there are replicated lines) do
{
        find the largest replicating span

        raytrace the pixels on the scanline at the middle of that span

        modify the rendered array element for that scanline to "raytraced"

        replicate the raytraced pixels to the empty spans below

        modify the pointers so that the largest replicating span found at
        the beginning of this iteration is replaced by two new formed spans
}
```

Fig. 1. Progressive raytracing algortihm

rendering process. The dark pixels represent the actually calculated pixels. The remaining light gray pixels are replicated from the dark ones in order to fill the gaps. In the next pass, one of the replicated lines is chosen and rendered. Then the gaps below it is filled this new line. This process continues until no replicated line is left (all pixels in the image are calculated).

The initial number of replicated lines between actually calculated lines determine how good the initial approximation will be. Moreover, allowing horizontal gapping in addition to vertical can improve the technique.

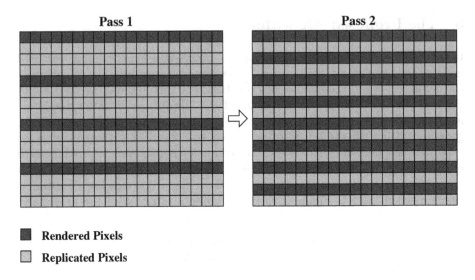

Rendered Pixels

Replicated Pixels

Fig. 2. Interlacing process

4 Implementation

4.1 Ray Tracer Implementation

We have based our progressive raytracer on a simple standard raytracer. This raytracer incorporates basic raytracer functionality in a modular way. Supported basic features include

- basic primitives like spheres, discs and polygons,
- reflections,
- refractions and
- texture mapping.

Although the basic raytracer also includes antialiasing routines, we removed them since antialiasing and filtering modifies pixel (or sample) color by a function of neighboring pixels (or samples). This is because progressive raytracing needs each pixel value to be calculated independently so that the pixel can be calculated in any order.

4.2 Graphical User Interface

Development of a Graphical User Interface (GUI), which will allow simultaneous management of display window and the raytracer, was the hardest part. In most of the classical window systems like Xwindows (with Toolkit), the interaction with the user is strictly event-based [15]. Since progressive raytracing requires a raytracer core running and calculating pixels while a GUI managing the display window, we created separate processes for each part. The overall structure of the processes can be visualized like in Fig. 3. These two processes are concurrent processes which could be assigned to parallel processors.

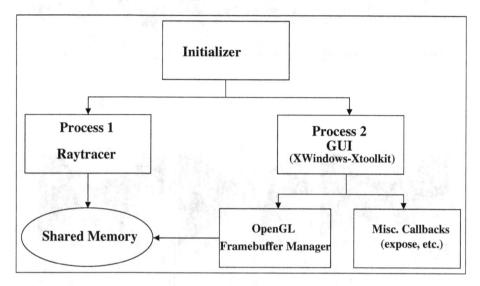

Fig. 3. Overall structure of the processes

The Initializer allocates necessary memory, prepares the execution environments and creates two new processes one of which is the actual raytracer and the other will be the GUI. Upon initialization, the raytracer and the GUI processes are allocated a shared memory which will hold the data. Raytracer, then reads the scene description and other peripheral data (like texture maps) and starts rendering and pushing data into the shared memory. The GUI, on the other hand, initializes the Xtoolkit and constructs the window to display the data calculated. Then, this window is associated with OpenGL[1] and callbacks are registered. One of these callbacks is a timeout mechanism which continuously generates expose callbacks in fixed intervals (2 seconds by default). Consequently, expose callback notifies OpenGL to refresh the screen from the shared memory.

5   Results

A series of raytraced images generated by our progressive raytracer is given in Fig. 4. As seen in the figure, there is no significant improvement in the image quality after 50 percent.

---

[1] OpenGL is a registered trademark of Silicon Graphics International, Inc.

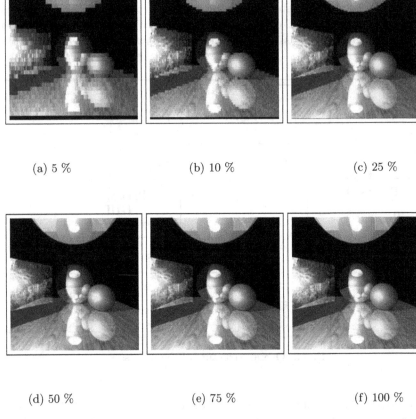

      (a) 5 %   (b) 10 %   (c) 25 %

      (d) 50 %   (e) 75 %   (f) 100 %

Fig. 4. Progressive generation of a raytraced image

6   Conclusions and Future Work

We implemented a progressive version of raytracing that allows us to generate
raytraced images in a progressive manner. This is useful since raytracing is a
time consuming process and the images are not seen until the intensities of all
pixels are generated. Sometimes it is necessary to see the low resolution versions
of the final image to get an idea about the final image and discontinue the image
generation process at the early stages if the final image will be useless. This is
especially useful for progressive transmission of images over the Internet. Future
work may include:

1. **Antialiasing:** As mentioned above, antialiasing techniques like oversam-
   pling or filtering are hard to incorporate into a progressive raytracer. Because
   in these methods, several samples affect the final pixel color and neighboring
   pixel can contribute into the final color. The final image can be computed
   in panels to implement antialiasing.

2. **Adaptive Refinement:** The computation power can be concentrated in parts displaying rapid changes in color (detail). This way faster convergence to the realism can be obtained.

3. **User Dictated Refinement:** A different user interface that will allow user to determine the parts to compute first can be developed, increasing the usability of the program.

Acknowledgments. This research is partially supported by an equipment grant from Turkish Scientific and Technical Research Council (TÜBİTAK) with grant number EEEAG 198E018. Thanks to Varol Akman for valuable discussions.

References

1. Cohen M.F. Chen, S.E., Wallace, J.R. and Greenberg, D.P., "A Progressive Refinement Approach to Fast Radiosity Image Generation", *ACM SIGGRAPH Conference Proceedings*, pp. 75-84, 1988.
2. CompuServe Inc., "GIF: Graphics Interchange Format", `http://www.daubnet.com/formats/GIF.html`.
3. Cook, R.L., Porter, T., and Carpenter, L., "Distributed Ray Tracing", *ACM SIGGRAPH Conference Proceedings*, pp. 137-144, 1984.
4. Glassner, A., "Space Subdivision for for Fast Ray Tracing", *IEEE Computer Graphics and Applications*, Vol. 4 No. 4, October 1984.
5. Glassner, A., *An Introduction to Ray Tracing*, Academic Press, New York, 1989.
6. Haines, E., "Progressive Ray Tracing and Fast Previews", Ray Tracing News, Vol. 10, No. 1, edited by Eric Haines, January 1997, `http://www.acm.org/tog/resources/RTNews/html`.
7. Hall, R.A., and Greenberg, D.P., "A Testbed for Realistic Image Synthesis", *IEEE Computer Graphics and Applications*, Vol. 3, No. 8, November 1983.
8. Kay, D.S., "Transparency, Refraction and Raytracing for Computer Synthesized Images", Master's Thesis, Cornell University, Jan. 1979.
9. Laur, D., and Hanranan, P., "Hierarchical Splatting: A Progressive Refinement Algorithm for Volume Rendering", *ACM SIGGRAPH Conference Proceedings*, pp. 285-288, 1991.
10. Levoy, M., "Volume Rendering by Adaptive Refinement", *The Visual Computer*, Vol. 6, No. 1, pp. 2-7, February 1990.
11. Maillot J-L. and Carraro, L. and Peroche, B., "Progressive Ray Tracing", *Proc. of Third Eurographics Workshop on Rendering*, pp. 9-20, Bristol, UK, May 1992.
12. Painter, J. and Sloan, K., "Antialiased Ray Tracing by Adaptive Progressive Refinement," *ACM SIGGRAPH Conference Proceedings*, pp. 281-288, 1989.
13. Pighin, F., Lischinski, D., and Salesin, D., "Progressive Previewing of Ray-Traced Images Using Image-Plane Discontinuity Meshing", *Proc. of 8th Eurographics Workshop on Rendering*, pp. 115-125, France, July 1997.
14. Raidl, G. and Barth, W., "Fast Adaptive Previewing by Ray Tracing", *Proc. of 12th Spring Conference on Computer Graphics*, edited by W. Purgathofer, pp. 247-255, Comenius University, Bratislava, Slovakia, June 1996.
15. Scheifler, R.W., Gettys, J. and Newman, R. *X Window System*, Digital Press, 1988.

16. Schlick, C. and Le Saëc, B., "A Progressive Ray Tracing based Radiosity with General Reflectance Functions", *Proc. of Eurographics Workshop on Photosimulation, Realism and Physics in Computer Graphics*, pp. 101-113, June 1990.

17. Wallace, J.R., Elmquist, E.A., Haines, E.A., "A Ray Tracing Algorithm for Progressive Radiosity", *ACM SIGGRAPH Conference Proceedings*, pp. 315-324, 1989.

18. Ward, G., "The RADIANCE Lighting Simulation and Rendering System", *ACM SIGGRAPH Conference Proceedings*, pp. 459-472, 1994.

19. Weghorst, H., Hooper, G, and Greenberg, D.P., "Improved Computational Methods for Ray Tracing," *ACM Transactions on Graphics*, Vol. 3, No. 1, pp. 52-69, 1984.

20. Whitted, T., An Improved Illumination Model for Shaded Display", *Communications of ACM*, Vol. 23, No. 6, pp. 343-349, June 1980.

Fuzzy Signatures for Multimedia Databases*

Václav Snášel

Department of Computer Science, Palacky University of Olomouc, Tomkova 40,
779 00 Olomouc, Czech Republic
vaclav.snasel@upol.cz

Abstract. In this paper we suggest a new method of information retrieval based on signature files which uses fuzzy sets and fuzzy logic for
signature construction and manipulation. We provide the definition of
fuzzy signatures and we discuss their possible applications.

1 Introduction

In the very near future, multimedia applications will be as routine as text applications today. Images and videos will be captured, manipulated, stored, searched, and reproduced much the way we manipulate text today. It is extremely
important to develop technologies for the management of large archives of visual
information. In particular, image and video data need to be organized efficiently
and fast query mechanisms are required to perform content-based retrieval of
the stored information. The most difficult part of the problem is to find feature
vectors that represent image archives as close as possible and data structures
that organize the feature vector space efficiently and thus speed up the search
process. In addition, a feature vector has to be computationally inexpensive to
facilitate query processing in real time.

Signature files are one of the methods used for full text retrieval. While some
authors suggest that their practical usage is limited to special cases in which
they outperform inverted files, the most commonly used method, we believe
that exploring the idea of signature files can lead to interesting new techniques
for retrieval of not only text documents, but also of other types of documents
mentioned earlier.

In the context of our search for new ways of looking at signature methods we
tried to apply a fuzzy approach to signature construction and manipulation. The
results were the definition of fuzzy signatures which we present in this paper.

2 Previous Work

NEC ART MUSEUM [9] is one of the early content-based image indexing systems
based on image edge features in the raw image domain. An edge map for each

---

* This work was done under grant from the Grant Agency of Czech Republic, Prague
No.: 201/00/1031

T. Yakhno (Ed.): ADVIS 2000, LNCS 1909, pp. 257–264, 2000.

image is stored in the database. The query retrieval system accepts hand-drawn sketches as queries. Edge map of the query image is matched with every edge map stored in the database by incrementally sliding the query image in 2-d over a fixed range. Each image is scored on how well it matches the query. Finally, all the images are sorted on their scores and the first one is presented to the user as a matching image. The method is obviously not scalable with respect to database size. Hence the query response time will linearly increase with the database size. IBM's Query By Image Content (QBIC) system [4] allows queries using multiple features based on image contents. These include color histogram, texture, shapes, and spatial relation- ship of objects. Each image in the database has multiple representation, one corresponding to each feature space. The query retrieval system supports comparison of each of the features separately. The calculation of multiple feature spaces is expensive. Also, each stored image is first compared against the query image and is discarded if not similar. VisualSEEk system [15] is another example of indexing on raw image data. Prior to queries, images in the database are processed once to extract regional color information or salient color regions. Along with color information, region sizes and spatial locations of regions are also extracted and are used as features to index images. To select the images matching with the query image, the color region and location information of the query image is compared against the same features of the all the images in the database. An alternative to indexing images based on raw-image data is the transform-domain based indexing. The main advantage expected from a transformation is effective characterization of local image properties. Moreover, since transformations are also used to compress image data, indexing based on compressed data increases the storage efficiency and performance of the multimedia systems.

3   Brief Description of Signature Files

Before we introduce the definition of fuzzy signatures, we would like to mention the basic principles of all signature methods.

Information retrieval methods are used to separate a subset of relevant documents within the finite set of all documents $D = \{D_1, D_2, \ldots, D_k\}$. The extracted documents are said to be relevant to a given query Q. A certain amount of work may be required before any query can be evaluated. During this phase, auxiliary data structures are created on disk. These structures enable later evaluation of any given query. In signature methods, signatures are auxiliary structures and are stored in a signature file on disk.

An ordinary signature is a bit string $s_1 s_2 \ldots s_n$, of a fixed length n. Signatures of all documents in a set are created and can be used for query evaluation as the signature of the query is compared with them. The lengths of the document signature S_i and the query signature S_Q are the same. Document D_i is relevant only if its signature S_i contains ones in all positions in which ones are encountered in the query signature S_Q.

An important issue is the way the signatures are created. There are two basic possibilities: superimposed coding and concatenation. Since superimposed coding is the only way actually used, we only describe this method. The signature of a document is created in the following way:

In the beginning, the bit string of signature S_i contains zeros in all positions. Let's assume that document D_i contains a finite number of distinct words $\{w_1, w_2, \ldots, w_l\}$. Then m positions in the signature bit string are selected for each word. If the bit string contains zeros in these positions, they are replaced with ones. These positions should be distributed evenly throughout the full length of the signature. They are typically chosen as a result of a hash function which takes the word w_j as an argument. The resulting effect is the same as if one signature S_{w_j} was created for each word w_j. This signature would contain up to m ones. The signature of the whole document would be the result of superimposing all signatures $S_{w_1}, S_{w_2}, \ldots, S_{w_l}$.

The query signature is created in the same way. The query contains a finite number of words which are used as arguments for the same hash function which was used for the construction of document signatures.

Another important issue is the organization of signatures in the signature file. It affects the efficiency of query evaluation. We will discuss possible organizations in the part devoted to fuzzy signature files.

4 Fuzzy Signatures

As hash functions are for signature extraction, collisions are bound to occur. This results in the selection of the same position p for the distinct words w_i, w_j. The bit in the position p, set to one, indicates the presence of the word w_p in the corresponding document. The word w_p belongs in the set M of all words, for which the hash function has the same result p. That is why the value in position p is the truth value of the statement that the corresponding document belongs to the set of all documents containing any word from the set M. This value is expressed using two numeric values, 0 and 1.

The ideal situation for signature methods occurs when each position in the signature corresponds to exactly one word. In this case, the set M contains just one word.

This interpretation of values in signature string positions is just one step from using fuzzy sets and fuzzy logic for signature extraction. Assuming that the values in signature positions represent degrees of membership in certain sets, we can extend these degrees using numbers from the interval $\langle 0, 1 \rangle$.

Definition 1. *The fuzzy signature F is a vector $(f_1, f_2, \ldots, f_n)$, where $f_i \in \langle 0, 1 \rangle \forall i = 1, 2, \ldots, n$.*

Provided that we have created fuzzy signatures for all documents in the set D and that we have the fuzzy signature F_Q of the query Q, we can use the operation of conjunction to find the relevant documents. The operation is defined in the same way as in fuzzy logic.

Definition 2. *The conjunction of fuzzy signatures F_i and F_j is the fuzzy signature*

$$F_i \bigwedge F_j = (f_{i_1} \wedge f_{j_1}, f_{i_2} \wedge f_{j_2}, \dots, f_{i_n} \wedge f_{j_n})$$

The operation $\wedge$ is defined for all elements of the fuzzy signature as

$$f_{i_r} \wedge f_{j_r} = \min\{f_{i_r}, f_{j_r}\}$$

In order to find all documents relevant to the given query Q we have to find all documents D_i which satisfy the formula $F_i \bigwedge F_Q = F_Q$. This actually means that for a document to be relevant to the query, all the elements of its signature must be equal to or greater than all the corresponding elements in the query signature.

The operation of disjunction can be defined in a similar way:

Definition 3. *The disjunction of fuzzy signatures F_i and F_j is the fuzzy signature*

$$F_i \bigvee F_j = (f_{i_1} \vee f_{j_1}, f_{i_2} \vee f_{j_2}, \dots, f_{i_n} \vee f_{j_n})$$

The operation $\vee$ is defined for all elements of the fuzzy signature as

$$f_{i_r} \vee f_{j_r} = \max\{f_{i_r}, f_{j_r}\}$$

These definitions of logical operations with fuzzy signatures reflect the common definitions of logical operations in fuzzy logic. The definitions can be found in [11].

We have described the method of evaluating a query by examining the fuzzy signatures of individual documents and the signature of the query itself, but we have not discussed how fuzzy signatures are actually constructed. We assume that the elements f_i of the string F are degress of membership of the feature M_i.

The application of fuzzy signatures is not limited just to text documents but that they could also be applied to multimedia databases. It is possible to use this method for describing geometrical objects. We could specify certain features of these objects then check for the presence of these features in each object. The elements of the fuzzy signatures would express our degree of confidence that a particular feature is present in a particular drawing.

Another area of fuzzy signature application is searching in a fingerprint database. Fingerprints are characterized by features identified in them see [6,14]. The elements of the fuzzy signature would again express our degree of confidence that a particular feature is present in the fingerprint.

5   Organization of Fuzzy Signatures

Another issue to be resolved is the way fuzzy signatures are organized in the signature file. The simplest way would be to store the fuzzy signatures in sequential order. This method is not efficient if the time required for query evaluation is concerned. That is why we have modified the data structure called S-tree, which is traditionally used to store ordinary signatures. In the next section we will sum up the specification of the original S-tree. Following that we will present the modification which enables to use this data structure to store fuzzy signatures.

5.1   Fuzzy S-Tree

We can obtain a data structure for the storing of fuzzy signatures by a modification of the S-tree.

Fuzzy signatures of documents will be stored in leaf pages rather than ordinary signatures. In the non-leaf pages there will be fuzzy signatures too but each fuzzy signature in a non-leaf page will correspond to another page at the lower level. These signatures are created as disjunctions of all fuzzy signatures in the corresponding pages. The operation of disjunction was defined in 4.

S-tree is a balanced tree which uses similar principles as the well known B-tree or its variation, the B^+-tree. S-tree is a data structure which allows to search for, insert and remove signatures.

Fuzy signatures are created using a feature extraction function applied to some objects in the database. They are then stored in the leaf pages of the tree. Each signature is accompanied by a link which points to the object described by the signature, or by the object itself. The pages at higher levels contain signatures too, but these signatures are created by superimposing all signatures in the pages of their corresponding successors. This means that each record in a non-leaf page has one whole page assigned at the lower level. In the non-leaf pages there are links to successor pages rather than database objects. Several rules similar to B-tree are defined, for the implementation of the above operations.

1. Each path from the root to any leaf has the same length h.
2. The root page contains the minimum of 2 records and the maximum of K records, except for the cases when it is a leaf page at the same time.
3. Each page except for the root page contains the minimum of k records and the maximum of K records.

k and K are constants.

The major advantage of the S-tree structure is the reduction of the number of signatures which must be searched during the evaluation of a query. In an ideal case, this number would be proportional to the height of the tree. However, this situation is very unlikely as we will explain in the section devoted to splitting of tree pages.

5.2   Operations Enabled by S-Tree

Searching. Searching is the most common operation. It gives the possibility to identify signatures which are relevant to a given query. The query is transferred into a query signature using the same hash function which was used for the extraction of object signatures. The query signature is used for navigation through the tree from the root to a leaf in the following way: All signatures In the current page are examined to identify all signatures which contain ones in all positions in which there are ones in the query signature.

There can be more than one signature which satisfies this condition. If the current page is a leaf page, objects described by the signatures identified are added to the result of the whole search. If it is not a leaf page, the search continues in the pages of all successors corresponding to the signatures identified.

Inserting a Signature. Insertion of a record is the next most commonly used operation in most applications. The record contains the signature of the inserted object and a link to the object, or the object itself. The record must be placed in a leaf page and the procedure of insertion must follow the defined rules. The algorithm of insertion starts in the root page again and continues down the tree until it reaches a leaf page. In each step, the algorithm selects one page from the successors of the current page. The signatures in the selected page must be as similar to the inserted signature as possible. To specify the similarity, several measures can be used. Two most commonly used methods are as follows:

1. Hamming metrics δ

$$\delta(S, S') = \gamma(S \vee S') - \gamma(S \wedge S')$$

   where S and S' are signatures of the page and of the inserted object, L is the length of the signatures
   $\gamma(S) = \sum_{i=1}^{L} s_i$ is the weight of the signature S
2. increase of weight ϵ

$$\gamma(S \vee S') = \gamma(S) + \epsilon(S, S')$$

$$\epsilon(S, S') = \gamma(S \vee S') - \gamma(S)$$

This measure is not commutative and therefore it is not a metric. Nevertheless, it proves to be more suitable for our purposes because we try to minimize the number of ones in the signature of the whole page as described in detail in the following paragraph.

Splitting of Pages. If the page selected for insertion of a new record is already full (i.e. it contains K records), then it is necessary to split this page into two new pages containing k and $(k + 1)$ records. The record of the original page is replaced in the page of the predecessor by two records of the newly created pages. In case the predecessor page is also full, the splitting continues until it eventually reaches the root page. If the root page is split, the height of the tree will increase by one. This is the only way how the tree can grow.

To preserve the logarithmical class of all operations, it is important that the number of pages which must be searched after the the current page is left is as close to one as possible. That is why we try to maintain the weight of all pages at minimum. The higher the number of ones in the signature of a page, the more likely it is that the signature will match the query signature even though its combination of ones resulted from superimposing several signatures with a lower weight.

It is our aim to find a way of splitting pages which would preserve low weight of the newly created pairs of pages while maintaining high Hamming distance between the pages in the pairs. There is no optimum algorithm at the moment and that is why heuristic approach is used. The following way of splitting the pages is suggested by Depisch in [7].

The signature with the highest weight is identified among the signatures within the original page, including the newly inserted signature. The record of this signature is marked as seed α and it is stored in the first of the two new pages. The record whose signature has the largest distance from α is marked as seed β and it is stored in the second page. The remaining records are divided between the new pages depending on whether their distance to seed α is larger than the distance to seed β.

Deletion of a Record. The operation of deletion is relatively rare and can be implemented in the following way. First we find the leaf page which contains the record to be removed. The record is deleted from the page. If the number of records in the page has not dropped below the constant k, the operation is completed. Otherwise it is necessary to reconstruct the whole tree to make sure that it satisfies the defined rules. This can be achieved by removing the whole affected leaf page and inserting its records back using the usual procedure for inserting records into the tree. The only exception is that the reinserted records must stay at the same level where they had been before they were removed from the tree.

Table 1. Comparison of different image indexing systems

System	Image Size	Database Size	Feature Vector Size	Search Time
LiangKuo [12]	192x128	2119	212 bytes	NA
WBIIS [18]	128x128	10,000	768 bytes	3.3 sec
QBIC [4]	100x100	1000	NA	2-40 sec
U. Wash [10]	128x128	1093	$O(m)$ bytes	47.46 sec
UDel [1,2]	512x512	1000	10 bytes	50 msec
Fuzzy Signatures	512x512	1000	200 bytes	20–50 I/Os

6   Conclusion and Future Works

The systems for indexing of pictures are compared in table 1. Fuzzy signatures, organized by S-tree, demand very small number of I/O operations. In future works, we want join creating of feature vector with compression, similarly to the method described in [1,2,4]. Implementation of the method is presented in [16].

The author thanks P. Zezula for his valuable comments.

References

1. E. Albuz, E. D. Kocalar, A. A. Khokhar. Scalable Image Indexing and Retrieval Using Wavelets, *SCAPAL Technical Report, 1998.*
2. E. Albuz, E. Kocalar, A.A. Khokhar. Scalable Image Indexing and Browsing using Wavelets, *IEEE Transaction on Knowledge and Data Engineering, 1998.*
3. E. Albuz, E. D. Kocalar, A. A. Khokhar. Vector-wavelet based scalable indexing and retrieval system for large color image archives, *ICASSP'99.*
4. J. Ashley, R. Barber, and M. Flickner. Automatic and semi-automatic methods for image annotation and retrieval in qbic. *In Proceedings in Storage and Retrieval for Image and Video Databases – III, volume 2420, pages 24–35, San Jose, CA, Feb 1995.*
5. R. Baeza – Yates and B. Riberio – Neto. *Modern Information Retrieval.* Addison Wesley 1999.
6. N.Cook. *Classifying fingerprints*, Addison-Wesley, 1995
7. U.Deppisch. S-tree: A Dynamic Balanced Signature Index for Office Retrieval. *Proc. of ACM Research and Development in Information Retrieval*, Pisa, Italy, Sept. 8-10, 1996, pp. 77-87.
8. C. Faloutsos. *Searching Multimedia Database by Content.* Kluwer Academic Publishers, 1996.
9. K. Hirata and T. Kato. Query by visual example. *Advances in Database Technology EDBT '92, Third International Conference on Extending Database Technology, Vienna, Austria, march 1992. Springer-Verlag.*
10. C.E. Jacobs, A. Finkelstein, and D. H. Salesin. Fast multiresolution image querying. *Proceedings of SIGGRAPH 95, 1995.*
11. G.J.Klir, St.U.Clair and B.Yuan. *Fuzzy set theory: foundations and applications*, Prentice-Hall, 1997.
12. K.C. Liang and C.C.J. Kuo. Wavelet-compressed image retrieval using successive approximation quantization (saq) features. *SPIE Voice, Video, and Data Communications, Dallas, Texas, november 1997.*
13. W.Y.Ma and B.S.Manjunath. Pictorial queries: Combining feature extraction with database search. *Technical Report 18, University of California at Santa Barbara, Dept. of Electrical Engineering, 1994.*
14. PCASYS – A Pattern – Level Classification Automation System for Fingerprints *NIST NISTIR 5647.*
15. John R. Smith and Shih-Fu Chang. Visualseek: a fully automated content-based image query system. *ACM Multimedia, 1996.*
16. V. Snášel. Fuzzy Signatures for Information Retrieval. *Workshop ISM 2000, accept*
17. P.Zezula and P.Tiberio. Storage and Retrieval: Signature File Access. *Encyclopedia of Microcomputers*, edited by Kent, A. and Williams, J.G., published by Marcel Dekker, Inc., 1995, Vol. 16, pp. 377-403
18. J. Z. Wang, G. Wiederhold, O. Firschein, and S. X. Wei. Wavelet-based image indexing techniques with partial sketch retrieval capability. *Journal of Digital Libraries, 1997.*

Evolutionary Prefetching and Caching in an Independent Storage Units Model

Athena Vakali

Department of Informatics
Aristotle University of Thessaloniki, Greece
E-mail: avakali@csd.auth.gr

Abstract. Modern applications demand support for a large number of clients and require large scale storage subsystems. This paper presents a theoretical model of prefetching and caching of storage objects under a parallel storage units architecture. The storage objects are defined as variable sized data blocks and a specific cache area is reserved for data prefetching and caching. An evolutionary algorithm is proposed for identifying the storage objects to be prefetched and cached. The storage object prefetching approach is experimented under certain artificial workloads of requests for a set of storage units and has shown significant performance improvement with respect to request service times, as well as cache and byte hit ratios.

Index terms: *data prefetching and caching, parallel storage units, object-based storage models.*

1 Introduction

According to [3], the amount of storage sold has been almost doubling each year, and storage demands are rapidly increasing due to the complexity and diversity of many current applications. Research has focused on minimizing the so called " I/O bottleneck". Here, we consider an object-based storage model and we propose a prefetching and caching approach in order to reduce data access times and improve data availability over a specified storage subsystem of a number of independent storage units.

Modern disk drives attributes and characteristics have been identified in [11] and their most significant performance factors have been indicated in [10,12]. Network Attached Storage systems [9,1] and *NASD* [4] have been introduced as new scalable bandwidth storage architectures with an object-based storage interface model.

File prefetching has been proven a quite effective technique for improving file access performance. In [7] an analytical-based prefetching mechanism is proposed and the prefetching approach has been proven quite effective while cache miss ratios have been reduced significantly. Recommendations on how to improve and benefit of file prefetching are pointed in [14]. Cooperative prefetching and

T. Yakhno (Ed.): ADVIS 2000, LNCS 1909, pp. 265–274, 2000.

caching is also discussed in [16], where the use of network-wide global resources support prefetching and caching in the presence of hints of future demands. Traditional caching in a distributed file system is discussed in [2].Finally, a web-based evolutionary model has been presented in [15] where cache content is updated by evolving over a number of successive cache objects populations and it is shown by trace-driven simulation that cache content is improved.

This paper presents a theoretical analytical model which introduces the idea of prefetching and caching of data objects collected by a number of independent storage units. The storage units are considered to store information in the form of the so-called *storage objects* which are variable sized data blocks. The prefetching is not based on continuous data allocation but on request frequency of the data storage objects. Storage objects are considered to formulate the individual members of a large "population" residing among a predefined set of parallel independent storage units. An initial process constructs a population of data storage objects to be prefetched on a local cache server from the various storage units. These data blocks populations are evolved such that their members are as frequently requested and efficiently retrieved as possible. The identification of the data objects to be prefetched is done by the introduction of an evolutionary-type algorithm based on the Genetic Algorithm idea. The model performs the request servicing by searching for requested data at the cache area first, then at the other storage units. The prefetching process is applied at regular intervals such that the populations are updated and confront with the requests access patterns. The remainder of the paper is organized as follows. The next section has the definition of the considered object-based storage model and of the storage units characteristics and parameters. Section 3 presents the prefetching and caching approach whereas the requests workload, the model's experimentation and results are presented in Section 4. Section 5 summarizes the conclusions and discusses potential future work.

2   The Object-Based Storage Model

Figure 1 shows the architecture and topology of the proposed storage model. A number of n clients requests data stored among k parallel independent storage units. A local server is considered to host the cache area which is contacted by the clients and a cache controller is assumed to handle and manage the caching and prefetching.

 – **The Cache :** A cache acts as a buffer area for storage of the most frequently requested data. The caching policy is based on the idea that when a user (client) requests a piece of data, the cache should be checked first. The cache area is modeled as an information table which contains information about the cached data block(s). Each row in the cache table has an index number which uniquely characterizes objects stored and is also accompanied by a number of attributes such as size, time of its being cached, storage unit it resides. The cache area has a limited predefined size and there is a specific retrieval

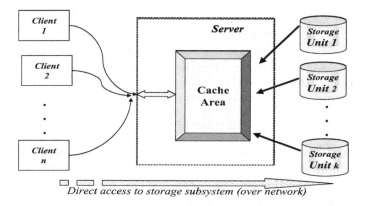

Direct access to storage subsystem (over network)

Fig. 1. The Object-based Storage Model

Table 1. The storage model parameters

parameter	description
id_i	object i identification (an index number).
s_i	object's i size, in KBytes.
t_i	time the object i was prefetched.
a_i	number of accesses since the last time object i was accessed.
cr	the cache area retrieval rate, in MBytes per second.
CS	the total cache size, in MBytes.
sr	the storage unit retrieval rate, in MBytes per second.
D	number of independent disks of the storage subsystem .
C	number of cylinders per disk in each storage unit.

rate for request servicing. The parameters associated with the considered object-based storage model are given in Table 1.

The most common performance metrics used for cache characterization are the cache-hit ratio and byte-hit ratio :

— *Cache hit ratio :* represents the percentage of all requests being serviced by a cache copy of the requested data, instead of searching the other original storage unit. Similarly, a *cache miss* is related to requested data not found in cache.

— *Byte hit ratio :* represents the percentage of all data transfered from cache, i.e. corresponds to ratio of the size of data retrieved from the cache area.

— **The Storage Unit :** We mainly concentrate on a multiple disks subsystem where storage units are considered as similar technology disk drives, with similar configuration requirements that can serve requests in parallel under a considered storage subsystem.(Disks parameters are given in Table 1). *Request servicing* is performed by accessing the storage unit which has the requested data and reading them from this drive. Disks serve requests in parallel in order to exploit the system's responsiveness.

The service time of a request in the disk mechanism is a function of the seek time (ST), the rotational latency (RL) and the data transfer time (TT). [12, 13]. The most widely used formula for evaluating the expected service time involves these time metrics and it is expressed by :

$$E[Disk\_Service\_Time] = E[ST] + E[RL] + E[TT] \tag{1}$$

where $E[ST]$ refers to the expected seek time, $E[RL]$ refers to the expected rotational delay and $E[TT]$ refers to the expected transferring time.

The following function has been used widely for the approximate evaluation of the seek time, which is a major performance factor :

$$Seek\_Time(dist) = \begin{cases} 0 & \text{if } dist = 0 \\ a + b \sqrt{dist} & \text{if } 0 < dist < cutoff \\ c + d \; dist & \text{if } dist \geq cutoff \end{cases} \tag{2}$$

where a, b, c, d and $cutoff$ are device-specific parameters and $dist$ is the number of cylinders to be traveled. The expected rotational delay is evaluated by $E[RL] \approx \frac{Revolution\_Time}{2}$ for randomly distributed requests. The transfer time depends on the amount of data to be transferred and is evaluated by $E[TT] = \frac{Request\_Size}{sr}$ under a constant sr disk drive retrieval rate.

3   The Prefetching and Caching Algorithm

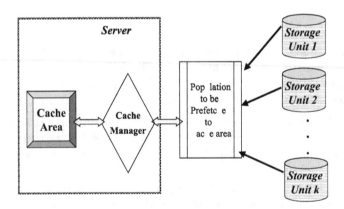

Fig. 2. The Prefetching and Caching Algorithm

Users/clients make requests which refer to data objects stored among the k storage units spread over parallel storage units. Each request refers to an arbitrary amount of data of a certain file. The file system divides the request into several block-sized segments, each served separately by the file system [14]. Several blocks could be grouped in a cluster or segment in order to define the "storage object" which is either cached or stored at a storage unit. Figure 2 depicts the proposed prefetching and caching process.

Definition 1 : The *storage object* is a group of logically sequential data blocks that are stored consequently on a disk. A number of KBytes corresponding to x data blocks defines the size of each stored object.

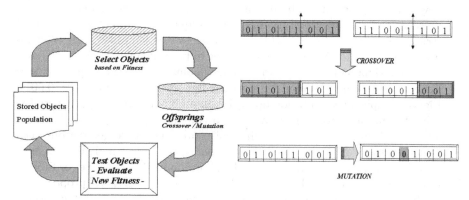

Fig. 3. The Genetic Algorithm Cycle

Fig. 4. GA operators: crossover and mutation

The Genetic Algorithm (*GA*) idea is applied in a considered population of the storage objects as defined above. The GA is used because of two main reasons : First, the basic idea of the GAs is based on the evolution of populations by the criterion "survival of the fittest" and the objects to be prefetched should be the fittest (i.e. the most frequently accessed in best retrieval rates) of the stored objects. Second, the GAs are applied to problems demanding optimization out of spaces which are too large to be exhaustively searched and all storage units have a huge amount of storage objects, impossible to be searched exhaustively in a realistic amount of time. Figure 3 depicts the cycle of a GA applied in a space of individual stored objects. In the present paper, the stored objects are modeled as the individuals considered for evolution. The individuals are assessed according to predefined quality criterion, called *objective or fitness function*. Two genetically-inspired operations, known as *crossover* and *mutation* are applied to selected cached objects (considered to be the population individuals) to successively create stronger "generations" of considered storage objects. The propose GA model follows the simple GA proposed in [5].

– **Encoding and Operators** : Each stored object individual must be identified according to a predetermined encoded string. The encode scheme is chosen such that the potential solution to our problem may be represented as a set of parameters. These parameters are joined together to form the encode string. In order to consider the identification of each stored object individual, the stored objects filenames are mapped to the integer values $1, 2, \ldots, O$ where O is the total number of objects to be prefetched in the are reserved for prefetching and caching. Parameters act_i, df_i, cr and s_i are the ones to guide the optimization problem, therefore they are included in the proposed encoding string. Each parameter is assigned a value and the presence of that parameter is signaled by the presence of that value in the ordered encode string. *Crossover* is performed between two stored object individuals ("parents") with some probability, in order to identify two new individuals resulting by exchanging parts of parents' strings. Figure 4 presents the crossover operation on an example of an 8-bit binary encoded string, partitioned after its 5th bit, in order to result into two new 8-bit individuals. *Mutation* is introduced in order to prevent premature convergence to local optima by randomly sampling new points in the search space. It randomly alters each individual with a (usually) small probability (e.g. 0.001). Figure 4 depicts the mutation operation in an binary 8-bit string where the 4th bit is mutated to result in a new individual.

The stored objects population will evolve over successive generations such that the fitness of the best and the average stored object individual in each generation is improved towards the global optimum. An objective (or fitness) function is devised based on the need to have a figure of merit proportional to the utility or ability of the encoded stored object individual. Our fitness function is the considered access frequency of each storage object.

- **The GA Prefetching and Caching Algorithm** Each population is formed by the most promising and strong storage objects of all considered storage units. Then, the standard operators defined above mix and recombine the encoding strings of the initial population to form offspring of the next generation. In this process of evolution, the fitter stored object individuals will create a larger number of offspring, and thus have a higher chance of "surviving" to subsequent generations. A pseudo-code version of the GA prefetching and caching algorithm follows :

```
initialize()
old_storage_pop <- initial objects population
evaluate_fitness(old_storage_pop)
generation <- 1
while (generation <= maxgen)  do
    par1 <- selection(popsize, fitness, old_storage_pop)
    par2 <- selection(popsize, fitness, old_storage_pop)
    crossover(par1,par2,old_storage_pop,new_storage_pop,p_cross)
    mutation(new_storage_pop, p_mutate)
    evaluate_fitness(new_storage_pop)
    statistical_report(new_storage_pop)
    old_storage_pop <- new_storage_pop
    generation <- generation + 1
```

In the above GA *maxgen* corresponds to the maximum number of successive generation runs, *popsize* is the stored objects population size, *fitness* is the stored objects fitness metric. Variables $par1$ and $par2$ define the parents chosen for the reform of each generation, p_cross, p_mutate are the probabilities for the crossover and mutation operators, respectively. The $old\_storage\_pop$ refers to the initial population in every GA cycle whereas the $new\_storage\_pop$ is the resulting population of each GA run.

4   Experimentation — Results

Table 2. The model's storage units parameters and their values

Secondary Storage Parameters	
Number of Storage Units Drives (k)	10
Seek time parameters (equation 2)	$a=3.24$ms $b=0.4$ms. $c=8.0$ms $d=0.008$ms. $cutoff=383.$
Number of Cylinders (C)	1936.
Rotational Speed	4002 rpm.
Data Transfer Rate (sr)	10 MB/sec.

Regarding the storage units, the model configuration considered, follows the disk drive configuration for the *HP 97560* disk drive which has been used in previous research [11,6]. The values for the parameters characterizing this particular disk drive are given in Table 2.

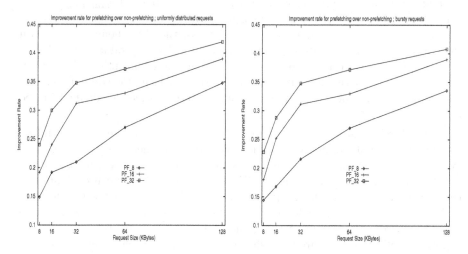

Fig. 5. prefetch/non-prefetch; req size   **Fig. 6.** prefetch/non-prefetch; req size

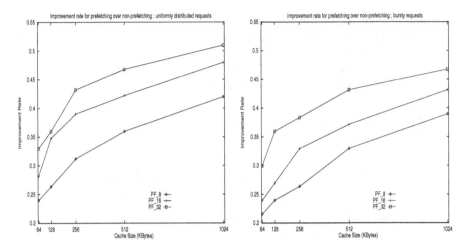

Fig. 7. prefetch/non-prefetch;cache size   **Fig. 8.** prefetch/non-prefetch;cache size

The proposed analytic model for prefetching and caching has been experimented under various request workloads. The workload is characterized by its arrival process, its request rate and its burstiness of arrivals. There were more than 100,000 requests generated for every execution cycle and the requests were randomly distributed among the considered storage units. The notations *PF_8*, *PF_16* and *PF_32* refer to the prefetching and caching algorithm applied on variable sized data blocks grouped into objects of sizes 8, 16 and 32 KBytes, respectively. The prefetching and caching algorithm follows the GA idea and the crossover and mutation probabilities values are $p_{crossover} = 0.6$ and $p_{mutation} = 0.001$ since these values are in the range of suggested representative trial sets for many GA optimizations [5,8]. The initial population for the GA scheme as applied in the considered storage units is generated by a randomly produced population. Furthermore, the typical non-prefetching approach of a conventional storage system has been experimented in order to serve as a basis for comparisons and discussion. The performance metrics evaluated are the metrics defined in Section 2.

The *PF_8*, *PF_16* and *PF_32* approaches were used for prefetching and caching under several requests workloads of various experimentation data sets. The average service time of the storage system has been evaluated by using the results for the service time of each request. The service times have been evaluated for the conventional Non-Prefetching as well as for the three proposed prefetching and caching algorithms, under varying request sizes and under varying cache sizes. The three proposed prefetching and caching algorithms, have been proven to be quite effective compared to the conventional Non-Prefetching strategy.

Figures 5 and 6 represent the percentage rate for the improvement in service times compared to the non-prefetching scheme, under request sizes of $8, 16, \cdots,$ 128 KBytes and under a cache size of 128 KBytes. Figure 5 refers to results from a workload of 112,000 uniformly ditributed requests and Figure 6 results from a similar 111,800 workload of more bursty requests. All of the prefetching and caching approaches have resulted in better performance metrics compared to the typical non-prefeching scheme. The benefits in service times are improved as the request sizes increase. This is explained since in a non-prefetching scheme, the larger the request size, the more search among storage hierarchies has to be done, resulting in worse service times. The *PF_32* approach is the most beneficial for the storage system since the service times improvement rates get to 42% as the request size increases. This shows that prefetching and caching favors the larger request sizes since the requested data can be accessed faster. Servicing requests by the cache area instead of contacting the original storage unit results in increasing of service times due to the slower characteristics of the disk storage units (ref. parameters Table 2).

Figures 7 and 8 represent the percentage rate for the improvement in service times compared to the non-prefetching scheme, under cache sizes of $64, 128, \cdots$, 1024 KBytes and under a request size of 32 KBytes. Figure 7 refers to results from a workload of 112,000 uniformly ditributed requests and Figure 8 from a similar 111,800 workload of more bursty requests. All of the prefetching and caching approaches have resulted in better service times compared to the typical non-prefeching scheme. The benefits in service times are improved as the cache sizes increase. This was expected since the data availability in cache is improved when having more space reserved for prefetching and caching. Therefore, the difference in service times between the proposed schemes and the conventional non-prefetching scheme will become more intense as the prefetching and caching area gets increased. Again, the *PF_32* approach is the most beneficial for the storage system since the service times improvement rates can get to about 48% as the request size increases. For example there is a 51% for *PF_32* under cache size of 1,024 KBytes of the first workload and there is a 46% for *PF_32* under cache size of 1,024 KBytes under the second workload. In conclusion, it has been shown by the experimentation that the GA-based prefetching and caching can be a quite effective approach in order to improve the request servicing process in a hierarchical storage model. The performance gain is significant, since service times can be improved at rates of 24%–48% value as the cache sizes and the request sizes increase.

5   Conclusions — Future Work

This paper has presented a study of applying the prefetching and caching idea to a theoretical model of parallel storage units. The proposed approaches for prefetching and caching were based on algorithms of the Genetic Algorithm idea, guided by an objective function in relation to objects frequency of access. The results have shown that the proposed prefetching and caching schemes could be very beneficial for the request servicing process since the improvement in this figure can get as high as 49% for certain request and cache sizes.

Further research should experiment the present scheme under a simulation model which could be based on a certain storage subsystem topology. This is a challenging issue since different storage systems have complex characteristics and requirements that need to be synchronized and parameterized. Furthermore, the use of traced workloads will help in identifying the demerit figures for the proposed model in order to use it in storage system implementations.

References

1. D. Anderson : "Network Attached Storage Research", *http://www.nsic.org/nasd/meetings.html*, Mar., Jun. 1998.
2. M. A. Blaze: Caching in Large-Scale Distributed File Systems, Princeton University, PhD thesis, Jan 1993.
3. G.A. Gibson, J.S. Vitter, J. Wilkes et al.: "Strategic directions in Storage I/O Issues in Large-Scale Computing", *ACM Computing Surveys*, Vol.28, No.4, pp.779–763, 1996.
4. G.A. Gibson, D. F. Nagle, W. Courtright II, N. Lanza, P. Mazaitis, M. Unagst and J. Zelenka : "NASD Scalable Storage Systems", *USENIX 1999 Extreme Linux Workshop* Monterey, California, Jun 1999.
5. D. Goldberg: *Genetic Algorithms in Search, Optimization, and Machine Learning*, Addison-Wesley, 1989.
6. D. Kotz, S.B. Toh and S. Radhakrishnan: "A Detailed Simulation Model of the HP 97560 Disk Drive", Department of Computer Science, Dartmouth College, Technical Report *TR94-220*, Jul 1994.
7. H. Lei and D. Duchamp : "An analytical Approach to file prefetching", *Proceedings of the 1997 USENIX Annual Technical Conference*, Anaheim, California, Jan 1997.
8. Z. Michalewicz: *Genetic Algorithms + Data Structures=Evolution Program*, 3rd edition, Springer-Verlag, 1996.
9. NSIC : National Storage Industry Consortium, *http://www.nsic.org/nasd*, 1999.
10. S.W. Ng: "Advances in Disk Technology — Performance Issues", *IEEE Computer*, Vol.31, No.5, pp.75–81, 1998.
11. C. Ruemmler and J. Wilkes: "An Introduction to Disk Drive Modeling", *IEEE Computer*, Vol.27, No.3, pp.17–28, 1994.
12. E. Shriver: "Performance modeling for realistic storage devices", *Ph.D. Thesis*, Department of Computer Science, New York University, May 1997.
13. E. Shriver, A. Merchant and J. Wilkes: "An Analytic model for disk drives with readahead caches and request reordering", *ACM SIGMETRICS'98*, Conference Proceedings, pp.182–191, Jun 1998.
14. E. Shriver, C. Small and K.A. Smith : "Why does file system prefetching work ?", *Proceedings of the 1999 USENIX Annual Technical Conference*, pp.71–84, Monterey, California, Jun 1999.
15. A. Vakali: A Web-based evolutionary model for Internet Data Caching, *Proceedings of the 2nd International Workshop on Network-Based Information Systems*, *NBIS'99*,IEEE Computer Society Press, Florence,Italy, Aug 1999.
16. G. M. Voelker et. al : "Implementing Cooperative Prefetching and Caching in a Globally-Managed Memory System", *ACM SIGMETRICS'98*, Conference Proceedings, pp.33–43, Jun 1998.

A Video Streaming Application on the Internet

Aylin Kantarcı[1] and Turhan Tunalı[2]

[1] Computer Engineering Department, Ege University, 35100, Bornova, İzmir, Turkey
kantarci@bornova.ege.edu.tr
[2] International Computer Institute, Ege University, 35100, Bornova, İzmir, Turkey
tunali@ube.ege.edu.tr

Abstract. A new software only streaming application is being developed for the Internet. The application is based on multithreaded architecture. RTP (Real-time Transport Protocol) is being used for the transmission of video data. The system is adaptive to dynamic network conditions. In this paper, performance results of the implementation are reported.

1 Introduction

Internet, the ideal network that connects all computers in the world, is a popular platform for many applications. First Internet applications focused on the integration of discrete data types such as graphics, text and pictures. With the developments in communication and computing technology, distribution of continuous media such as video and audio that demand more resources than other media has gained popularity.

Streaming is a technique to distribute video and audio data over the Internet. Prior to streaming, continuous media files were downloaded onto client's disk as a whole before the playback began. Streaming overlaps transmission and playback and thus eliminates high waiting times and large disk space requirement that occur in downloading approach. Although downloading yields better quality, the quality produced in streaming is within acceptable levels.

Developing a streaming application entails the cooperation in many technologies such as compression, communication and computation. The choice of the compression algorithm to be used depends on the application type being developed. Video conferencing applications require real-time compression in which frames should be quickly captured and sent to the network. Therefore, compression is required to be performed by hardware and compression quality is not required to be very high. Applications such as video on demand deal with stored data and hence compression is performed off-line. In this situation, compression may be performed by software and a compression approach producing high quality may be preferred. MPEG standard is a good choice for the compression of stored video. MPEG has many versions. MPEG-1 was developed for the storage of videos on CDs. It allows encoding bit rates upto 1.5 Mbps and frame rates upto 30 fps. Maximum frame resolution being supported is 352x288. MPEG-2 was developed for high quality TV applications. It allows encoding bit rates in the orders of 4-6 Mbps and frame rates upto 30 fps. Maximum supported resolution is 720x480. MPEG-4 has been

T. Yakhno (Ed.): ADVIS 2000, LNCS 1909, pp. 275-284, 2000.

developed for Internet applications. It is intended for applications that require video conference quality. It supports low encoding bit rates between 4.8 - 64 Kbps and a frame rate of 10 fps [1].

Internet is a best effort shared datagram network. Although current protocols transmit conventional data such as text and graphics, they fail in continuous media traffic that requires real-time transmission and high bandwidth. Therefore, existing protocols should be supported with new protocols that took into consideration the requirements of multimedia data. IETF (Internet Engineering Task Force) has developed RTP protocol for the transmission of multimedia data over the Internet. RTP is mostly used with UDP. It allows many typed of payloads such as MPEG, JPEG, CellB, H.261, H.263, etc. It provides time stamping and sequence numbers for the ordering of packets. It is implemented as part of the application, not of the OS Kernel. It has multicast provision. It offers no reliability mechanism. RTP is accompanied with another protocol, RTCP, for QoS monitoring and congestion control. Participants periodically send each other RTCP packets that include transmission statistics such as jitter and loss rate [2].

First multimedia research was performed in networking field. Later, it was understood that end systems play important roles in the performance of multimedia applications. Conventional operating systems have low clock resolution and frequent context switches to kernel that result in unpredictability an delays. POSIX 4 real-time extensions allow finer clock granularity and performs many functions at user level, eliminating costly context switches [3]. Multithread programming paradigm has changed the traditional view of process that is composed of many processes that communicate and share data through shared memory, message queues and pipes. Multithreading divides the tasks of an application into threads that share the same address space and therefore it eliminates costly IPC mechanisms. Additionally, multithreading allows parallel execution of I/O and CPU bound parts of an application [4].

Our goal in this study is to build a software only video streaming application for the Internet by using multithreading and POSIX 4 real-time extensions on a general purpose operating system. There are existing VOD systems for the Internet. The first well known example is the *Continuous Media Player* of Berkeley University [5]. CM Player has a process based architecture. It uses UDP for video transmission and TCP for the exchange of control data. It has its own heuristics for QoS based adaptation. Another software only solution is the player of Oregon University [6]. This player is different from Berkeley's player in that it uses a toolkit based software feedback approach for QoS control and synchronization instead of an ad-hoc one. *Vosaic* of Illinois University [7] extends the architecture of the WWW to encompass video and audio. Vosaic is built on a protocol called VDP (Video Datagram Protocol) similar to RTP. VDP is different from RTP in that it has built-in fault tolerance for video transmission. Fault tolerance is achieved by the retransmission of lost and delayed packets. There are also commercial protocols such as Real Player, XingMPEG for real time video transmission. Those systems are all based on traditional process concept and require costly IPC mechanisms. Our player takes the advantage of multithreading by allowing parallel execution of I/O and CPU bound threads to maximize the display frame rate and minimize the loss rate.

In this paper, we extended performance results of our previous papers [8, 9, 10, 11] on the streaming application that we developed for the Internet. The paper is organized as follows. Section 2 introduces our system. Experimental results are given in Section 3. Section 4 is the conclusion.

2 The Developed System

Research on VOD systems evolves in two directions. The first direction deals with the storage and retrieval issues of videos while the second direction concentrates on the transmission of videos. We focused on the second direction in our studies.

Fig. 1 illustrates our VOD system. The system has a client-server architecture. An important feature of our system is that it is a software only application developed by using multithreading and POSIX 4 real time extensions. More detailed information can be found in our previous papers [8, 9, 10, 11].

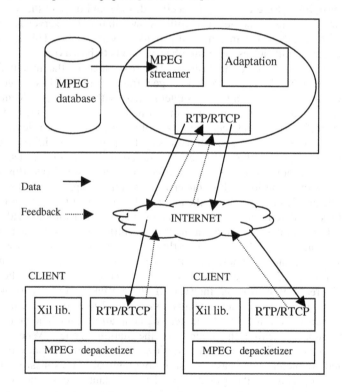

Fig. 1. The architecture of our streaming application

The server accepts connection requests. A thread and an RTP/RTCP connection is created for each request. Main responsibility of the server thread is to stream the vides to clients over RTP connection. Streaming is based on the rules stated in the related internet draft [12]. These rules aim at maximizing the amount of decodable data at the receiver. Control information between the server and a client is exchanged via TCP.

The server schedules packet delivery based on the file size statistics of the videos. Once a video is added to the database, the streaming module is executed and the number of packets to deliver the video is determined. This value is stored in a file. During actual transmission, packet interval between consecutive packets is calculated by dividing the video duration by the number of packets.

Another function of the server is to perform quality of service control by collecting RTCP reports from receivers. RTCP reports give statistics about packet loss rate. Packet loss rate is the indicator of the congestion in the network. Congestion may be short or long termed. Short and long term congestion are discriminated by using a low pass filter. Short term congestion is ignored while rate adaptation is performed by dropping frames from the GOP sequence in case of a long term congestion. Frame dropping is performed in levels by considering the dependencies among different frame types and the smoothness of the video streams. At each adaptation level, more frames are dropped from the GOP sequence.

The responsibility of the client is to receive, decode and display the video frames and to monitor the state of the network with its own status. The server is notified by the client and responds to the information provided by the client. This alleviates the burden of the server and causes larger number of clients to be served.

The client works in a pipeline fashion. The first stage of the pipeline receives packets from the network. The second stage gets frames from the first stage, decodes and displays them. Each stage is implemented with a thread. Multithreaded implementation of the receiver resulted in parallelism and maximized the display frame rate. A buffer is used between the stages of the pipeline. The role of the buffer is to order the incoming packets and to compensate the effects of different execution rates of the two stages of the pipeline. A flow control algorithm is used to keep enough frames in buffers to prevent buffer underflows and overflows. Buffer overflows result in frame loss while underflows result in gaps during display. Flow control module notifies the server via the control channel about the state of its buffers. The client does not determine the rate at which the server transmits the video. It only keeps track of the number of frames in the buffer. Buffer space is divided into buffer occupancy zones: Stability, overflow, underflow, overflow_warning, underflow_warning, danger zones. Stability zone keeps from 0.85 to 1.15 minutes of video. If more than 1.5 minutes of video exists in the overflow zone, the server is notified to increase transmission interval. If the buffer occupancy is within the underflow zone, that is the buffer contains less than 0.2 min. of video, the server is notified to decrease the frame rate by increasing adaptation level. The goal is to keep the buffer occupancy within the bounds of the stability zone. When the buffer occupancy is in the overflow_warning/underflow_warning, danger zones, the server is again notified to increase/decrease transmission interval. Transmission interval is changed by multiplying the theoritical interval by constant values. The constants are 1.2, 1.1, 0.9 and 0.8 for overflow, overflow_warning, underflow_warning and danger zones, respectively.

3 Experimental Results

The system we developed has been tested on Ege University Gigabit Ethernet Campus LAN (Fig. 2). The server is a Sun ULTRA 10 workstation. Clients are Sun ULTRA 1 and Sun ULTRA 5 workstations. The video database consists of MPEG-1 videos encoded at two different resolutions and five different bit rates (100 Kbps, 200 Kbps, 500 Kbps, 1000 Kbps and 1500 Kbps). As stated in [13], bit rates of 200 Kbps and lower are suitable for the streaming of MPEG-1 videos over the Internet. Although higher bit rates give better quality, they lead to heavier traffic on the network and more processing cost at the server and client. We included videos encoded at higher bit rates than 200 Kbps to observe the behavior of our system during the transmission of higher quality videos. In the following subsections, we will present collected performance results pertaining to our system.

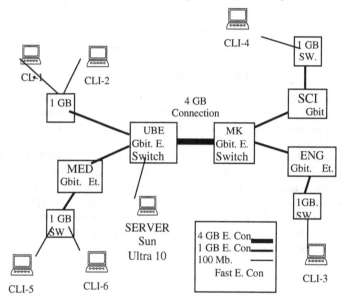

Fig. 2. The test bed of our system

3.1 System Overhead and System Capacity

Server overhead is important in order to determine the number of streams that can be supported successfully without affecting each other. Less server overhead means more clients can be served concurrently.

Server overhead is mainly composed of two components: CPU time spent for packetization and CPU time spent for RTP library processing at the server site. We measured these costs at different encoding bit rates and at different frame rates. Theoretical number of concurrent clients is calculated by dividing the video duration by the total server overhead. Fig. 3 shows the number of clients that can be supported simultaneously. As seen from Fig. 3, system capacity increases as the encoding bit rate and frame rate decreases.

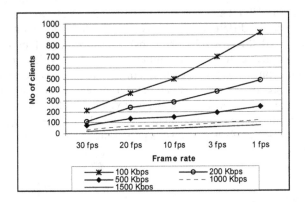

Fig. 3. System Capacity

Client overhead is mainly the processing time spent for the display of videos. Display process has two steps: decoding and dithering. Our experimental findings show that decoding has negligible contribution to client overhead while dithering consumes most of the time.

Time consumed during display is the most important contributor to the performance of the client. If the frame rate is r, then each frame should be processed and displayed at intervals of 1/r. For example, at 30 fps, XIL library costs should be less than 1/30 secs = 33 milliseconds. We measured that XIL library displays a frame in about 28 milliseconds at resolution 320x240 and in about 9 milliseconds at resolution 160x120. Therefore, it can be concluded that videos can be displayed at full frame rate on the clients' screen.

3.2 Packet Scheduling

Graphs in Fig.4 illustrate the behavior of our packet scheduling scheme which starts transmission under the assumption of constant packet intervals between consecutive packets. Initial packet interval is computed by dividing the video duration by the number of packets required to send the video.

As the transmission progresses, flow control module forces the server to change the packet interval. At the encoding bit rate of 100 Kbps, buffer occupancy enters into overflow zone and the server is asked to increase the sending interval. After the buffer occupancy re-enters into stability zone, transmission interval is reset to its initial value. At 200 Kbps, no change in transmission interval is required, because buffer occupancy remains in the stability zone. At 500 Kbps, it has been observed that the transmission interval is settled into a stable value as the transmission progresses.

At 1000 and 1500 Kbps, frequent changes in packet interval have been observed. Interaction with packet scheduling module and flow control module increases. Server is forced to change the frame rate frequently, since the buffer occupancy level repeatedly falls into underflow zone.

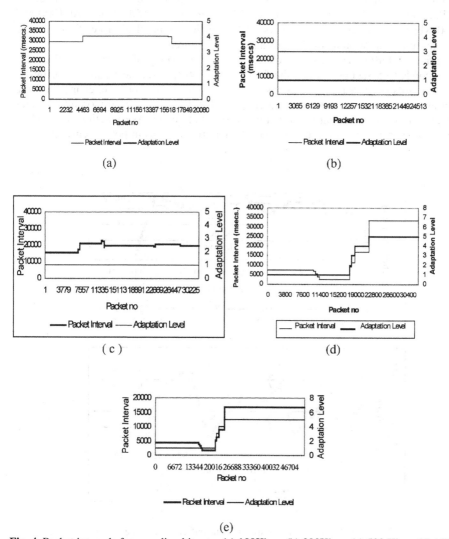

Fig. 4. Packet intervals for encoding bit rates (a) 100Kbps, (b) 200Kbps, (c) 500 Kbps, (d) 1000 Kbps and (e) 1500 Kbps and interaction with our rate adaptation module.

3.3   Flow Control and Buffer Management

Fig. 5 shows the performance of our flow control module. For the cases where the encoding bit rate is 100 Kbps and 200 Kbps, the input rate to the buffer and output rate from the buffer are nearly the same. Therefore, buffer occupancy is within the stability zone. As bit rate is further increased, consumption rate becomes greater than the input rate. At 500 Kbps, as soon as the buffer level falls below the stability zone, the server is forced to increase its rate. As soon as the increased rate causes the occupancy to enter overflow zone, the server is forced to reduce its rate. Therefore, oscillations within the stability zone have been observed at 500 Kbps. At 1000 and 1500 Kbps, buffer level quickly falls into underflow zone. Therefore, frequent rate changes and adaptation level changes have been required.

Currently, buffer occupancy is checked after the display of every 10 GOPs. This corresponds to 5 seconds where the frame rate is 30 fps. As the frequency of buffer level controls is decreased, it may take longer to detect a buffer level change. On the contrary, increases in control frequency enable us to detect important changes quickly. However, frequent controls may also initiate unnecessary actions by leading to the misinterpretation of transient changes in buffer level as important changes.

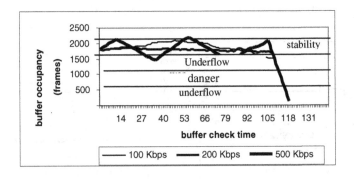

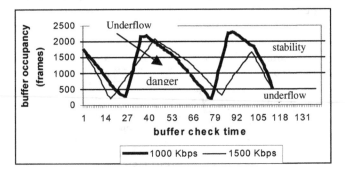

Fig. 5. Buffer occupancy graphs for our buffer management module

3.4   QoS Control with RTCP Statistics

Clients regularly send RTCP reports which provide feedback on the status of the network to the server. Among the fields of RTCP reports, loss statistics and jitter statistics fields are related to the quality of the transmission.

Loss statistics are reported in two fields: Loss rate and cumulative number of lost packets. Loss rate statistics are the indicator of the short term losses and transient congestion. Cumulative number of lost packets field keeps the number of lost packets since the beginning of the transmission. In other words, they provide information on the long term packet loss.

Since our current test bed contains limited number of clients and our LAN is built upon Gigabit Ethernet Backbone, packet loss is negligible. We tested our rate adaptation module by using artificial packet intervals and illustrated the operation of our rate adaptation module in Fig 6.

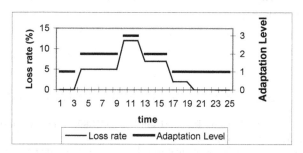

Fig. 6. Loss rate vs. Rate adaptation

Jitter field shows the statistical variance of the inter arrival times between consecutive packets. It is the second indicator of transient congestion after loss rate statistics. Future packet losses can be predicted by analyzing jitter statistics. A sudden increase in jitter may be the indicator of a future packet loss and the loss may be avoided if the necessary precautions are taken in time. Fig. 7 shows the jitter statistics for encoding bit rates 200 and 1000 Kbps. For 1000 Kbps case, we observed a sudden increase in jitter was accompanied by an increase in loss statistics.

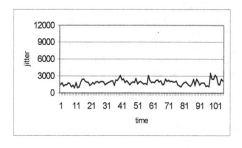 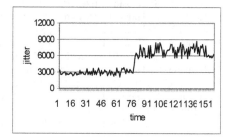

Fig. 7. Jitter statistics for bit rates of 200 Kbps and 1000 Kbps

4 Conclusion

The design, implementation and evaluation of a distributed real-time MPEG video player has been presented. Our experimental results demonstrate that our packet scheduling, flow control and feedback based QoS control modules are effective and the system can be employed on Campus LANs successfully. Our next goal is to include PCs to our test bed and to obtain more experimental results, especially on loss rate statistics. We also plan to measure the performance when the number of concurrent clients increases. Our other goals are to embed our system into WWW and to add audio to our system.

References

1. Fluckinger F.: Understanding Networked Multimedia: Applications and Technology, Prentice Hall (1995).
2. Schulzrinne H., Cosner S., Frederic R., Jacobson V.: FC1889: RTP: A transport protocol for real-time applications, (1996).
3. Gallmeister B. O., Programming for the real world: POSIX 4, O'Reilley & Associates, USA (1995).
4. Lewis B., Berg D. : Multithreaded Programming with Pthreads, Sun Microsystems, California (1998).
5. K. Mayer-Patel, L. A. Rowe: Design and performance of the Berkeley Continuous Media Toolkit, SPIE Proceedings of ACM SIGCOMM'97, Cannes, France (1997).
6. Cen S., Pu C., Staehli R., Cowan C., Walpole J.: A Distributed Real-Time MPEG Video Audio Player, Proceedings of NOSSDAV'95, Durham, New Hampshire (1995).
7. Chen Z., Tan S. M., Campell R. H., Li Y.: Real-Time Video and Audio in the World Wide Web, Proceedings of the 4th International World Wide Web Conference, Boston, Massachusetts, (1995).
8. Kantarc• A., Tunal• T.: Transmission of stored MPEG-1 videos over the Internet, Computer Networks Symposium, BAS2000, Ankara, Turkey (2000).
9. Kantarc• A., Tunal• T.: Design and Implementation of a VoD system for the Internet, Packet Video 2000 Conference, PV2000, Sardinia, Italy (2000).
10.Kantarc• A., Tunal• T.: The Design and Implementation of a Streaming Application for MPEG Videos. IEEE International Multimedia Conference and Expo (2000).
11.Kantarc• A., Tunal• T.: Real-time Transmission of Stored MPEG-1 Videos over the Internet, 4th CSCC, 2000 WORLD MULTICONFERENCE, Athens, Greece (2000).
12.Hoffman D., Fernando G., Goyal V., Civanlar M.: RFC2250: RTP payload format for MPEG1/MPEG2 video (1998).
13.DVmpeg User's Guide, Darim Multimedia and System (1997).

What Do Hyperlink-Proposals and Request-Prediction Have in Common?

Ernst-Georg Haffner[1], Uwe Roth[1], Andreas Heuer[1],
Thomas Engel[1], and Christoph Meinel[1]

[1] Institute of Telematics, Bahnhofsstr. 30-32,
D-54292 Trier, Germany
{Haffner, Roth, Heuer, Engel, Meinel} @ti.fhg.de

Abstract. This paper focuses on fundamental similarities between *proposing links for hypertexts* and *predicting user-requests*. It briefly outlines the theoretical background of both categories of problems and, as an example, explores common implementation strategies for handling them. Even though there are some important differences between link-proposals and prediction of requests, we believe that in regarding these categories as special cases of a more general and abstract mathematical model, improvements on the one side will also result in advantages for the other.

1 Introduction

Due to the breath-taking growth of the World Wide Web (WWW), the need for high quality hypertexts is rapidly increasing, and finding appropriate links is one of the most difficult of tasks. Ultra-modern online authoring systems[1] that provide possibilities to check link-consistencies and administrate link management should also propose links in order to improve the usefulness of the HTML-documents.

Another major problem of today's Internet applications - and, at first glance, an entirely different one to finding hyperlinks - is the performance of Client/Server communication: servers often take a long time to respond to a client's request. There are several strategies to overcome this problem of high user-perceived latencies; one of them is to predict future requests. This way, time-consuming calculations on the server's side can be performed even before a special request is being made. If the server is "sure" that certain documents will soon be requested, the associated data can be sent to the client in advance (or can be pre-fetched by the client) even while the user is unaware of this process.

The two problem categories discussed here do not seem to have much in common. In this paper, we mean to show that there are certain, similar, solution strategies to take care of both problems. Therefore, we will first have a closer look at hyperlink-proposals (section 2). Then, we will present a prediction scenario and outline advanced strategies to foresee future user-requests on a statistical base (section 3). A comparison and an abstraction of both methodologies will be highlighted in section 4. Finally, a summary and an outlook on future events will be presented in section 5.

---

[1] For instance: *Microcosm* [12], or *Daphne* [28]. Multilingual approaches are sketched in [13].

T. Yakhno (Ed.): ADVIS 2000, LNCS 1909, pp. 285-293, 2000.
© Springer-Verlag Berlin Heidelberg 2000

2 Hyperlink-Proposals

The theory of hyperlink-research already has a long history. Hyperlink-research has been carried out ever since the introduction of the World Wide Web-service to the Internet. Kaindl et. al. present a compact outline of the progresses made so far and draw some conclusions [19].

As its final aim, link retrieval research strives to achieve the automatic generation of hyperlinks. Several systems were built to perform link-proposals. A very promising description of Chang's *HieNet* can be found in [7]. Due to various problems in finding links for hypertexts Allan distinguishes between three major characteristics of link-types: *manual, automatic* and *pattern-matching* [1]. Every effort in the Hyperlink-Research can thus be categorized. We will focus on retrieving "automatic" links on a statistical base with approved methods (e.g. [27] or [25]).

It is a very complex task to measure the quality of hyperlink-proposal algorithms. Cleary and Bareiss mention as important factors the *recall*, the share of appropriate proposals of all good links and the *precision*, the share of appropriate proposals of all proposals [6]. Often, the quality of proposals in detail is only measurable by human experts. The algorithm for the proposal of hyperlinks in this paper is based on the idea of case-based reasoning [20] with some improvements concerning efficiency and complexity.

In order to be able to propose hyperlinks for texts on a statistical base [5] without any further use of a semantic model, one has to build up a database for possible link targets first (see also [24] for another insight regarding this topic). Given a straightforward case, the targets of hyperlinks are simply the documents of the hyperlink-management system. In this paper, we mean to present a hyperlink-proposal-system where all links that are part of any hypertext can be proposed no matter whether the target document is part of the system or not. Storing the possible hyperlinks is not enough, though. Of major importance is the fact that the system must store the relationship between the text and the associated links. This step is called a *learning process*. Therefore, the database can also be called a *knowledge base*. The quality of this knowledge base depends on the quality of the learning process: how can the information of the texts be combined with appropriate hyperlinks?

One possibility would be the *human teacher*. A person or a group of persons could derive the important information of the document *manually*. This process is very cost- and time consuming and is not feasible in practice. Most of today's learning algorithms conquer the problem of high quality knowledge retrieval by extracting some information automatically on the basis of advanced heuristics. The learning process itself evaluates the relevance of the extracted information. At first, the algorithm treats a text with hyperlinks as if it did not contain any link and derives the relevant information. Next, it proposes one or more hyperlinks and compares the result to the hyperlinks that the document in fact contains. By using this method, the learning process can be carried out without any human teacher. A disadvantage arises from the strong relationship between the quality of the learning process and the quality of the initial documents. Furthermore, only those hyperlinks can be proposed that are already known to the system and at the very beginning of the learning process there are no links to be proposed at all.

After this learning step has been taken, hyperlinks for new documents can be proposed by using the knowledge base. This phase is also called the *classification phase*. The part of retrieving the relevant information is just the same as in the learning phase, and link-proposals can be calculated. In fact, a learning component can also be found in this step. In general, the user of the hyperlink-management system accepts or rejects a proposal of the classification algorithm and thus plays the role of a human teacher. The learning algorithm can - just as in the learning phase - compare the proposal of the system with the reaction of the human user and thus adapt the relevance values. Figure 1 shows the described ideas in a graphical form.

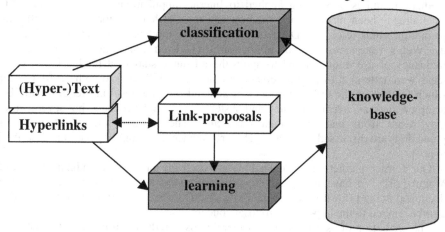

Fig. 1. Principle structure of a statistical-based hyperlink-proposal algorithm

A straightforward implementation of such a hyperlink-proposal algorithm could be realized by using mathematical objects such as vectors or matrixes. If the entire knowledge of the link-proposal systems were to be stored in a *relevance-matrix R*, every column would represent all the existing text attributes and every row would stand for a possible link-proposal. Furthermore, a text (without links) in the classification process could be modeled by an attribute vector t where every element corresponds to an - existing or not - textual property. The classification process would then be a multiplication of the relevance matrix with the vector and the resulting vector r would contain the probabilities for all existing links in the knowledge-base whether they are appropriate as hyperlink of the new text or not (1).

$$r = R \cdot t. \tag{1}$$

The learning process has to guarantee that a given resulting vector r', that represents the really existing hyperlinks in the text t, would just be the result in a classification step of t. Therefore, the relevance-matrix R has to be adapted. With other words, the learning process applies a function f on R so that the following equation holds true (2):

$$r' = f(R) \cdot t. \tag{2}$$

In practice, the learning and the classification steps are combined so that the system is able to make proposals even though the knowledge base is still rather small. A description of an implementation and an evaluation of this principle can be found in [17].

In section 4, we will come back to the main ideas of this approach and we will show that these algorithms can be improved by drawing parallels for the - at first glance - quite different field of request-prediction.

3 Request-Prediction

The notion of predicting future events originates from compiler construction (branch prediction) and is now being applied to Internet applications [18]. Many attempts have already been made to find the best and most efficient algorithms for fulfilling prediction needs (e.g. [2]). Discrete Markov-Chain-models were used as a basis for the Web's first algorithms of prediction [21]. Their conception is to store the frequency of user-requests and to apply the adequate statistical model. Later, these ideas were extended to a continuous chain-approach [22] and to path profiling [26] which focuses on the order of document demands and the resulting request path. By using predictions, average latency and system load may be reduced but several risks may result from inaccurate data prediction. The negative effects of incorrect predictions are discussed in [9] and [4]. Performance modeling in general is described in [10].

Our former prediction-approach [14] is based on an idea of Padmanabhan and Mogul [23]. We have improved this straightforward approach to model time and document aging [15]. It is within the focus of this paper to verify that these ideas can also be applied to the hyperlink-proposal concept.

The prediction of user-requests in general aims at reducing user-perceived latency. The first concept models a group of requests as a *session*. Even though the session cannot be justified on the basis of the standard WWW protocol HTTP[2], it is very important to group several requests together. In general, a session is regarded as a time period of about 30 minutes [26]. During such a session, a user can request several documents (or other kinds of data packets). The main goal of prediction aims at foreseeing some of the upcoming requests during the same session on the base of the requests that have already been made. Therefore, the prediction-algorithm generates *relative probabilities* for future document requests on the basis of the *relative frequencies* of requests in the past. Certainly, there are also some other approaches, but the perhaps most straightforward one simply stores the actual user-requests at first and then calculates probabilities for future request wishes. The former process is a (rather simple) kind of *learning phase*, while the latter one can be called a *classification phase*.

If we do not take care of maintaining the correct request order, a straightforward mathematical conception of a session will lead to a vector s. Every request of a document corresponds to an element of s. The size of the vector corresponds to the number of documents that should be part of the prediction algorithm (*predictable documents*). Details about finding the appropriate documents and criterions whether prediction should be made can be found in [14] and [16].

To store the relative frequencies of requests, e.g. the number of requests for every document depending on the requests of all the other documents, we need - at least - a quadratic matrix with the dimension of the number of predictable documents. We call

---

[2] **Hypertext Transfer Protocol**

this matrix the *Memory Matrix M*. In the case where the request order is irrelevant, the matrix *M* is also symmetric. For efficiency, the relative probabilities are not stored directly in *M*, but they are calculated from the relative frequencies at the moment when they are needed. If the function g retrieves the relative probabilities from the according frequencies, the classification of the (not completed) session s^n to the predicted final session s' can be described as (3):

$$s' = g(M) \cdot s^n. \qquad (3)$$

Again, the learning process can compare the results of the classification step (s') to the real user behavior that leads to the (completed) session s. The function g must be adapted by using a heuristic function h so that the following equation will be true (4):

$$s = h \circ g(M) \cdot s^n. \qquad (4)$$

Additionally, the Memory Matrix M must be adapted to M' after every new request during a user session so that it still represents the relative frequencies (5).

$$M' = (M_{ij}') = \begin{cases} M_{ij} + 1 : \text{iff } s_i = 1 \wedge s_j = 1 \\ M_{ij} : \text{otherwise} \end{cases} \qquad (5)$$

A very critical point in generating request-prediction is to take care of the different costs. Not only the system costs for the prediction itself must be taken into account, but also the network load and the server load for wrongly predicted documents. These considerations are very important especially for making pre-fetches. In this paper, cost-aspects are not part of the focus. Detailed information about this topic can be found in [3] and [8].

4 Abstraction and Generalization

Closer analysis of the categories *finding hyperlink-proposals* and *predicting user-requests* leads to astonishingly similar results. The core function of both solution-algorithms is the storage and classification of information and the completion of partially given information in a foreseeing-step by calculating probabilities which are based on statistical data. Moreover, mathematically speaking, existing information can be stored in a (dynamically adapted) matrix, and the classification step is executed by multiplying a vector with this matrix. Differences arise when the matrix and the vector elements are modeled. When searching for hyperlinks, this is a very difficult task: in this case, extracting appropriate keywords to retrieve important text attributes for later classification steps is rather delicate. Too many keywords will result in a very large matrix and thus in a long duration of the process of calculation. Too few or even false keywords, though, may lead to a wrong classification and inappropriate link-proposals. In the case of predicting user-requests, the modeling of a session-vector is a very simple and straightforward task. Nevertheless, calculation of costs for incorrectly predicted and pre-fetched data is highly complex. And even though wrongly proposed hyperlinks are somehow disturbing, wrongly predicted data can cause enormous network and system load and thus presents a much higher danger than the one mentioned before.

Table 1 shows a short and abstract characterization of the main tasks in predicting user-requests and proposing hyperlinks for texts as presented in the sections 2 and 3. In both cases, the core data structure is a matrix that stores the entire knowledge. The major difference is in the methods and the complexity of modeling the problem in order to gain vectors.

Table 1. Similarities between hyperlink-proposal and request-prediction algorithms

Phase	Step	Hyperlink-Proposals	Request-Prediction
Learning phase	Retrieval of knowledge	Keyword extraction of hypertexts and attached links	Different document requests during the same session (with or without taking care of the order)
	Storing of knowledge	Vectorization of keyword attributes and storage of values in a relevance matrix (idea of case-based reasoning [20])	Transforming sessions into (mostly binary) vectors and storing their values in a memory matrix
Classification phase	Recognition of information	Generating attributes (keywords) from texts (without considering link information)	Interpreting the first client request as ignition for calculation of relative frequencies
	Calculation of (likely) candidates	Multiplying the relevance matrix with the new attribute vector and thus getting link-relevance-probabilities → generating link-proposals	Multiplying the memory matrix with the current session-vector and thus getting relative probabilities for other documents to be requested soon → generating request-prediction

Still, the question remains: what is the advantage of recognizing similarities between link-proposals and request-prediction? Obviously, knowledge of the one can improve evolution steps of the other.

We will try to illustrate this through the following example. In [15], an advanced prediction model is shown where time and document aging is applied[3]. The evaluation of the usefulness of this approach is also confirmed. A time function f_t is applied to change the elements of the memory matrix M to model document aging (6) (corresponds to equation 4).

$$s_t = h \cdot g \cdot f_t(M) \cdot s^n. \qquad (6)$$

But how can algorithms for the proposal of hyperlinks be improved by this perception? Time and aging also play an important role in the latter task as the relevance of former attributes changes. Analogously, the same time function can be used to change the relevance matrix in order to improve the learning and classification phases of the hyperlink-proposal algorithm (7).

$$r_t' = f \cdot f_t(R) \cdot t. \qquad (7)$$

The usefulness of this idea is currently being evaluated by the authors and the first results are very promising. Figure 2 shows briefly the improvement-results for the introduction of the timing factor to the area of hyperlink-proposing algorithms after the initial tests.[4]

---

[3] General information on the topic of document aging can also be found in [11].

[4] More detailed results on that special topic will follow soon.

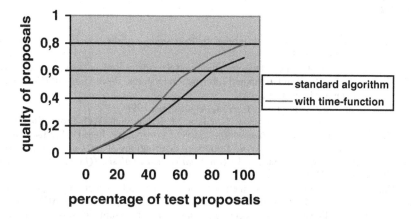

Fig. 2. Improvements of hyperlink-proposal efficiency by using time-functions to model document aging

The quality of hyperlink-proposals rises for higher degrees of knowledge-base load. Using time-factor modeling the highest qualities can be increased about 10 percent above the standard values. Thus - presented as an example - ideas of prediction can help to improve the efficiency of proposing hyperlinks.

5 Summary and Outlook

In this paper, we presented two different problem categories of special modern Internet applications. The task of proposing hyperlinks for texts should help the author of a hypertext to improve the quality of his/her work by adding additional link information automatically. The prediction of requests aims at reducing the user perceived latency while waiting for an answer from the server.

Both problems - even though very different at a first glance - could be proved as similar during a closer analysis. We presented two mechanisms to solve these problems by learning and classifying and we elaborated their common properties.

With this knowledge, improvements on the one side should help to make progress on the other side as well. As an example of this idea we briefly outlined how to apply the concepts of modeling time and document aging of request-prediction to the area of proposing hyperlinks. The first results were very promising.

In the near future, besides evaluating exactly the advantages of time-modeling for the area of proposing hyperlinks, we will try to find several further aspects of the one side to improve the other and vice versa. Perhaps, it makes sense to use a single algorithm to solve both problem categories. Thus, we could try to find further areas that fulfill the same pre-conditions as prediction and link-proposals. Even though the comparison of different topics of modern Internet applications is very useful and advantageous, we should not forget that differences always remain and that the most difficult parts of problem aspects are, as a rule, quite unique.

References

1. J. Allan. Automatic hypertext link typing. In *Proceedings of the Seventh ACM Conference on Hypertext (Hypertext '96)*, 1996
2. A. Bestavros, Speculative Data Dissemination and Service to Reduce Server Load, Network Traffic and Service Time in Distributed Information Systems, *Proceedings of ICDE'96*, New Orleans, Louisiana, March 1996
3. P. Cao, S. Irani, Cost-Aware Proxy Caching Algorithms, 1998
4. R. Cáreres, F. Douglis, A. Feldmann, G. Glass, M. Rabinovich, Web Proxy Caching: The devil is in the details, *Workshop on Internet Server Performance*, June 1998, Madison, WI
5. R. J. Chitashvili, R. H. Baayen. Word frequency distributions of texts and corpora as large number of rare event distributions. In *Quantitative text analysis, (Quantitative linguistics, Vol. 52)*, 1993. WVT Trier
6. C. Cleary, R. Bareiss. Practical methods for automatically generating typed links. In *Proceedings of the Seventh ACM Conference on Hypertext (Hypertext '96)*, ACM, 1996
7. D. T. Chang. HieNet: A user-centered approach for automatic link generation. In *Proceedings of the Fifth ACM Conference on Hypertext (Hypertext'93)*, ACM, 1993
8. E. Cohen, B. Krishnamurthy, J. Rexford, Improving End-to-End Performance of the Web Using Server Volumes and Proxy Filters, 1998
9. M. Crovella, P. Barford, The Network Effects of Prefetching, IEEE Infocom 1998
10. S. Erickson, D. Yi, Modelling the performance of a Large Multi-Tiered Application, American Management Systems, AMS Center For Advanced Technology, 1999
11. J. Griffioen, R. Appleton, The design, Implementation, and Evaluation of a Predictive Caching File System, CS-Department University of Kentucky, CS-264-96, 1996
12. W. Hall, H. Davis, G. Hutchings. *Rethinking Hypermedia: The Microcoms Approach.* Kluwer Academic Publishers, 1996
13. A. Heuer, E.-G. Haffner, U. Roth, Z. Zhang, T. Engel, C. Meinel. Hyperlink management system for multilingual websites. In *Proceedings of the Asia Pacific Web Conference (APWeb '99)*, 1999. http://www2.comp.polyu.edu.hk/~apweb99/
14. E.-G. Haffner, U. Roth, T. Engel, Ch. Meinel, A Semi-Random Prediction Scenario for User Requests, *Proceedings of the Asia Pacific Web Conference*, APWEB99, 1999
15. E.-G. Haffner, U. Roth, T. Engel, Ch. Meinel. Modeling Time and Document Aging for Request Prediction - One Step further. *Symposium on Applied Computing*, SAC2000, Como, Italy, 2000
16. E.-G. Haffner, U. Roth, T. Engel, Ch. Meinel. Optimizing Requests for the Smart Data Server. *Applied Informatics*, IASTED AI2000, Innsbruck, Austria, 2000
17. E.-G. Haffner, A. Heuer, U. Roth, T. Engel, Ch. Meinel. Advanced Studies on Link-Proposals and Knowledge-Retrieval of Hypertexts with CBR. *Proceedings of the International EC-Web Conference*, ECWeb2000, Greenwich, United Kingdom, LNCS, Springer-Verlag, 2000
18. Z. Jiang, L. Kleinrock, An Adaptive Network Prefetch Scheme, IEEE J Sel Areas Commun, 1998
19. H. Kaindl, S. Kramer. Semiautomatic generation of glossary links: A Practical Solution. In *Proceedings of the Tenth ACM Conference on Hypertext (Hypertext '99)*, ACM, 1999
20. J. Kolodner, D. Leake. A tutorial introduction to Case-Based Reasoning. In *Case-Based Reasoning*. AAAI Press, the MIT Press, 1995
21. Achim Kraiss, Gerhard Weikum, Vertical Data Migration in Large Near-Line Document Archives Based on Markov-Chain Predictions, Proceedings of the 23rd VLDB Conference, 1997
22. A. Kraiss, G. Weikum, Integrated document caching and prefetching in storage hierarchies based on Markov-chain predictions, The VLDB Journal, Springer-Verlag, 1998
23. V.N. Padmanabhan, J.C. Mogul, Using Predictive Prefetching to Improve World Wide Web Latency, Computer Communication Review, 1996

24. F. J. Ricardo. Stalking the paratext: speculations on hypertext links as second order text. In *Proceedings of the Ninth ACM Conference on Hypertext (Hypertext '98)*, ACM, 1998
25. J.H. Rety. Structure analysis for hypertext with conditional linkage. In *Proceedings of the Tenth ACM Conference on Hypertext (Hypertext '99)*, ACM, 1999
26. Schechter, Krishnan, Smith, Using path profiles to predict HTTP requests. In *Proceedings of the World Wide Web Conference*, 1998
27. J. Tebbutt. Finding links. In *Proceedings of the Ninth ACM Conference on Hypertext (Hypertext '98)*, ACM, 1998
28. Z. Zhang, E.-G. Haffner, A. Heuer, T. Engel and C. Meinel. Role-based Access Control in Online Authoring and Publishing Systems vs. Documentation Hierarchy. In *Proceedings of the SIGDOC '99*, ACM, 1999

The Valid Web: An XML/XSL Infrastructure for Temporal Management of Web Documents

Fabio Grandi and Federica Mandreoli

CSITE-CNR, Dip. di Elettronica, Informatica e Sistemistica,
Università di Bologna, Italy, email: {fgrandi,fmandreoli}@deis.unibo.it

Abstract. In this paper we present a temporal extension of the World Wide Web based on a complete XML/XSL infrastructure to support valid time. The proposed technique enables the explicit definition of temporal information within HTML/XML documents, whose contents can then be selectively accessed according to their valid time. By acting on a navigation validity context, the proposed solution makes it possible to "travel in time" in a given virtual environment with any XML-compliant browser; this allows, for instance, to cut personalized visit routes for a specific epoch in a virtual museum or a digital historical library, to visualize the evolution of an archaeological site through successives ages, to selectively access past issues of magazines, to browse historical time series (e.g. stock quote archives), etc. The proposed Web extensions have been tested on a demo prototype showing, as application example, the functionalities of a temporal Web museum.

1 Introduction

A great deal of work has been done in recent years in the field of Temporal Databases (TDBs) [15,9,26,7,16]. Due to this effort, a large infrastructure (namely data models, query languages, index structures, etc.) has been developed for the management of data evolving in time, for which successive versions need to be maintained rather than being overwritten or discarded by destructive changes. However, research interests on temporal information have been almost focused on highly structured data (e.g. relational or object-oriented), and, for instance, no textual data or less structured multimedia documents have been diffusely considered for temporal extensions so far.

On the other hand, the World Wide Web (WWW, W3 or Web [2]) is a large distributed collection of hypertextual and multimedia irregular documents, currently formatted according to the HTML standard [20], available on-line on the Internet. The new emerging standard for publishing documents over the Web, which has been recently recommended by the W3C Consortium [25], is the eXtensible Markup Language (XML [18]), which has been designed to overcome the main limitations of HTML (and is also suitable to describe semistructured data). The browsing of XML documents can be best enjoyed by means of provided *stylesheets*: in particular, the eXtensible Stylesheet Language (XSL [19]) allows the

T. Yakhno (Ed.): ADVIS 2000, LNCS 1909, pp. 294–303, 2000.

definition of XML document transformation rules [30] and the specification of formatting semantics. Although the functionalities of the Web, including markup language potentialities and browser capabilities, have been greatly increased in recent years, scarce attention has been so far devoted to the *temporal* aspects, whereas Web documents may also contain intrinsically temporal (i.e. *historical*) information.

Our proposal concerns the introduction of valid time into the Web to support the management of historical information, borrowing the basic timestamping and temporal selection techniques introduced in TDB theory. To this purpose, in Sec. 2, we will define a complete XML/XSL infrastructure to embody temporal information into Web documents and put temporal navigation and query facilities at Web user's disposal. The solution we propose does not require changes in the current Web technology as it is based on XML and related standards. The adoption of a suitable XML schema for document timestamping and a provided XSL stylesheet for temporally selective document processing will enable any XML-compliant browser, like Microsoft Internet Explorer 5 (Ie5 [24]), to support temporal documents. The proposed Web extensions have also been tested on a prototype implementation [5] which will be briefly described in Sec. 3. Our implementation also largely exploits another powerful Web technology supported by Ie5: the Document Object Model (DOM [17]), which is an application programming interface (API), which allows dynamic manipulation of HTML and XML documents (e.g. via scripting languages like JavaScript [23]). Conclusions can finally be found in Sec. 4.

2   Integrating Valid Time into the Web

In this Section we outline our proposal concerning an XML/XSL infrastructure for the definition and use of valid-time temporal Web documents. Generally speaking, the adoption of valid time is aimed at allowing the management of historical information (past, present or future). Historical information must explicitly be coded within the Web documents, to be selectively accessed during temporal browsing. To this purpose, distinct parts of a Web document can be timestamped with their own validity during the document creation.

2.1   Defining Valid Temporal Documents

The addition of valid time to Web documents we propose is based on the extensions of the XML markup language [18] with timestamping tags. The functionalities of the new tags can be fully specified by means of suitable XML schemas and stylesheets, without requiring modifications to the Web browsers supporting XML (like Ie5).

In particular, our proposal consists of the addition of a new XML tag, <valid>, to define a *validity context*. The validity context is used to assign a

```
<?xml version="1.0" ?>
<Schema xmlns="urn:schemas-microsoft-com:xml-data"
        xmlns:dt="urn:schemas-microsoft-com:datatypes">
   <AttributeType name="from" required="yes" dt:type="date" />
   <AttributeType name="to" required="yes" dt:type="date" />
   <ElementType name="validity">
      <attribute type="from" minOccurs="1" maxOccurs="1" />
      <attribute type="to" minOccurs="1" maxOccurs="1" />
   </ElementType>
   <ElementType name="valid" content="mixed">
      <element type="validity" minOccurs="1" maxOccurs="*" />
   </ElementType>
</Schema>
```

Fig. 1. The `ValidSchema.xml` XML schema.

specific time pertinence to a piece of a document to be used for temporally-selective manipulation. The contents embraced in a validity context can be of any allowed kind (namely text, graphic elements, and any other XML structure including nested <valid> elements). The timestamps can be specified in a validity context by means of <validity> tags, which allow the definition of a temporal interval through its boundaries (i.e. the values of the from and to attributes of the <validity> XML element). In general, multiple intervals can be used: in this case, the timestamp is defined as the union of all the validity intervals specified; formally, the timestamp is a *temporal element* as defined in the BCDM temporal data model [8]. For instance, the following code:

```
<valid>  <!-- definition of a validity context -->
   <validity from="1980-01-01" to="1985-12-31" />
   <validity from="1995-01-01" to="2000-12-31" />
         This is text <b>valid from 1980 to 1985</b>
         but also <b>valid from 1995 to 2000</b>...
</valid>
```

defines a validity context whose validity is [1980–1985] ∪ [1995–2000]. The time constants are specified according to the ISO 8601 format [3], corresponding to the XML date data type. The introduction of validity contexts conforms with the XML schema, named `ValidSchema.xml`, displayed in Fig. 1. Such a schema should be included in any XML temporal document (in the way usual for XML-Data [27] schemas), in order to ensure syntactic checks on the well-formedness of temporal documents to be automatically effected by the XML-enabled browers[1]. The adoption of an XML schema instead of a Document Type Definition (DTD [12]) is due to its flexibility and extensibility, in addition to the availability of predefined data types. Unlike a DTD, an XML schema is based on an open content model, and thus it can be applied to any kind of documents (irregular data), also containing XML elements defined by other schemas or even not defined anywhere. In this way, our proposal concerns a way of adding the timestamping facility to generic XML documents, also just containing plain HTML code. Any existing HTML-based Web site can thus easily be made temporal, by adding XML <valid> timestamps to its documents in accordance to the ValidSchema.xml schema above. Therefore, the purpose of our proposed timestamping schema is (at least) twofold:

- it can be used to make temporal "legacy" Web sites (featuring HTML multimedia documents), in order to support the representation of historical information, and enabling a temporally selective navigation with respect to information validity;
- it can be used to make temporal the emerging deployment of XML for data representation and exchange on the Web, in order to support the management of temporal structured or semistructured data, and enabling the utilization of functionalities as developed by temporal database research (e.g. TSQL2-like temporal query languages);

2.2   Temporal Browsing and Navigation

The default valid time used for the Web navigation is usually the whole time range, enabling to view the full contents of documents, which corresponds to the basic use of non-temporal (standard) documents. For a selective temporal navigation, the time range can be reduced by the user (e.g. via some browser facility). It does not seem too restrictive to adopt a time *interval* as validity context, since also most queries in valid-time databases are commonly based on the comparison with an interval. Once set up by the user, such validity interval is known by the Web client as a *navigation validity context*. When temporal documents are processed, only the parts whose valid timestamp *overlaps* the navigation validity context are effectively taken into account and displayed. The same is automatically applied when new documents are retrieved by following a link, enabling a full-fledged temporal navigation.

In our proposal, the valid-time selection relies on the adoption of an XSL stylesheet [19], named Valid.xsl, to perform a dynamic filtering of the document contents according to the navigation context. Such stylesheet embeds XSL

---

[1] This also applies to the date constants specified as from/to attribute values. Unfortunately, the Ie5 parser and DOM do not currently support the date type and, thus, do not effect correctness checks on date values when applying the XML schema.

```
<?xml version="1.0" ?>
<xsl:stylesheet xmlns:xsl="http://www.w3.org/TR/WD-xsl">
<xsl:template> <!-- identity transformation template -->
   <xsl:copy>
      <xsl:apply-templates select="@*|*|comment()|pi()|text()" />
   </xsl:copy>
</xsl:template>
<xsl:template match="valid"> <!-- recursive selection template -->
   <xsl:choose>
      <xsl:when test="validity[condition on from and to attribute values]">
         <xsl:copy>
            <xsl:apply-templates select="@*|*|comment()|pi()|text()" />
         </xsl:copy>
      </xsl:when>
      <xsl:otherwise>
         <xsl:apply-templates select="*//valid" />
      </xsl:otherwise>
   </xsl:choose>
</xsl:template>
</xsl:stylesheet>
```

Fig. 2. The XSL `Valid.xsl` stylesheet.

transformations [30] and adopts the XML path language [28] facilities implemented in Ie5 (see [29] for reference). The definition of the stylesheet can be seen in Fig. 2: the first part consists of a simple identity-transformation template, whereas the second part is devoted to the temporal selection of the contents of validity contexts. The processing of the new XML `<valid>` element causes the output of the element contents when a validity selection condition (involving the document element timestamp) is verified. For instance, if the condition has the form "`@from[. $le$ '1999-12-31'] and @to[. $ge$ '1999-01-01']`", each `<valid>` element whose validity overlaps year 1999 is included in the stylesheet output: the selection condition matches any `<validity>` element where the `from` attribute value is $\leq 1999/12/31$ and the `to` attribute value is $\geq 1999/1/1$.

The particular structure of the selection template causes the execution of a test involving the navigation context and all the `<validity>` timestamps found in the current `<valid>` element. The conditional processing uses the `xsl:choose` instruction which provides for an `xsl:otherwise` case (not supported by the `xsl:if` XSL element), in order to recursively look for nested validity contexts. The `xsl:when` instruction is activated if at least one of the intervals (corresponding to a `validity` element) belonging to the timestamp satisfies the selection condition. The `xsl:otherwise` instruction is activated only when none of the timestamps of the current `<valid>` environment satisfies the selection condition.

When the navigation validity context is changed by the user during his/her navigation, such a condition should dynamically be changed in the stylesheet. If the stylesheet is then re-applied to the document, also the document visualization changes to reflect the user's action. For instance, in our prototype implementation, which is based on Ie5, the stylesheet is actually changed to update the selection condition and re-applied to the document by means of the DOM methods' functionalities (see [4] for details). In this way, the change is actually effected on the stylesheet copy loaded (as an XML document object) in the main memory space managed by the browser or on a copy of the stylesheet cached on a local disk of the machine on which the browser is running.

Notice also that more complex temporal selections than the simple overlap could be implemented by applying a different condition or even by defining a different filtering template in the stylesheet. Therefore, also sophisticated temporal query and reasoning facilities could easily be added to a temporal Web site by means of appropriate direct management of the Valid.xsl temporal selection condition. As a matter of fact, as it is based on the Ie5 XSL support (with Microsoft XSLT [30] and XPath [28] extensions) our proposal represents a straightforward temporal extension of the XQL query language [13]; extension that has been designed as an application of XQL itself. For instance, in a related application [4], we showed how XSL filters can be used to support all the TSQL2-like selection predicates [14] over XML temporal data.

3   A Reference Application: The Temporal Web Museum

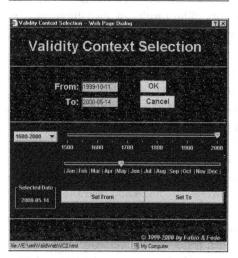

Fig. 3. Selection of a validity context.

The proposal outlined in this paper concerns the addition of the valid-time dimension to (HTML/XML) Web documents and is finalized to support temporal navigation in virtual environments which are sources of historical information. An extremely appropriate example of such an environment is a Web museum, where temporal selective browsing allows the definition of personalized visit paths through centuries and artistic or historical periods within the museum collections. In order to plan a visit, we can act on valid time selection to change the historical period of interest. For instance, we can choose the High Renaissance period, by selecting the validity range 1495–1520. Hence, we may start our virtual visit entering some virtual hall or gallery: only the temporally relevant paintings or sculptures would be present; by changing the validity context we could see some works vanish and some different works materialize. For example, in a hall dedicated to the Italian High Renaissance, we could view the evolution of the painting styles of Leonardo da Vinci, Raphael, Michelangelo and Titian and, say, have a look to works contemporaneous to the Mona Lisa picture.

The interest for museum applications on the Internet is constantly growing. This is shown by the increasing number of available museum sites and by the development of a specific discipline [1], with dedicated journals (e.g. *Archives and Museum Informatics*) and conferences (e.g. *Museums and the Web*). The "Web Museum" [11], authored and maintained by Nicholas Pioch, was one of the very first to open and is probably the most popular virtual museum on-line. It is basically a collection of image data representing famous paintings, heterogeneous as to their origin, which can accessed, for example, via an artist or a theme index. In order to test our proposal, we realized a temporal version of a subset of the Web Museum pages and developed a Web environment for the temporal browsing of its collections. The prototype, named "The Valid Web" [5], consists of a Web site which can be browsed with Ie5 (see Fig. 4). The pages of the site are organized in two frames. A small service frame in the bottom part of the window contains all the required controls to deal with the user interactive specification of the validity context to be used for temporal navigation, including the visualization of the current validity context. All the controls are implemented as JavaScript functions. A larger frame, occupying almost all the browser window space, is used to display temporal documents, that is the results of the temporally selective filtering effected by the `Valid.xsl` stylesheet on timestamped XML documents. The results of such filtering is a plain HTML document which is then rendered by the browser as usual.

In general, the valid-time selection implies the choice of an interval. This can be done by an independent choice of the two time points representing the interval bounds. The selection of each interval bound can be based, for instance, on a graphic *scrollbar* or *slider* for analog fine selection of a time-point (at a given granularity level). In our prototype implementation, time-points are dates (i.e. the granularity level is the day) and the selection of an interval can be effected by means of a Java JFC/Swing applet [22,6], which contains two graphic sliders: the former to select the year and the latter to select the day of the year (see Fig. 3). The former slider, for the user's convenience, has a 500-year range, which can be changed (from 0–500 to 2000–2500) by means of a *multiple-choice menu* available next to the year slider. Assume we have to fix a date, say 1996/3/7. We can start by choosing the year 1996 with the former slider (with the default range 1500–2000 set) and then choose the March, 7 date with the latter slider. The chosen value can then be assigned to the From or To interval bound by means of the corresponding "Set" *button*. However, also editable input fields

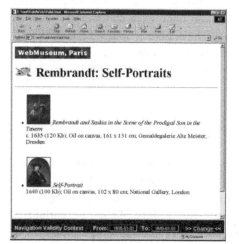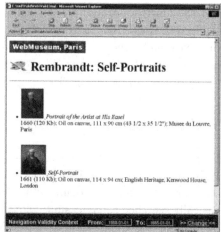

Fig. 4. Temporal navigation of the Web Museum.

for direct typing of a valid time value (in the "YYYY-MM-DD" string format) are always available in the dialog window containing the running applet. The communication between the applet and the JavaScript control functions in the calling service frame (e.g. to return the selected validity context) is managed by means of the LiveConnect package [10] supported by the Java Plug-in 1.2.2 [21]. Once the navigation validity context choice is confirmed by the user, the temporal selection over the currently displayed document is automatically re-executed by means of the DOM mechanism described in the previous Section. Furthermore, in order to enable a full-fledged temporal navigation, each time the user changes the displayed document in the usual way (e.g. by following a link), the current validity context is automatically "inherited" by the newly loaded page, if also the new document is a temporal XML one. This behaviour is forced in our prototype thanks to the dynamic HTML facilities supported by Ie5. In fact, we used a slightly modified `Valid.xsl` stylesheet with respect to Fig. 2. The actual stylesheet implements a dynamic callback mechanism by inserting some JavaScript code in the preamble of the processed document. Such a script provokes, on load of the document by the browser, the immediate activation of the temporal selection functionality: as when the validity context is changed, the `Valid.xsl` filter is updated on the fly (to include the overlap with the current validity context as selection condition) and then re-applied to the displayed document by means of DOM method calls.

For example, Fig. 4 shows two snapshots of the navigation of a sample page containing Rembrandt's self-portraits. The full page contains seven pictures, dating from 1629 to 1669. Fig. 4 shows on the left the page when the validity context has been set to [1635–1640] and only two pictures are visible (the third and the fourth one) and on the right the same page with the validity context changed to [1660–1665] and two other pictures have been displayed (the fifth and the sixth one). The current navigation context is always visible in the bottom service frame of the window. The "Change" command on the right is a link that activates the applet of Fig. 3.

4   Conclusions

In this paper we outlined an approach for the integration of valid time into the Web. We addressed the issues concerning the extensions required to document structure and format, document processing and temporal browser usage. We basically sketched some feasible solutions and gave an idea of their potentialities by means of the application to a Web Museum. The solution we proposed relies on simple XML-based extensions: the introduction of a new <valid> markup tag with the definition of an XML schema for the creation of temporal documents, and the use of an XSL stylesheet for selective filtering of temporal documents according to a user-defined navigation validity context. We also developed a prototype temporal Web site, accessible with Ie5, which implements the proposed solution and provides tools for the management of the validity context with a friendly user interface. Also legacy HTML-based Web sites can easily be made temporal by converting them into XML documents and adding the required timestamps to the historical multimedia information they contain. Our proposal is fully compatible with the nowadays Web technology and can be enjoyed by millions of users all over the world.

Notice that the adoption and use of the time dimension is a powerful weapon to improve the selectivity of certain information searches over the net, which could also enhance the usefulness and functionality of already available navigation aids (e.g. Web search engines). The setting of a suitable validity coordinate in browsing a Web site containing historical information (e.g. a digital library or a virtual museum, but also an archive of newspaper issues or stock quotes) would improve the search quality: only relevant information would be displayed, instead of being immersed in a mass of temporally heterogeneous and non-pertinent stuff. This is made possible by the temporal semantics, which is carried by the new timestamping XML tags and allows the selection of information pieces in a temporally marked multimedia document via the evaluation of even complex temporal predicates. This is very different indeed from the textual search for matching date strings available with a search engine.

Future work will be devoted to the extension of the presented insfrastructure to include more sophisticated temporal representation and retrieval facilities with the addition, for instance, of a TSQL2-like support [14] for temporal indeterminacy, multiple granularities and calendars.

References

1. Archives & Museums Informatics, www.archimuse.com.
2. T. Berners-Lee, R. Cailliau, A. Lautonen, H.F. Nielsen, A. Secret, "The World Wide Web," *Commun. of the ACM* 37(8), 1994.
3. Date and Time Formats, W3C Note, www.w3.org/TR/NOTE-datetime.
4. F. Grandi, F. Mandreoli, "The Valid Web: it's Time to Go...", TR-46, TimeCenter Tech Rep, 1999, *available from* www.cs.auc.dk/research/DP/tdb/TimeCenter/.
5. F. Grandi, F. Mandreoli, "The Valid Web ©", *Proc. of Software Demonstrations Track at the EDBT'2000 Intl. Conference*, Konstanz, Germany, March 2000.
6. Java Foundation Classes (JFC), java.sun.com/products/jfc/.
7. C.S. Jensen, J. Clifford, R. Elmasri, S.K. Gadia, P. Hayes, S. Jajodia (eds.) *et al.*, "A Consensus Glossary of Temporal Database Concepts - February 1998 Version," in O. Etzion, S. Jajodia and S. Sripada (eds.), *Temporal Databases - Research and practice*, LNCS N. 1399, Springer-Verlag, 1998.
8. C.S. Jensen, M.D. Soo, R.T. Snodgrass, "Unifying Temporal Data Models via a Conceptual Model," *Information Systems* 19(7), 1994.
9. N. Kline, "An Update of the Temporal Database Bibliography," *SIGMOD Rec.* 22(4), 1993.
10. LiveConnect, in *JavaScript Guide*, Ch. 5, Netscape Communications, developer.netscape.com/docs/manuals/communicator/jsguide4/livecon.htm.
11. N. Pioch, "The Web Museum," www.cnam.fr/wm/.
12. Prolog and Document Type Declaration, in *Extensible Markup Language (XML) 1.0*, W3C Recommendation, www.w3.org/TR/REC-xml#sec-prolog-dtd.
13. J. Robie, J. Lapp, D. Schach, "XML Query Language (XQL)," *Proc. of QL'98*, Boston, MA, Dec. 1998, www.w3.org/TandS/QL/QL98/pp/xql.html.
14. R.T. Snodgrass (ed.) *et al.*, *The TSQL2 Temporal Query Language*, Kluwer Academic Publishers, Boston, Massachussets, 1995.
15. M.D. Soo, "Bibliography on Temporal Databases," *SIGMOD Rec.* 20(1), 1991.
16. A. Tansel, J. Clifford, V. Gadia, S. Jajodia, A. Segev, R.T. Snodgrass (eds.), *Temporal Databases: Theory, Design and Implementation*, Benjamin/Cummings Publishing Company, Redwood City, CA, 1993.
17. The Document Object Model (DOM) Home Page, www.w3.org/DOM/.
18. The Extensible Markup Language (XML) Res. Page, www.w3.org/XML/.
19. The Extensible Stylesheet Language (XSL) Res. Page, www.w3.org/Style/XSL/.
20. The HyperText Markup Language (HTML) Home Page, www.w3.org/MarkUp/.
21. The Java Plug-in Home Page, java.sun.com/products/plugin/
22. The Java Technology Resource Page, java.sun.com.
23. The JavaScript Resource Page, developer.netscape.com/tech/javascript/.
24. The Microsoft Internet Explorer Home Page, microsoft.com/windows/Ie/.
25. The World Wide Web Consortium (W3C) Home Page, www.w3.org/.
26. V.J. Tsostras, A. Kumar, "Temporal Database Bibliography Update," *SIGMOD Rec.* 25(1), 1996.
27. XML-Data, W3C Note, www.w3.org/TR/1998/NOTE-XML-data.
28. XML Path Language (XPath) Version 1.0, W3C W. Draft, www.w3.org/TR/xpath.
29. XSL Developer's Guide, msdn.microsoft.com/xml/xslguide/.
30. XSL Transformations (XSLT) V 1.0, W3C Working Draft, www.w3.org/TR/xslt.

Application of Metadata Concepts to Discovery of Internet Resources

Mehmet Emin Kucuk[1], Baha Olgun[2], and Hayri Sever[2]

[1] The Department of Library Science
Hacettepe University
06532 Beytepe, Ankara, Turkey
[2] The Department of Computer Engineering
Hacettepe University
06532 Beytepe, Ankara, Turkey

Abstract. Internet resources are not yet machine-understandable resources. To address this problem a number of studies have been done. One such a study is the Resource Description Framework (RDF), which has been supported by World-Wide Web (WWW) Consortium. The DC (Dublin Core) metadata elements have been defined using the property of extensibility of RDF to handle electronic metadata information. In this article, an authoring editor, called H-DCEdit, is introduced. This editor makes use of RDF/DC model to define contents of Turkish electronic resources. To serialize (or to code) a RDF model, SGML (Standard Generalized Markup Language) has been used. In addition to this work, a possible view of RDF/DC documents is provided using Document Style Semantics and Specification Language (DSSSL) standard. H-DCEdit supports use of Turkish language in describing Internet resources. Isite/Isearch system developed Center for Networked Information Discovery and Retrieval (CNIDR) organization with respect to Z.39.50 standard is able to index documents and allows one to query the indexed terms in tagged elements (e.g., terms in RDF/DC elements). In the scope of our work, the localization of this Isite/Isearch system is completed in terms of sorting, comparison, and stemming. The feature of supporting queries over tags provides basis for integrating H-DCEdit authoring tool with Isite/Isearch search engine.

Keywords: Discovery of Internet Resources, Web mining, RDF/DC authoring editor.

1 Introduction

As it is witnessed by everybody, the Internet has been growing rapidly since its inception in December 1969 and we are experiencing a boom in electronic information availability on the Internet. 1996 statistics showed that the number of Web sites were doubled in 100-125 days. Again, 1997 statistics proved that the number Web sites doubled in every 4 months. According to a survey of the Internet Valey Inc., in January 1998, there were 29.67 million hosts on the Internet with 2.5 million domains and 2.45 million Web sites [2]. In year 2000, it

T. Yakhno (Ed.): ADVIS 2000, LNCS 1909, pp. 304–313, 2000.

was estimated that 50 million computers are connected to the Internet, there are 304 million Internet users and 200 million Internet documents and 7 million Web sites are available on the Internet (Nua surveys. [online] http://www.nua.ie/ [2000, May 5]).

Obviously, organizing this huge amount of information sources and retrieving the desired information are not easy matter. Although the Internet and its most commonly used tool WWW represent significant advancement to retrieve desired information, we are experiencing some retrieval and information dissemination problems on the Internet. The major search engine companies have often claimed that they can keep up with the size of the Web, that is, that they can continue to index close to the entire Web as it grows. The reality however shows different story: according to a survey on the search engines, Northern Light covers 16covers 15covers 2.5that the entire Web is not covered and indexed by the engines [5].

In addition to the coverage, precision and recall rate are the other problematic area. Recall and precision factors of engines frequently involve precision factors of much less than 1 percent. For example, a search of the WWW using search engine ANZWERS on acronym "IETF" (which stands for Internet Engineering Task Force) retrieved 100,888 match on 12 April 2000, 91,017 match on 5 August 1999, 896,354 match in 1998. Every Web page which mentioned the IETF in an incidental way retrieved by this search. This example illustrates that search engines can return a lot of irrelevant information because they have no means (or very few means) of distinguishing between important and incidental words in document texts [3].

1.1   What Is Metadata ?

There are many definitions of the term metadata. Literally meaning *information about data,* the most simplified and referred definition is *data about data.* One recognizable form of metadata is the card catalogue in a library; the information on that card is metadata about a library material (e.g., book, journal, proceeding, etc.) Regardless of the context being used for publishing metadata, the key purpose remains the same which is to facilitate and improve the retrieval of information. In an environment such as the traditional library, where cataloging and acquisition are the sole preserve of trained professionals, complex metadata schemes such as MARC (Machine Readable Cataloging) are, perhaps, acceptable means of resource description [6].

1.2   Metadata Standards

There are variety of metadata standards such as Content Standards for Digital Geospatial Metadata, Encoded Archival Description, Text Encoding Initiative, Categories for the Description of Works of Art, Dublin Core (DC) and etc. Some standards have been developed to describe and provide access to a particular type of information resource, such as geospatial resources. Other standards such as DC, have been developed to provide a standard way of describing a wide range of different types of information resource, allowing these diverse types to be retrieved through a single searching process.

There will always be variety of metadata standards. However, DC is the most commonly used and promoted metadata standard since it aimed all types of documents in any subject. The DC (http://purl.org/dc/) metadata element set was developed during 1995 and 1996 and has been supported by the W3 Consortium. DC Element Set includes 15 data elements which are listed as Title, Author or Creator, Subject and Keywords, Description, Publisher, Other Contributors, Date, Resource Type, Format, Resource Identifier, Source, Language, Relation, Coverage, and Rights Management.

1.3   Tools for Metadata Creation

Metadata creation tools can be categorized as editors and generators. The first type, loosely labelled as *editor,* allows the creation of metadata by providing a template for entering new metadata content. The supporting software places the content into the HTML tag, which may then be *cut* and *pasted* into a document [1].

The second type, loosely labelled as *generator,* extracts metadata from existing HTML- encoded documents and places the content into the HTML tag. The generators have a choice of outputs: for example, some produce HTML v3.2, or HTML v4.0 and some generate XML for immediate inclusion in a repository (Meta Matters: Tools [online] http://www.nla.gov.au/meta/tools.html [2000, July 25]).

2   H-DCEdit

DC element set was translated into 22 languages including Turkish [2] [10]. However, translating DC element set into a different language without supporting this particular language by a metadata creation tool does not make a significant improvement in identifying and retrieving the documents in this particular language. As a response to the need to improve retrieval of electronic information in Turkish, H-DCEdit Software based on RDF/DC[3] standard was developed as a part of KMBGS (Kaşgarlı Mahmut Bilgi Geri-Getirim Sistemi) research project at the Department of Computer Engineer, Hacettepe University in 1999.

Our objection with the H-DCEdit editor is to define the metadata content of Internet sources using RDF model and DC elements. As shown in Figure 1, the

---

[1] It is important to note that the metadata language is an internal representation used by the system, not for end users to specify directly. This is similar to PostScript, which is used to specify page layouts for printers. Users do not generate PostScript directly, instead they use a graphical application to automatically generate it [9].

[2] Currently, Turkish version of DC has only Turkish counter-parts of some DC elements.

[3] RDF/DC stands for Resource Description Framework on top of which the controlled vocabulary of Dublin Core has been defined.

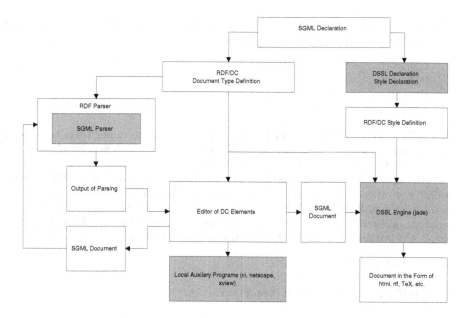

Fig. 1. Functional View of H-DCEdit System Model

H-DCEdit editor utilizes the SGML/XML [4] concrete syntax notation with the version of ISO 8879. The functionality of H-DCEdit editor depends on SGML declaration including name spaces of XML. We have used core RDF structs excluding the rules for statements about statements, where the predicate field of <subject, predicate, object> triple concerns with creation of sets or bags objects of the subject. The components filled with gray color in Figure 1, indicate names of freeware modules, which are available for Unix platform and incorporated into implementation of H-DCEdit editor whose components are explored in the rest of this section.

The SGML declaration was designed in such a way that it is not only consistent with the reference concrete syntax, but also supports use of Turkish characters. The eight-bit character set of document is based on two character sets: ISO646 and ECMA-128 (ISO Latin 5) for the positions of 0-127 and 128-255, respectively. In the RDF model, XML namespaces are used for schema specification as well as for control of vocabulary.

An element name is associated with a XML namespace; hence, its structure and correct interpretation become possible to software agents (or robots), even if namespaces are nested in describing an Internet resource[5]. This kind of associations in XML is denoted in the form of <schema:element name> and

---

[4] Since, as discussed later, SGML's type declaration was designed in consistent with some XML notation, it is called SGML/XML, instead of SGML.

[5] any object that has a valid URL address is called Internet resource.

becomes part of document type definition by introducing the following clause to the NAMING section of SGML declaration.

```
LCNMSTRT     ""
UCNMSTRT     ""
LCNMCHAR     "-.:"
UCNMCHAR     "-.:"
```

The LCNMSTRT and UCNMSTRT articles define lower and upper letter to be used for naming characters. By adding the characters of -.: into both articles respectively, these characters are conceived as naming characters.

The document type definition of RDF/DC notation is provided using SGML declaration. On other words, both XML compatibility and Turkish character support is also provided to RDF/DC notation for the reason that XML is used as a serialization language of the RDF/DC model that is, in turn, represented by acyclic and directed graph. Critical sections of RDF/DC document type definition is given as follows.

```
<!--    RDF Elements -->

<!ELEMENT rdf:RDF - -  ( rdf:Description )* >
<!ATTLIST rdf:RDF
    xmlns:rdf CDATA "http://www.w3.org/RDF/"
    xmlns:dc  CDATA "http://purl.org/DC/"
>
```

This recursive coding (Kleene closure of <rdf:Description>) provides building blocks for document content enclosed by <rdf:RDF> and < /rdf:RDF> tags. The attributes of <rdf:RDF> element are listed as <xmlns:rdf> and <xmlns:dc> with contents of http://www.w3.org/RDF/ and http://purl.org/DC/, respectively. The content model of <rdf:Description> would be any property as shown below.

```
<!ENTITY % property "ANY">
<!ELEMENT rdf:Description - - %property;>

<!--    DC Elements     -->

<!ENTITY % dccontent "(#PCDATA)">

<!ELEMENT DC:TITLE          - -    %dccontent;  >
<!ELEMENT DC:CREATOR        - -    %dccontent;  >
<!ELEMENT DC:SUBJECT        - -    %dccontent;  >
<!ELEMENT DC:DESCRIPTION    - -    %dccontent;  >
<!ELEMENT DC:PUBLISHER      - -    %dccontent;  >
<!ELEMENT DC:CONTRIBUTOR    - -    %dccontent;  >
<!ELEMENT DC:DATE           - -    %dccontent;  >
<!ELEMENT DC:TYPE           - -    %dccontent;  >
<!ELEMENT DC:FORMAT         - -    %dccontent;  >
<!ELEMENT DC:IDENTIFIER     - -    %dccontent;  >
<!ELEMENT DC:SOURCE         - -    %dccontent;  >
<!ELEMENT DC:LANGUAGE       - -    %dccontent;  >
<!ELEMENT DC:RELATION       - -    %dccontent;  >
```

```
<!ELEMENT DC:COVERAGE       - -    %dccontent;  >
<!ELEMENT DC:RIGHTS         - -    %dccontent;  >

<!ATTLIST DC:DATE
     year       CDATA  #IMPLIED
     month      CDATA  #IMPLIED
     day        CDATA  #IMPLIED>

<!ATTLIST DC:RELATION
      type  CDATA #IMPLIED
      resource  CDATA #IMPLIED >
```

This parametric property provides us to define DC elements as properties of RDF elements, but at the same time, allows possible extensions in terms of new element definitions that might be needed in the future. The attribute list of ¡rdf:Description¿ is declared as follows.

```
<!ATTLIST rdf:Description
          ID          NMTOKEN     #IMPLIED
          about       CDATA       #IMPLIED
          aboutEach     CDATA        #IMPLIED
          bagID       NMTOKEN     #IMPLIED
>
```

The above attributes describe (or identify) the rdf objects as discussed in[7]. Let us give a simple RDF model of an Internet resource in Figure 2. Here, the source object, sgml.cs.hun.edu.tr is described by RDC/DC elements in the form of <source, dc elements, value>. For this example, the output of HDC-Edit is given in the Appendix (GUI based user interfaces were developed using MOTIF on Unix platform. Because of the size limitation of the paper, we refer the reader to home page of KMBGS Project for appearances of these interfaces [8].

The RDF/DC parser component incorporates with the SP package (SGML parser) via API (application program interface) to parse an SGML document complying with type definition of RDF/DC. The compilation process yields an intermediate output that is in suitable form to the RDF/DC editor whose browser module in turn displays that output.

DSSSL standard contains three main sections, namely transformation, style, and query. The HDC- Edit uses the style utiliy of DSSL engine which was embedded into the SGML processing system. Note that the SGML declaration and document type definition is also valid for the DSSSL engine. The SGML document generated by RDF/DC editor can be translated to well-known languages like RTF, TeX, HTML along with its CSS (cascaded style sheet).

3   Application

In this section, we introduce a traditional way of exploiting metadata. Specifically we use Isite/Isearch search engine, which is a freeware software developed by CNIDR (Center for Networked Information Discovery and Retrieval). Database of information items given in Figure 4 consists of RDF/DC documents generated by H-DCEdit editor (strictly speaking this is not a true picture, because RDF/DC documents were

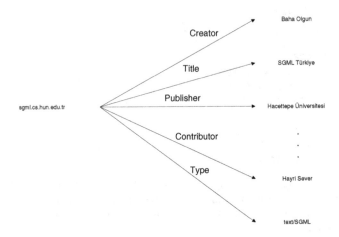

Fig. 2. Content representation of a simple Internet resource by RDF model

translated to HTML using DSSSL engine of SGML processing system). This back-end database can be populated by remote Internet resources using a Web crawler. Note that the choice of Isite/Isearch is twofold. The first one is that it allows users to query over contents of tagged elements, which is something we need for DC elements. The other is that it operates on Z.39.50 networked information retrieval standard whose support to DC model (in addition to MARC records) is imminent.

As a simple application we have cataloged our small departmental library using H-DCEditor and indexed using Isite system. We encourage the reader to take a look at the query interface and view source of documents located at http://ata.cs.hun.edu.tr/~km/gerceklestirim.html. We will not go through details of the integration of Isite/Isearch system with SGML processing system, but, instead, give a brief description of this search engine along with the design of a Web crawler.

3.1   Web Crawler

A Web crawler has been designed and implemented to retrieve documents from the World Wide Web and create a database. The implementation has been done in JAVA to effectively utilize the power of its platform independence, secured access and powerful networking features.

As seen in Figure 3, the Web crawler agent is composed of three tightly-coupled submodules, namely Downloader, Extractor, and Parser. Initially the downloader starts with a seed (root) URL and then navigates Web catalog by a breadth-first expansion using the queue for URLs to be traversed. It retrieves Web documents via hypertext transfer protocol (http) and in the process, passes the HTML document to the extractor. The detection of bad links are handled by the downloader by trapping the HTTP status code returned by the HTTP server of the URL to be visited. The hyperlink extractor detects the hyperlinks in the web documents, extracts them and passes

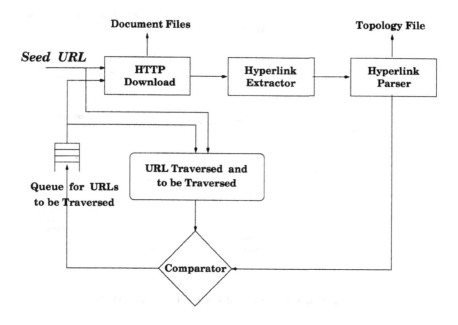

Fig. 3. The View of Web Crawler

them to the parser for further processing. It searches for HTML tags of the form
$< a \quad href = "http : // \cdots" >$ or $< frame \quad src = " \cdots " >$. The parser converts these
relative URLs to absolute URLs following the Internet Standards (RFC 1738, 1808)
drafted by the Network Working Group. No parsing is necessary for absolute URLs.
The host name (more specifically, the domain name of the network host) of each of
these URLs is then compared with the host name of the seed URL and only those URLs
whose host name match with that of the seed URL are added to the queue. Care should
be taken so that any new URL that is added to the queue is not a repetition of any of
the URLs that has been already traversed or will be traversed. Thus the Web crawler
retrieves all the web documents within the site specified by the root URL. URLs added
by the parser to the queue are restricted to certain specific types only and any URL to
postscript (.ps), image (.gif, .jpeg, .jpg, etc.), compressed (.gz, .tar, .zip, etc.) files
are not added.

3.2   Retrieval Engine

Functional diagram of descriptor-level retrieval is shown in Figure 4. The retrieval of
documents using the keywords and matching them to descriptors is called descriptor-
level retrieval. For descriptor-level document/query processing we use the Isite/Isearch
system. Isite database is composed of documents indexed by Iindex and accessible by
Isearch. Isite/Isearch allows one to retrieve documents according to several classes of
queries. The simple search allows the user to perform case insensitive search on one
or more search elements, e.g. the use of titles/headers in matching search terms with
document terms. Partial matching to the left is allowed. The Boolean search allows

312   M.E. Kucuk, B. Olgun, and H. Sever

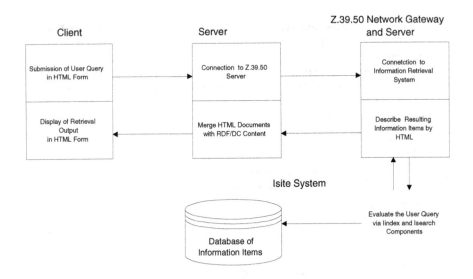

Fig. 4. Functional View of H-DCEdit System Model.

the user to compose a two-term query where the two terms are related by one of the Boolean operators AND,OR, and ANDNOT. "Full Text" is the default search domain unless the user selects a particular element for a term from the term's pull-down menu. The advanced search form accepts more complex Boolean queries that are formed by nesting two-term Boolean expressions. To narrow a search domain from *Full Text* to a single search element, the term is prefixed with the element name and a forward slash, e.g., DC:TITLE/metadata. Format and size of the information items listed in a retrieval output may be specified by choosing target elements from a pull-down menu. Isearch is capable of performing a weighted search that employs the inverse document frequency of terms.

As a final note, we have made some patching to Isite/Isearch to enable indexing of Turkish documents (in terms of stemming, sorting and comparison) as well as use of Turkish characters in expression of user queries[1]. A new module of stemming based on G. Duran's graduate thesis was added, which can be considered as a supplementary module to Isite/Isearch[4].

Acknowledgment. This work is partly funded by T.R. State Planning Organization under the project number of 97K121330 (please, refer to the home page of KMBGS Project at http://www.cs.hacettepe.edu.tr/~km for detailed information).

Appendix: The Output of HDC-Edit for the Example in Figure 2

```
<!DOCTYPE RDF:RDF PUBLIC "-//Baha Olgun//DTD RDF and DC//EN">
<RDF:RDF xmlns:rdf="http://www.w3.org/RDF/"
```

```
        xmlns:dc="http://purl.org/DC/">
<RDF:Description about="http://sgml.cs.hun.edu.tr">
<DC:IDENTIFIER>http://sgml.cs.hun.edu.tr</DC:IDENTIFIER>
<DC:CREATOR>Baha Olgun</DC:CREATOR>
<DC:TITLE>SGML T&uuml;rkiye Kullan&iwhdot;c&iwhdot;lar&iwhdot;</DC:TITLE>
<DC:SUBJECT>SGML T&uuml;rkiye</DC:SUBJECT>
<DC:DESCRIPTION>SGML T&uuml;rkiye Web Sayfas&iwhdot;</DC:DESCRIPTION>
<DC:PUBLISHER>Hacettepe &Uuml;niversitesi</DC:PUBLISHER>
<DC:CONTRIBUTOR>Hayri Sever</DC:CONTRIBUTOR>
<DC:RIGHTS>Her Hakk&iwhdot; Sakl&iwhdot;d&iwhdot;r</DC:RIGHTS>
<DC:TYPE>text</DC:TYPE>
<DC:FORMAT>text/sgml</DC:FORMAT>
<DC:LANGUAGE>tr</DC:LANGUAGE>
<DC:DATE year=2000 month=Jan day=01></DC:DATE>
<DC:RELATION resource=http://www.cs.hun.edu.tr type=IsReferencedBy>
</DC:RELATION>
</RDF:Description>
</RDF:RDF>
```

References

1. Akal, F. Kavram Tabanlı Türkçe Arama Makinası. Yuksek Muhendislik Tezi. March 2000. Fen Bilimleri Enstitusu, Hacettepe Universitesi, 06532 Beytepe, Ankara, Turkey.

2. Chowdhury, G.G. The Internet and Information Retrieval Research: A Brief Review. *Journal of Documentation*. March 1999, 55 (2): 209-225.

3. Connolly, D. Let a thousand flowers bloom. *In interview section of IEEE Internet Computing*. March-April 1998, pp. 22-31.

4. Duran, G. Gövdebul: Türkçe Gövdeleme Algoritması. Yuksek Muhendislik Tezi. Fen Bilimleri Enstitusu, Hacettepe Universitesi, 06532 Beytepe, Ankara, Turkey, 1997.

5. Lawrence, Steve and C. Lee Giles. Searching the World Wide Web. *Science*. 3 April 1998, 280: 98-100.

6. Marshall, C. Making Metadata: a study of metadata creation for a mixed physical-digital collection. *In Proc. of the ACM Digital Libraries'98 Conf.*, Pitsburgh, PA (June 23-26, 1998), pp. 162-171.

7. Miller, E. An introduction to the resource description framework. *D-Lib Magazine*. May 1998.

8. Olgun, B. Dublin Core Ustveri Elemanlari Editoru. Yuksek Muhendislik Tezi. Fen Bilimleri Enstitusu, Hacettepe Universitesi, 06532 Beytepe, Ankara, Turkey. 1999. This thesis is also available at http://www.cs.hacettepe.edu.tr/~km.

9. Singh, N. Unifying Heterogeneous Information Models. *ACM Comms*. May 1998, 41(5):37-44.

10. Weibel, S. The state of the Dublin Core metadata initiative. *D-Lib Magazine*. April 1999.

Evaluation of the Object-Orientation
of a Multi-agent System

Bora İ. Kumova

Dept. of Comp. Eng., Dokuz Eylül University, 35100 İzmir, Turkey
kumova@cs.deu.edu.tr

Abstract. In this work, we discuss the multi agent system AgentTeam from a software engineering perspective. AgentTeam is a framework for agent-based flexible management of distributed relational data stored in heterogeneous databases. In order to fit the inherently distributed application environment of the network, it is designed in an object-oriented fashion. It consists of domain-dependent system concepts and partially domain-independent software design concepts. The software design concepts are evaluated with respect to heterogeneity, scalability, and re-useability and discussed in detail on the prototype CourseMan.

1 Introduction

With the rapid growth of the network, the network capability of applications becomes irremissible. Furthermore, some systems presume a network, in order to be applicable, such as computer supported co-operative work (CSCW), networked information retrieval (NIR) [14], distributed database management (DDBM), or manufacturing systems. It is widely known that systems operating in a network environment should itself be inherently distributed. To achieve this requirement, multi-agent systems (MAS) have been proposed, extensively discussed in the literature, and successfully applied to various domains [25], [22], [5], [20], [11]. Since, related research projects' prototype systems are ad-hoc implementations that usually do not comply to well-known software design principles, such as [2], [9] the software design of related MASs is not sufficiently discussed in the literature. However, a detailed discussion of the software concepts of as MAS can explain the design concepts, visualise system structures, and can encourage software designers to utilise the agent design paradigm [17], [19]. Besides the agent design paradigm, at a lower conceptual level, object-orientation has been identified as a further key design approach for distributed systems [23], [27]. The aim at this work is to offer software engineers re-useable design concepts for designing a new system in form of a MAS.

We have designed the MAS framework AgentTeam for flexible DDBM. Where, flexible means that additionally to deterministic DDBM non-deterministic functionality is included [15]. Both, framework and the prototype CourseMan are object-oriented [16]. We evaluate the object-orientation of the software design concepts with respect to the software properties heterogeneity, scalability, and re-

T. Yakhno (Ed.): ADVIS 2000, LNCS 1909, pp. 314–323, 2000.

useability, which are critical for MASs. With heterogeneity we mean that a concept can be used to represent objects of a given application domain with slightly different functionality. Scalability means that a concept can be used multiple times inside a specific design. Re-useability means that a concept can be used in the design of systems of different application domains.

First, the AgentTeam framework is introduced, thereafter the class structure, object structures, state transitions, and the module structure of the CourseMan prototype. After a discussion of the work, we conclude with some final remarks.

2 Design Concepts of the AgentTeam Framework

AgentTeam have been already introduced elsewhere [10], [12], [13]. The basic concepts of the application domain and the basic domain-independent software design concepts are summarised here and explained briefly.

2.1   Application Domain Concepts

The framework AgentTeam is designed for flexible management of relational and distributed data stored in heterogeneous databases. Below, some basic concepts of this domain are listed.

Data Model: Data to be managed must be available in structured form, such as relational, object-oriented, or logical (knowledge base data). Data can be distributed over different sites.

Data Management: A DDB can consist of multiple shared databases stored in local autonomous DBMSs. Involved DBMSs need not to be aware of each other or of AgentTeam. The system can manage multiple DDBs at the same time [15].

Semantic Data Control: Consists of view management, data security, and semantic integrity control.

Distributed Query Processing: Depending on requested data and its location, a user transaction is transformed into one or more queries. Therefore, a query is processed and optimised principally distributed by the involved agents, but locally by the related DBMSs.

Distributed Transaction Management: The execution of a transaction is planned and broken down to queries at client site. A query is scheduled and executed at server site. Distributed concurrence control is implemented at server site.

User Model: Co-operating users can dynamically define a DDB and the degree of data consistency. Since, the system is flexible against data consistency, co-operating sites can be scaled to a large number.

2.2   Software Design Concepts

We introduce our software design concepts here in form of general abstractions, which are intended to be domain-independent.

Class: A class consists of data and functionality that, as a whole is a software abstraction over a group of real or imaginary objects.

Class Relationship: Class relationships define structural interdependencies between classes. We differentiate between three types inheritance, has, and using. The inheritance relationship, where data and methods of more general classes can be used by more specific classes. The has relationship, where a class can create instances of an another class. The using relationship, where a class can access instances of a class already created by another class. Class relationships are used to represent the static structures of a system.

Task: A task defines pre- and post-conditions. If the pre-conditions are satisfied, then some processing is started and continued, until the post-conditions are satisfied. A task represents some consistent functionality.

Instance: An instance is a sample object of a class. Usually, all instances of a class have the same structure, but different values assigned.

Object: An object stands for the group of all possible instances of a class. It is used to represent the dynamic structures of a class.

Object State: Object states document consistent states in the behaviour of an object. They visualise the dynamic structures of an object. A state transition is caused by a task.

Module: A module combines one or more objects and appears as a process started by the operating system.

Agent: An agent is a module that represents meta knowledge of its environment, which may consist of other modules. Therefore, it usually appears as an entity that autonomously acts in its environment. We define an intelligent agent as an agent with logical behaviour and communication capabilities.

Agent Knowledge Representation: This is the knowledge of the application domain represented in the behaviour of an agent.

Agent Communication Language (ACL): An ACL is a language, in which an agent can communicate with other entities of its environment. Sample ACLs are KQML [21], ARCOL [24], AgentCom. We have introduced the model of a bilingual agent that results from separating non-logical from logical behaviour of an agent [18].

Agent Communication Structure: These are module inter-dependencies inside the environment of a specific agent. They represent the possible messages between modules.

Agent Co-operation Structure: These are semantics defined over communication structures. They define how some communication structures need to be used to accomplish a task.

Agent Co-ordination Structure: A control structure is necessary for each co-operation structure. It synchronises conflicting messages inside a task or between different tasks.

Agent Behaviour: Agent behaviour is an abstraction that includes and generalises all above mentioned agent concepts. Sharing the behaviour of an agent can facilitate inter-agent communication [8].

Agent Property: An agent property qualifies a behavioural attribute. All agent properties constitute its behaviour. Agent properties are discussed in literature mainly from a logical perspective [6].

We believe that the above concepts are relative steady against heterogeneity, principally scalable to a large number, and are highly re-useable, since they have been built independently from any domain-specific concept. However, the application of a software concept to different domains can introduce different semantics, which is the next concern of our discussion. In the next chapters, above software concepts are interpreted in the context of DDBM.

3 Class Structure of CourseMan

A class diagram is used to show the existence of classes and their relationships in the design of the system. A single class diagram represents a view of the class structure of a system [1]. It visualises static structures of a system. CourseMan consists of various object-oriented classes, following classes and interrelationships were developed (*Figure 1*).

SemanticNode: Provides basic functionality to construct and modify a semantic node, and to find its attributes. It is used to store attributes, values, and procedures of an object.

SemanticNet: It creates and uses *SemanticNode* and provides the functionality to construct, store, and modify a semantic net, and to find nodes. It maintains a list of all stored templates, where each template represents a group of concepts, and a list of all stored concepts. A semantic net is the basic knowledge representation form of the system.

Template: A template is a semantic net that defines the skeleton for a group of semantic nets. It represents a class.

Instance: An instance is a semantic net structured according the related template. It represents an instance of the related template.

FIPA ACL: The agents use a simplified ACL that complies with FIPA ACL [4], to communicate at the logical level of their behaviour.

AgentCom: Implements the agent communication language AgentCom [18]. Uses *Template* and *Instance* to wrap between data represented in AgentCom in form of attribute-value pairs and its semantic net representation. AgentCom can also be viewed as an application programming interface (API) to access knowledge base objects. Agents use AgentCom to communicate on the non-logical level of their behaviour.

InferenceMechanism: Uses *Template* and *Instance* and implements mechanisms to evaluate the state of an instance that is the semantic net with the current focus. It determines firing conditions and executes related procedures.

Agent: This is a super class that forces its sub-classes to create one instance of *InferenceMechanism*. For each AgentCom message an *AgentCom* instance is created. It is further an abstract class that forces its sub-classes to implement some general agent properties, such as the bilingual communication capability.

UserAgent: It is a sub-class of Agent. It creates an *InferenceMechanism* object to evaluate *Template* and *Instance* objects. It can wrap between semantic net data and data to be displayed in the user interface. It creates one or more *TaskAgents* to execute user transactions and other tasks.

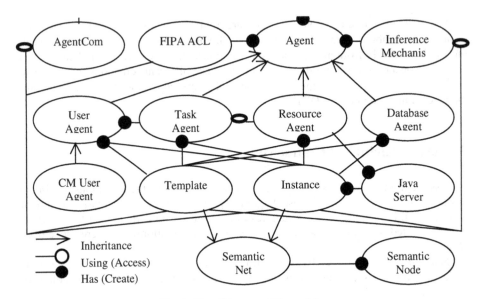

Fig. 1. Class Diagram of CourseMan

CMUserAgent: This is the CourseMan User Agent. It implements a default user interface for the CourseMan domain. It is a sub-class of *UserAgent*.

TaskAgent: It is a sub-class of Agent. It creates an *InferenceMechanism* object to evaluate *Template* and *Instance* objects. It uses remote created *ResourceAgents*. If the local machine is determined to serve data of a DDB, then it creates one local *ResourceAgent*. Divides a user transaction into one or more sub-transactions and establishes for each one an AgentCom session, which are stored in a list. However, only one session can be active at a given time

JavaServer: When started on server site, it invokes the local persistent *ResourceAgent*, if existent, otherwise a new *ResourceAgent* is created.

ResourceAgent: It is a sub-class of Agent. Creates one or more *DBAgents* and brokers AgentCom messages between them and *TaskAgents*. This class also implements the client site stub and the server site skeleton. It runs as a demon process that listens on Java RMI's standard TCP/IP port for incoming AgentCom messages.

DatabaseAgent: It is a sub-class of Agent. It creates an *InferenceMechanism* object to evaluate *Template* and *Instance* objects. It can wrap between semantic net data and SQL data.

4   Object Structures of CourseMan

An object diagram is used to show the existence of objects and their relationships in the design of a system. A single object diagram is typically used to represent a scenario [1]. It visualises static and dynamic structures of a system.

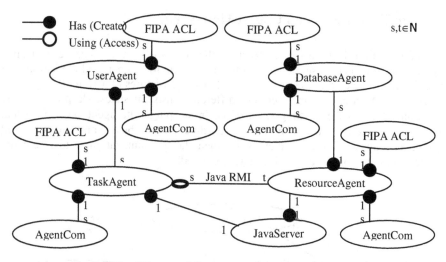

Fig. 2. Object Diagram of CourseMan (Communication Structure)

The MAS CourseMan consists mainly of four agent types, which communicate with each other in a well-defined communication structure (*Figure 2*). Most of the relationships between objects have variable cardinality, where a concrete cardinality depends on the state of an agent (*see next chapter*). Each agent type can create one or more instances of FIPA ACL and AgentCom, in order to communicate with another agent. The role of a task agent is to determine for a given task the necessary co-operation structures and to co-ordinate the communication. Where, co-ordination structures are determined by finding *ResourceAgents* and *DBAgents* necessary to accomplish a task. A simplified object structure is depicted in (*Figure 3*).

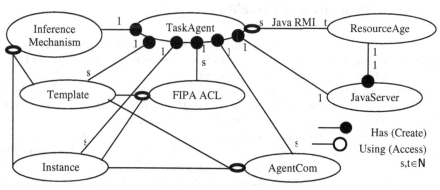

Fig. 3. Object Diagram of a CourseMan Task Agent

5 State Transitions of a CourseMan Agent

A state transition diagram is used to show the state space of an instance of a given class, the events that cause a transition form one state to another, and the actions that result form a state change [1]. It visualises dynamic structures of a system.

A CourseMan agent can enter four different consistent states: sleeping, waiting, communicating, and evaluating (*Figure 4*). In sleeping state agent stays persistent, until requested to do a task. If no task is scheduled, then a busy-wait state is entered. If data is requested from another agent, then the communication state is entered. While a computation the evaluation state is entered.

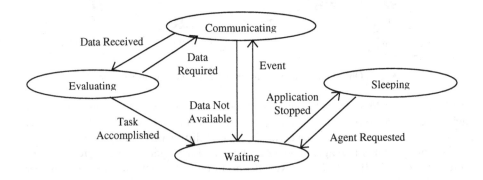

Fig. 4. State Transition Diagram of a CourseMan Agent

6 Module Structure of CourseMan

A module diagram is used to show the allocation of classes and objects to modules. A single module diagram represents a view of the module architecture of a system [1]. It visualises static and dynamic structures of a system.
The whole CourseMan system consists of two modules, one for a client site, and one for a server site. Scalability of modules is unrestricted for both, clients and servers (*Figure 5*). The whole code is written entirely in Java [7] scripts.

On client side, the user agent is embedded in an HTML page as an applet. At run-time, the user agent dynamically creates one or more task agents, which establish the network connection to resource agents at server sites. To receive messages from database servers asynchronously, the user agent creates a server side skeleton on the client that listens on the standard port of the RMI system. All programmes running at client site are compiled to Java applets.

On server side, the resource agent is introduced to the operating system as a service programme to be started at boot time. The resource agent is written as a Java application. With the first request to a database, it creates a database agent for that database. A database agent can access a database over a related ODBC configuration, by using Java's JDBC-ODBC Bridge. All programmes running at server site are bound to one Java application.

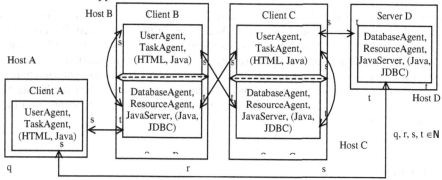

Fig. 5. Module Diagram of CourseMan (All Possible Configurations)

All agents implement the AgentCom language interface to communicate with each other. Communication between agents that are compiled to one code is synchronous. Communication between agents over the network can be established optionally in asynchronous mode.

7 Discussion

The debate on whether an application is an agent or not is subject of current research. Our definition is simplified, compared to others [3], but aimed at giving designers general concepts than complicated definitions.

The design of a bilingual agent is aimed at separating logical communication and behaviour from non-logical communication and behaviour. The goal of this approach is to find common logical communication semantics that are independent from any application domain. On the other hand, domain-specific behaviour that cannot be represented in logical form is implemented separately and communicated in AgentCom [18].

AgentTeam is designed for managing relational distributed data. Since, agent-base NIR systems have similar architectures [14], the re-useability of the whole framework in a variation is considered in a current project.

In the visualisation of the concepts we were motivated by widely used notations for object-oriented design [1], [26].

Finally, in this paper, we have not discussed implementation concepts. Nevertheless, it may be interesting to mention that scalability in CourseMan is due to the management of objects in thread and list.

8 Conclusion

We have discussed the object-oriented design of AgentTeam, which is a framework for multi-agent-based flexible DDBM. We have further discussed its prototype implementation CourseMan. We have evaluated the object-oriented and domain-independent design concepts with respect to heterogeneity, scalability, and re-useability. Our objective was to enlighten a sample MAS from different viewpoints, to visualise the inherently object-oriented system architecture, to show the highly re-useable concepts, and to encourage designers to utilise these concepts in other application domains.

References

Referenced URLs were valid by July 05, 2000 !
[1]   Booch, Grady; 1994; "Object-Oriented Analysis and Design - With Applications"; The Benjamin/Cummings Publishing Company, Redwood City, California
[2]   Cohen, Philip R.; Cheyer, Adam; Wang, Michelle; Baeg, Soon Cheol; 1994; "An Open Agent Architecture", *AAAI Spring Symposium*; pp. 1-8, March
[3]   Duego, Dwight; Oppacher, Franz; Ashfield, Bruce; Weiss, Michael; 1999; "Communication as a Means to Differentiate Objects, Components and Agents"; *TOOLS 30*; IEEE Computer
[4]   Foundation for Intelligent Physical Agents; 1999; "Agent Communication Language", FIPA 99 Draft 2-1999; http://www.fipa.org/spec/fipa99spec.htm
[5]   Genesereth, Michael R.; Keller, Arthur M.; Duschka, Oliver M.; 1997; "Infomaster: An Information Integration System"; *ACM SIGMOD*
[6]   Goodwin, Richerd; 1994, "Formalizing Properties of Agents", *Journal of Logic and Computation*, Volume 5, Issue 6
[7]   1999; "JDK 1.2 Documentation"; Sun Microsystems Inc., Palo Alto
[8]   Katoh, Takashi; Hara, Hideki; Kinoshita, Tetsuo; Sugawara, Kenji; Shiratori, Norio; 1998; "Behavior of Agents Based on Mental States"; ICOIN'98; *IEEE Computer*
[9]   Knapik, Michael; Johnson, Jay; 1998; "Developing Intelligent Agents for Distributed Systems: Exploring Architectures, Technologies, & Applications"; McGraw-Hill; New York
[10]   Kumova, Bora İ.; 1998; "System Specification for an Example Distributed Database"; MSc; D.E.U., Dept. of Comp. Eng., http://cs.deu.edu.tr/~kumova
[11]   Kumova, Bora İ.; Kut, Alp; 1998, "A Communication Model for Distributed Agents", *ISCIS'98*, http://cs.deu.edu.tr/~kumova, *Concurrent Systems Engineering Series*, Volume 53, IOS Press, Amsterdam
[12]   Kumova, Bora İ.; Kut, Alp; 1999, "Agent-based System Co-operation", *PDP'99*, Madeira, http://cs.deu.edu.tr/~kumova
[13]   Kumova, Bora İ.; 1999; "The Agent Model of AgentTeam"; *TAINN'99*; İstanbul
[14]   Kumova, Bora İ.; Alpkoçak, Adil; 1999; "A Survey on Networked Information Retrieval Systems"; *ISCIS'99*
[15]   Kumova, Bora İ.; 2000; "Flexible Distributed Database Management with AgentTeam"; *IASTED-AI'00*, Innsbruck; http://www.iasted.com/conferences/2000/austria/ai.htm
[16]   Kumova, Bora İ.; 2000, "Object-oriented Design and Implementation of the Multi-agent System AgentTeam", *ECOOP'00*, Cannes, http:// ecoop2000.unice.fr
[17]   Kumova, Bora İ.; 1999; "Object-oriented Design Concepts for Client-Broker-Server-based Multi-agent Systems"; *TAINN'00*; İzmir; http:// www.eee.deu.edu.tr/ tainn2000

[18] Kumova, Bora İ.; 2000; "The Agent communication Language AgentCom and its Semantics"; *IASTED-ASC'00*, Banff; http:// www.iasted.com/conferences/2000/banff/asc.htm

[19] Kumova, Bora İ.; 2000; "Software Design Patterns for Intelligent Agents"; *SSGRR'00*, lAquila; http://www.ssgrr.it/en/conferenza/index.htm

[20] Martin, David; Cheyer, Adasm J.; Moran, Douglas B.; 1999; "The Open Agent Architecture: A Framework for Building Distributed Software Systems"; *Applied Artificial Intelligence*, vol. 13, pp. 91-128

[21] Mayfield, James; Labrou, Yannis; Finin, Tim; 1995; "Evaluation of KQML as an Agent Communication Language"; *Intelligent Agents II -- Agent Theories, Architectures, and Languages*; Springer-Verlag, Berlin, 1996

[22] Nodine, Marian; Chandrasekara, Damith; 1998; "Agent Communities"; technical report, MCC

[23] Orfali, Robert; Harkey, Dan; Edwards, Jari; 1996; "The Essential Distributed Objects Survival Guide"; John Wiley Sons, New York

[24] Sadek, D.; Bretier, P.; Panaget, F.; 1997; "ARCOL agent communication language and MCP, CAP and SAP agent's cooperativeness protocols"; proposal, France Télécom

[25] Sycara, Katia; Decker, Keith; Pannu, Anandeep; Williamson, Mike; Zeng, Dajun; 1997; "Distributed Intelligent Agents"; *IEEE Expert*; Dec 96, http://www.cs.cmu.edu/~softagents

[26] Unified Modelling Language Version 1.3; 1999; "Notation Guide"; *Rational Software Corporation*; Santa Clara

[27] Vitek, Jan; Tschudin, Christian (eds.); 1996; "Mobile Object Systems: Towards the Programmable Internet"; *MOS '96*; Linz, Austria, July 8-9

Development of a Component Oriented
Garment Design Library

Ender Yazgan Bulgun[1] and Alp Kut[2]

[1]Dokuz Eylül University Department of Textile Engineering
Bornova 35100, Izmir, Turkey
ender.bulgun@deu.edu.tr
[2]Dokuz Eylül University Department of Computer Engineering
Bornova 35100, Izmir, Turkey
alp@cs.deu.edu.tr

Abstract. In today's garment industry manufacturers are not able to withhold the designs of the models that are created consequently, manufacturers have to spend considerable time presenting the model designs to new customers. For this reason, a software package has been developed to facilitate from customers with the help of this program, garment companies will be able to establish archives of their various models on the computer. Also, it will be possible to design new models more efficiently by using the computer rather than doing technical drawings by hand.

1 Introduction

In the past few years, computer usage has become wide spread in many sectors, including the textile and garment sector. In garment manufacturing the CAD (Computer Aided Design) system is used for model design and efficient cutting. In planning and production, the CIM (Computer Integrated Manufacturing) system is of great use. Along with these systems, the rapid development of communication systems and the internet will enable even greater use of computer technology in the garment industry.

In order to show new designs to their customers the garment companies are in the need of a large model library containing current collections made according the latest fashion trends and color schemes. Generally the design models are presented to customers with the help of the stylist's draft drawings and also fashion shows. At this stage it is time for the customer to make decisions. By taking in to account their desired models, sizes and counts the customers will place their orders.

In garment companies designers have much work to do they must closely follow the year's trends, color schemes, fabrics and styles. They must also know which products are sold to which stores by their clients. In short, the designers are in strong need of a large model library at hand . In today's garment manufacturing the technical drawings of the models and production information are stored and archived on paper. This makes it difficult to reuse past productions. However by storing this information on computers, a more rapid business environment would be made available for producers and customers. Also rapid access to the production information as well as

T. Yakhno (Ed.): ADVIS 2000, LNCS 1909, pp. 324–330, 2000.

to the model library would be made possible by the internet for both the employees and the customers.

The purpose of this study is to provide an efficient method for garment manufacturers to present their models to prospective customers by means of a large computer based model library.

This model library would contain technical drawings and production parameters for the various designs. By choosing the selections underneath the models in the data library the customers would be able to create their own designs. In this way there would no longer be a need to make drawings for new models on paper; instead technical drawings of models will be made over the computer with the aid of the selections. Also this program could also be used in custom made tailoring which has been drawing more wide spread in the past few years.

2 Former Work

In today's garment industry there are software containing various design models, CAD programs that allow designers to produce drawings over the computer and CIM systems that calculate costs of products.

"Prostyle" is the program developed for garment design by Lectra. This program allows for the making of aesthetic and technical drawings. With the aid of a scanner the program also allows for the placing of different fabrics over the drawings. The cost of row and supplemental material can be calculated using CIM's software package program called "Graphi-Cost". Thus, the total cost of the product can be found using this program. The program also allows the inventory control. By creating forms from all the information documents can be made available by the designer [8].

"The Lectra Catalog" a new program seen in the June 2000 IMB-Koln Clothing Machinery Fair was created to enable customers to choose their desired clothing over the computer. Photographs of the various models and fabrics are run thru a scanner and downloaded onto a computer. One is able to first select a fabric and then see their chosen model with this fabric. This program may also be used with software made for custom made clothing [8].

The Gerber company provides a large model library for the designer and any drawing possibility using their "Gerber Artworks Designer" software. In the latest version of the software, one can find a more enriched model library for the technical drawings, a new drawing menu for curved drawings and many more selections for the sewing appearances. With the "Accu-Mark Spectrum" package program information involving production, management and costs can be stored on a computer. Information about the desired models may be found and costs may be calculated. The information may also be changed frequently to remain current [5].

"Gerber Accu-Mark Made-To-Measure (M.T.M.) system" is one of the latest in the CAD systems. With the aid of this software, clothing made to fit the personal sizes of customers can be produced. Pictures and information about different models such as shirts, jacket, outerwear and pants have been stored on this program. Using this software the customer may choose the model they desire over the computer and place an order according to their size [5].

With Assyst company's "AssyForm" program, one can easily find information and graphics related the production. With "AssyCost", one can make cutting plans and calculate costs of products [2].

One of the software packages developed by Infotech company is "Sample Preparation and Model Archiving Software". This program contains information about the fabric, color, printing, embroidery and labeling, accessory specialties, pictures modeling studio and sewing instructions, a table of sizes, historical cost analysis and pricing, production information, customer orders and payment information [7].

Investronica company is working with Kopperman to produce design software. In the suggested design program technical aesthetic drawings can easily be made. The new version of the program contains an extended model library [6].

Bulgun E. and Baser G., have created a package intended to facilitate order processing for garment companies a specially clothing designers. With the aid of this program, companies will be able to create computer archives of their production and design parameters. They will also be able to quickly create data of new products using the data of similar models that have already been made [3].

3 Material Method

In this section, information about selected product, methods for this study and parameters related with production are given. The developed software is also introduced.

3.1 Sample Selection and Creating Technical Drawing

In this study, shirt is selected as a sample product for demonstrating software. At first several shirt manufacturers have examined and the most preferred shirt models have determined and finally the technical drawings have created. The details as sewing formats, pocket, collar, and type of yoke were formed very easily in technical drawing. Then the model of shirt were divided into pieces such as front, back, arm, collar, cuff. And finally, technical drawings were made for several models of each piece. Gerber "Artworks Designer" software were used for preparing technical drawings.

3.2 Defining Production Parameters

After completing technical drawings, the related production parameters were defined. These parameters were used for calculating unit costs. According to the type of fabric, the unit fabric consumption and cost, the accessory amount and cost, the operation time period, and the amount of sewing thread were calculated. Also, the following formula was used to calculate the cost per unit.

Shirt cost per unit = Fabric cost per unit + Accessory cost per unit +
Gross labor cost + Operating cost per unit +
Management cost per unit + Marketing cost per unit +
Financial cost per unit + Amortization cost per unit

Since the labor costs, financial expenditures, amortization expenditures, and business
expenditures vary from company to company, these factors should be counted as
additional parameters in the software [3].

3.3   Structure of Developed Software

Our software for handling Garment Design Library has four different parts:
− A database for storing and retrieving Garment Library
− A user-friendly interface for designing new models using reusable components
− A mechanism for remote accessing to our library
− An interface for entering several parameters to calculate unit cost of any model

Component Library Database. Our library contains technical drawings related with
both complete models and also their basic components. Fabric measurement for
related model , accessories list and quantity, job flow sequence, operation period and
unit cost are also located in the same database. This database is placed in a MS SQL
Server database under Windows NT operating system. Some important attributes of
our tables is shown in Table 1.

User Interface. This part is the most important component of our software that
allows communication between Component Library and users. This part represents
the following functions :

− Retrieving a model or a model's component and displaying related technical
  drawing on the screen
− While a model is located on the screen, the interface will retrieve the model's
  selected part from Library database and shows all possible component drawings in
  a new window.
− While a model is on the screen and also the component window is visible, the
  interface allows to user for selecting a new part from the window and places the
  newly selected item onto old part of model on the screen. This operation can be
  repeated for several parts and also several times.
− The interface also allow to turn model for reaching back part of any model and
  component
− After completing the changing operation the interface will store the new design
  with a new identical information.
− Our software's one of the future work will be to change original component's
  shapes to generate more components with this interface only for the special
  authorized users .

This part of software is implemented using Borland Delphi programming language. The reason for selecting Delphi as an implementation tool is to generate easy to use application and also to port this application over the internet.

Table 1. The Important Table Structures for Component Library

MAIN

Field Name	Field Type	Description
c_id	Integer	Unique id for all components
c_type	Boolean	True for complete items, False for basic components
c_name	String	Name of component or complete item
c_description	String	Description of an item

RELATION

Field Name	Field Type	Description
main-id	Integer	Unique id for all complete models
component-id	Integer	Unique id for subcomponents
c_description	String	Description of an item

COST

Field Name	Field Type	Description
main-id	Integer	Unique id for all complete models
Fabric	Real	Fabric cost for this model
Accesory	Real	Accesory cost for this model
Labor	Real	Labor cost for this model
Operating	Real	Operating cost for this model
Marketing	Real	Marketing cost for this model
Financial	Real	Financial cost for this model
Amortization	Real	Amortization cost for this model

Remote Access Mechanism. Our software's interfaces can be used both inside of company users via LAN and outside of company users via Internet. For this reason a communication layer is designed for handling remote access operations in a secure way. This part supports retransfer operations for incomplete transactions [9].

Cost Accounting. This part of our software takes necessary parameters such as workmanship cost, financial expenses, amortization expenses, management expenses, marketing expenses for calculating unit cost. This part of software is also implemented with Borland Delphi programming language [1],[4].

3.4   An Example Screen View of Our Software's User-Interface Part

At the beginning program starts with the technical drawings of trousers, shirt, dress, jacket etc. for selecting necessary model. After selecting one of the these model, selected model placed on the screen and user can select any component of the model (Fig. 1.1) Selected component changes its color and the component's all alternate figures come into a new window. Then the user must select a new shape for selected component of current model (Fig. 1.2). Finally newly selected item placed onto model on the screen.(Fig. 1.3)

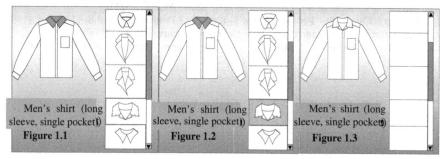

Fig. 1. An Example Shirt Model Screen's "Garment Design Library" Software. The user can make changes any part of selected model in an order one by one. The change model operation is shown in Figure 1 completely.

4 Conclusion

Today's garment industry manufacturers must produce and sell lists of models in very short period. Especially during the taking order period they must offer a good quantity of models to allow customers to design their own models. This will decrease the period of taking orders. In this study, a software was developed for archiving previously created models order by different types. This program also has utilities for users to create new designs using the previously created model components. The users of this software can be designers of any garment manufacturer company. Customers of companies can also use this software to create their own designs via the internet. This will cause a shorter period for taking offers for companies and also shorter decision making time for customers.

In this study shirts were selected as a sample product and all of the models and components technical drawings related with shirts were entered to "Garment Design Library ". Customers can create lots of new model designs any time these new models can also be stored to "Garment Design Library" via this software.

This software was also used for calculating unit costs with related production parameters such as fabric cost, accessory cost, management cost, labor cost, etc. Adding user defined model components will be future studies. With this option to create unlimited quantities of model parts will be possible.

References

1. Anon: Borland Delphi 5 Developers Guide, Borland Press, USA (1999)
2. Assyst Catalogues: Assyst GmbH, Henschering 15a, D-85551 Kirchheim, (2000)
3. Bulgun, E., Baser, G.: Development Of Garment Design Software. Fen ve
4. Mühendislik Dergisi, Cilt: 1, Sayi:4 (Accepted for printing) Bornova (2000)
5. Cantu, M., Gooch, T., Lam, F.J : Delphi Developer's Handbook. Sybex Inc, Alameda, CA, 94501 (1998)
6. Gerber Catalogues: Gerber Scientific Company, Connecticut, (2000)
7. Investronica Catalogues: C/Tomàs Bretòn 62 28045 Madrid (2000)
8. Info Tech Catalogues, Pera Bilgisayar Sistemleri ve Danismanlik Ltd.Sti, Istanbul (1998)
9. Lectra Catalogues: Image Force, Imp. Delteil, Bordeaux,-Pessac (2000)
10. Tanenbaum, Andrew S.: Distributed Operating Systems, Prentice Hall, USA (1997)

Using Wrapper Agents to Answer Queries in Distributed Information Systems

Luciano Serafini[1] and Chiara Ghidini[2]

[1] ITC–IRST, 38050 Pantè di Povo, Trento, Italy
serafini@irst.itc.it
[2] Dept. of Computing & Mathematics, Manchester Metropolitan University
Manchester M1 5GD, United Kingdom
C.Ghidini@doc.mmu.ac.uk

Abstract. In this paper we investigate an architecture for an agent-based integration of information, where both the centralised (or mediator-based) and decentralised approaches, studied in literature, may coexist. We provide a formal model for the architecture and we specify a set of basic tasks that must be supported by all wrapper agents in order to support query answering. A first implementation of these tasks, which is correct and complete w.r.t. the formal model, is also presented.

1 Introduction

Several reference architectures for information integration have been proposed in the literature [8,6,2]. In most of them the key role is played by a mediator, that provides an homogeneous and global model of the different information sources, and enables global applications to access the different information sources in a transparent way. Mediators suitably distribute queries to the relevant information sources, and provide answer integration. In this approach the applications associated to each single information source cannot access information that resides in the other sources. As a consequence users of the single information sources don't have any advantage from the integration process.

In this paper we investigate a more "cooperative" approach, in which each information source directly interacts with the others as it were a mediator. In this approach each source can obtain new information by directly querying the other information sources. This is made possible by using an multi-agent architecture, where the main role, in providing a correct interaction among different sources, is played by wrapper agents. In this architecture wrappers not only must provide their usual functionalities, but also they have to perform some of the functions usually fulfilled by the mediator. The architecture we use is described both at an informal level (Section 2) and at a formal level (Section 3). The latter is based on Distributed First Order Logics, a formalism described in [7]. A set of basic tasks that must be supported by wrapper agents is also presented. The distributed algorithm for query answering is described in Section 4 and a comparison with similar work is presented in Section 5.

T. Yakhno (Ed.): ADVIS 2000, LNCS 1909, pp. 331–340, 2000.

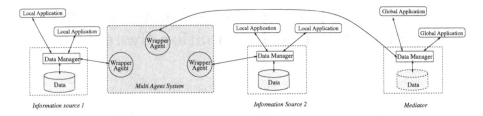

Fig. 1. High level structure of a distributed information system

2   The Architecture

The architecture we use is based on multi-agent paradigm and is depicted in
Figure 1. In this architecture, the key role is played by the wrapper agents associated to each information source. In a nutshell, every time a local application
submits a query Q to its data manager, the wrapper agent must collect the answer by suitably joining local data relevant to Q and data contained in other
sources relevant to Q. Notice that, in this architecture, a mediator can be considered an empty information source on a global schema which can access the
data of all the other information sources. Notice also that, differently from mediator based architectures, where the information flows from the sources to the
mediator, in our approach the information flows between sources.

In this paper we concentrate on query answering in a "read only" scenario,
that is a scenario where operations do not modify the data in the information
sources. In our approach wrappers not only must provide an interface between an
information source and other wrappers in order to answer a query, but they also
have to act as mediators. They must be able to distribute queries to other wrappers and to collect and integrate the results into the schema of their information
sources. More in detail, the agent wrapper W_i (associated to the information
source S_i) must support the following functionalities:

1. **Query propagation**. In receiving a query Q, W_i is able to establish (by
   itself or by asking some broker agent) whether there is an information source
   S_j containing information relevant to Q. W_i is able to send a sub-query Q'
   to the wrapper W_j of S_j. Q' can be expressed in the schema of S_i or in that
   of S_j, depending on the capabilities of W_i to do schema mapping.
2. **Query interpretation**. If W_i receives a query Q' from W_j and Q' is in the
   schema of S_i, then W_i submits Q' to S_i. Otherwise W_i tries to translate Q'
   into a corresponding query Q in the schema of S_i and it submits Q to S_i.
3. **Result propagation**. W_i collects the result R of a query Q that was submitted to S_i by another wrapper W_j and is able to send R back to W_j. Likewise
   query propagation, R can be sent to W_j using the data semantics of either
   S_i or S_j, depending on the capability of W_i to map different data semantics.
4. **Result interpretation**. If W_i receives from W_j the result R of a query that
   it previously submitted to W_j, and R is not expressed in the data semantics
   of S_i, then W_i tries to map R into a corresponding R' expressed in the data
   semantics of S_i.

5. **Local query planning.** W_i is able to perform query planning for queries expressed in the schema of S_i. This functionality can be provided by the information manager itself, which usually comes with some algorithms of query planning and optimisation.
6. **Local query submission.** W_i is able to submit a query to S_i and obtain the results of it.

Functionalities 1–4 require that W_i knows how to map queries among different schemata and results among different data semantics. This knowledge might be decided a priory by the designer of the system, or learned by the wrappers. In this paper we don't address this aspect. Note also that we do not force a wrapper to support 1–4 w.r.t. all the other wrappers in the system, and for all the information contained in its information source. Indeed W_i might be interested in communicating only with a proper subset of wrappers, and in exchanging data concerning a subset of expressions in the schema of its own information source.

3   Background

Models for the information sources. Abstractly enough, an information source is representable as a set of explicit information plus an inference machinery, which allows to deduce new information from the explicit information. This approach enables us to model information source as *definite deductive databases* [1]. In the following we recall some basic notions which are used throughout the paper.

A *first order relational language* L is a first order language defined on a set **dom** of individual constants, called *domain*, and a set **R** of *relation symbols* or predicates with a fixed arity. We use the terms *(ground) atom, (ground) literal, positive/negative literal* and *(ground) formula* in the usual way. A *deductive rule* is a formula of L of the form $h \leftarrow t$ where h (the head of the rule) is a literal and t (the body of the rule) is a conjunction of literals. A deductive rule is *definite* if the body is composed of positive literals. A *deductive database* DB is composed of an *extensional part* DB_E and an *intensional part* DB_I. DB_E is a set of ground literals and DB_I is a set of deductive rules. A deductive database is *definite* if DB_I contains only definite deductive rules.

The model for a definite deductive database DB is the *least Herbrand model* of DB on the domain $\mathbf{dom'} \subseteq \mathbf{dom}$ of the constants occurring in DB. This definition is based on the following assumptions on the data contained in DB:

CWA: all the information that is not true in DB is considered to be false;
DCA: there are no objects in the universe but those explicitly mentioned in DB;
UNA: distinct constants occurring in DB necessarily designate different objects.

CWA, DCA, and UNA must be reconsidered when a database DB is integrated into a distributed information system. CWA is not acceptable anymore. Indeed a database DB can obtain information which is not derivable from DB_E using the rules in DB_I, simply by querying other databases. Moreover, this new information might contain constants not explicitly mentioned in DB, invalidating also DCA. We substitute CWA and DCA with the following assumption:

D-CWA: all the information that is not true in DB and is not derivable by querying other databases in the information system is considered to be false;

D-DCA: there are no other objects in the universe than those in DB or those that can be contained in an answer from another database.

UNA says that DB never receives any information from another database concerning the equality of different constants. Even if this can be, in general, not true (e.g., by querying an astronomical database we might obtain *morning-star=evening-star*) we assume here that databases never submit queries containing the equality predicate to other databases, and therefore we retain UNA.

From this analysis, it follows that we cannot define a model for a definite deductive database DB in a distributed information system as it were a stand alone database. At this stage we can only define a set of *local models*. They correspond to the possible states of DB, depending on how DB is integrated with other databases and on the specific states of these other databases.

Definition 1 (Local model). *A* local model *of a definite deductive database DB, with language L on domain* **dom**, *is the set of Herbrand models for DB whose universe is contained in* **dom**.

Models for distributed information systems. Let $I = \{1, \ldots, n\}$ be a finite set of indexes, each denoting an information source. The first component of a definite deductive multiple database is a family $\{DB_i\}_{i \in I}$ (hereafter we drop the index $i \in I$) of definite deductive databases. L_i and $\mathbf{dom}_i$ denote the language and the domain of DB_i, respectively. The second component is a set of *deductive bridge rules*. Roughly speaking they are inter-databases deductive rules. More formally, a deductive bridge rule from i to j is an expression of the form:

$$j : h \leftarrow i : t \tag{1}$$

where h is a literal in the language of DB_j and t is a conjunction of literals in the language of DB_i. In this context, the procedural interpretation of (1) is that the query h, expressed in the language of DB_j, can be rewritten in terms of the subqueries in t, in the language of DB_i. A deductive bridge rule is *definite* if its body is composed of positive literal. As we do not assume the existence of a common domain, the third component of a definite deductive multiple database is a set of *domain relations*. A domain relation from i to j is a relation $r_{ij} \subseteq \mathbf{dom}_i \times \mathbf{dom}_j$. r_{ij} represents the capability of translating query results from $\mathbf{dom}_i$ into $\mathbf{dom}_j$. The procedural interpretation of a pair $\langle d, d' \rangle$ in r_{ij} is that the object d, which is in the result of a query to DB_i, can be translated into the object d' in the domain of DB_j. Deductive bridge rules and domain relations can be jointly used to define a predicate in DB_j from predicates in DB_i.

Definition 2 (Definite deductive multiple database). *A definite deductive multiple database MDB is a tuple* $\langle \{DB_i\}, \{br_{ij}\}, \{r_{ij}\} \rangle$, *where each* DB_i *is a definite deductive database, each* br_{ij} *is a set of definite deductive bridge rules from i to j and each* r_{ij} *is a domain relation from i to j.*

A definite deductive multiple database is a distributed information system where each information source is modelled as a definite deductive database. Each br_{ij} models the mapping from the schema S_i into the schema S_j. br_{ij} can be known by either W_i, or W_j, or both. The capability of W_i to propagate a query to W_j depends on the fact that W_i knows the bridge deductive rules in br_{ij}. Similarly, W_j is able to interpret a query received from W_i only if it knows br_{ij}. Analogously W_i (W_j) is able to perform result propagation (result interpretation) only if it knows the domain relation r_{ij}.

A *model* M for MDB is a family $\{M_i\}$ of local models of $\{DB_i\}$ such that for any deductive bridge rule $j : h \leftarrow i : t$ in br_{ij}, if $M_i \models t(d_1, \ldots, d_m)$ and $r_{ij}(d_k, d'_k)$, $(1 \leq k \leq m)$, then $M_j \models h(d'_1, \ldots, d'_m)$. This general definition does not take into account D-CWA, D-DCA, and UNA. Such assumptions are captured by the notion of minimal model. A model $\{M_i^*\}$ for MBD is a *minimal model* if it is a subset of any other model $\{M_i\}$ for MDB. A definite deductive multiple database has a *least model* which is the intersection of all the models of MDB. This is due to the fact that the intersection $\{M_i \cap M'_i\}$ of two models $\{M_i\}$ and $\{M'_i\}$ is still a model for MDB. Details about the construction of the least model for an MDB can be found in [3].

4   Distributed Algorithm for Query Answering

Every time an agent W_i receives a query Q, either from a local application or from some wrapper W_k, it activates a procedure, whose main steps are:

1. **Query interpretation.** If Q is not in the language of S_i, then W_i translates Q into a query Q' in the language of S_i.
2. **Local query planning.** W_i generates (asks its data manager for) a query plan Π for Q. A query plan contains information on the sub-queries that will be scheduled by the data manager in order to answer Q.
3. For each sub-query p in Π, W_i identifies the information sources S_j that are relevant to p. Then it retrieves answers to p by executing:
   3.1. **Query propagation.** W_i tries to rewrite p in some query Q_j in the language of S_j. If it succeeds, then it submits Q_j to W_j and waits for an answer; otherwise it submits p to W_j.
   3.2. **Result interpretation.** If the result R returned by W_j is still in the language of S_j, then W_i translates R into an equivalent result R'.
4. **Local query execution.** W_i asks the data manager of S_i to execute the query plan Π on the local data of S_i extended with the new data R'. Let R'' be the result of the execution of Π.
5. **Result propagation.** If Q was submitted by another wrapper W_k, then W_i communicates R'' to W_k, possibly in the language of W_k.

Steps 1–5 disregard a number of problems that must be considered in real implementations. First, the detection of infinite loops that can raise from circular (or recursive) sub-query calling. Second, the language that W_i uses to submit a query (return a result) to a wrapper W_j. To face these issues we assume that:

```
MQ(id : queryid, Q :query) → setoftuples
var:
   upd: database;
   p: predicate;
   ans: setoftuples;
   qp: setofpredicates;
begin
   qp := QP(Q);
   qp := qp \ {p | p appears in hist[id]}
   forall p ∈ qp do
      hist[id, p] := ∅;
      upd[p] := ∅;
   end;
   repeat
      hist[id] := hist[id] ∪ upd;
      forall p ∈ qp do
         forall i:p ← j:t ∈ br_ji do
            Send-Message(msg1);
            msg2 = Get-Message();
            upd[p] := upd[p] ∪ r_ji(Content(msg2))
         end;
      end;
   until (upd ⊆ hist[id]);
   ans := LQ(DB ∪ hist[id], Q);
   remove [id, Q'] from hist;
   return (ans);
end
```

```
msg1 = query-ref(
          :sender i
          :receiver j
          :content ({x | t(x)})
          :protocol (fipa-request)
          :language (language of j)
          :reply-with (query-id))

msg2 = inform-ref(
          :sender j
          :receiver i
          :content (<a setoftuples>)
          :protocol (fipa-request)
          :language (language of j)
          :in-reply-to (query-id))
```

$QP(Q)$: computes the query plan for Q
Send-Message(msg): sends the message msg
Get-Message(): waits for a new message
Content(msg): extracts the content of msg
$r_{ji}(X)$: computes the set $\{\langle x'_1, \ldots, x'_n \rangle |$ $\langle x_1, \ldots, x_n \rangle \in X \text{ and } \langle x_k, x'_k \rangle \in r_{ji}\}$;
$LQ(DB \cup hist[id], Q)$: computes the result of Q on the database DB extended with the content of $hist[id]$

Fig. 2. Procedure for distributed query answering

a. each information source S_i is a definite deductive database DB_i;
b. wrappers support only conjunctive queries without negation;
c. the query plan for a query Q submitted to DB_i is the set of relevant predicates to compute Q in DB_i. This set can be naively computed by collecting all the predicates occurring in the rules of DB_i used to answer Q;
d. wrappers accept only queries expressed in their own language and provide answers in their own language (i.e., query interpretation and result propagation are not necessary);
e. W_i's knowledge about query propagation is represented by a set of definite deductive bridge rules of the form $i : h \leftarrow j : t$, while W_i's knowledge about result interpretation is represented by a set of domain relations r_{ji};
f. the agents communication language is based on FIPA performative [5].

The procedure MQ associated to each wrapper W_i is shown in Figure 2 and is executed whenever W_i is requested to answer a query. Since wrappers should be able to manage multiple queries in parallel, W_i supports a multi-thread computation and MQ is executed on a new thread. In order to answer different external queries at the same time each wrapper must assign a new identifier id to every new query Q submitted by a local application. id is then used by all the wrappers to identify the sequence of sub-queries generated by the distributed algorithm in order to answer Q. Finally, each wrapper W_i is associated with a set of data modifiable by all the processes of W_i executing MQ: $hist$ associates a query identifier id to the informations which has been retrieved by W_i from the other sources, in answering Q; DB is the information available in the database associated to W_i; br_{ij} and r_{ji}, contain the knowledge of the W_i about query propagation and result interpretation, respectively.

We describe MQ by mean of an example. Consider a multiple database containing the following two databases without internal deductive rules:

$$DB_1 = \{S(1), P(2), P(3)\} \quad br_{12} = 2\!:\!R \leftarrow 1\!:\!S \wedge P \quad r_{12} = \{\langle 1, b\rangle, \langle 2, c\rangle, \langle 3, b\rangle\}$$
$$DB_2 = \{R(a), R(d)\} \quad br_{21} = 1\!:\!S \leftarrow 2\!:\!R \quad r_{21} = \{\langle a, 2\rangle, \langle b, 3\rangle, \langle c, 3\rangle\}$$

When an external query R is given in input to the database DB_2, the wrapper W_2 assigns a new identifier id to R and then calls the function $MQ(R, id)$. The sequence of steps generated in answering R is shown in Figure 3.

Proofs of correctness and completeness with respect to the fix point semantics for definite deductive multiple databases and proof of termination on all inputs are given in [3]. We have to admit that the basic algorithm is quite naive. Nonetheless there is a number of simple modifications that can lead up to optimised versions of MQ. First, MQ does not take advantage of caching results of sub-queries generated from the initial query R with identifier id. Given that the result of a query monotonically increases, every time a wrapper W_j is asked for a (sub)query P with query identifier id from wrapper W_i, W_j can answer back only the set Δ of new answers to p. In figure 3 this would imply that, e.g., data a and d are sent from W_1 to W_2 only once instead of six times. Second, in MQ a wrapper W_i sends the same sub-query to a wrapper W_j at least twice, and in general until no new results are calculated by W_j. It is easy to notice that whenever W_j executes MQ with $qp = \emptyset$, no other sub-queries can be generated and the answer set calculated by W_j depends only from the local query $LQ(DB \cup hist[id], Q)$. In this case W_i may disregard sending the same sub-query to W_j for the second time, because no new results can be calculated by W_j. In the example in figure 3 this would imply that, e.g., query R is submitted from W_2 to W_1 three times instead of six. Third, a wide number of optimisation methods used by mediators may be used by the wrappers as well. For instance, each W_i may cache the answer of a query Q up to a certain amount of time, and use it in answering Q, instead of running MQ.

5 Related Work and Conclusions

In the last few years significantly many attempts have been devoted to define systems for integrating multiple information sources. Most of them are based on the notion of *mediator* [10]. According to the classification of agent-based architectures for database federations proposed in [9], all these systems adopt a, so-called, *centralised approach*. Local applications access only local information sources, and global applications access all the information sources through a global mediator which provides the global schema of the federation. On the contrary, in the *decentralised approach*, no global mediator is provided. Every information source is associated to a wrapper that provides the interface for both local and global applications and communicates directly with the other wrappers.

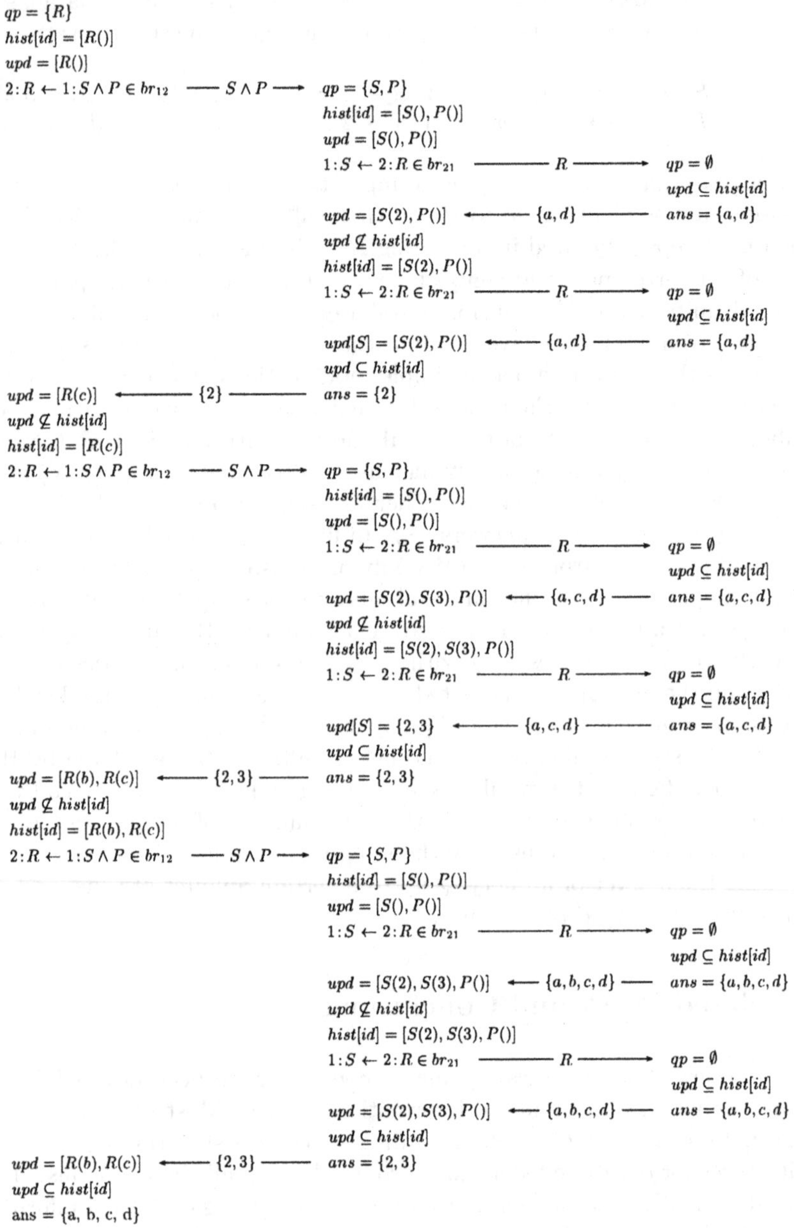

$qp = \{R\}$
$hist[id] = [R()]$
$upd = [R()]$
$2:R \leftarrow 1:S \wedge P \in br_{12}$ —— $S \wedge P$ ⟶ $qp = \{S, P\}$
                                                        $hist[id] = [S(), P()]$
                                                        $upd = [S(), P()]$
                                                        $1:S \leftarrow 2:R \in br_{21}$ ———— R ⟶ $qp = \emptyset$
                                                                                                $upd \subseteq hist[id]$
                                                        $upd = [S(2), P()]$ ⟵ $\{a, d\}$ —— $ans = \{a, d\}$
                                                        $upd \not\subseteq hist[id]$
                                                        $hist[id] = [S(2), P()]$
                                                        $1:S \leftarrow 2:R \in br_{21}$ ———— R ⟶ $qp = \emptyset$
                                                                                                $upd \subseteq hist[id]$
                                                        $upd[S] = [S(2), P()]$ ⟵ $\{a, d\}$ —— $ans = \{a, d\}$
                                                        $upd \subseteq hist[id]$
$upd = [R(c)]$ ⟵ $\{2\}$ —— $ans = \{2\}$
$upd \not\subseteq hist[id]$
$hist[id] = [R(c)]$
$2:R \leftarrow 1:S \wedge P \in br_{12}$ —— $S \wedge P$ ⟶ $qp = \{S, P\}$
                                                        $hist[id] = [S(), P()]$
                                                        $upd = [S(), P()]$
                                                        $1:S \leftarrow 2:R \in br_{21}$ ———— R ⟶ $qp = \emptyset$
                                                                                                $upd \subseteq hist[id]$
                                                        $upd = [S(2), S(3), P()]$ ⟵ $\{a, c, d\}$ —— $ans = \{a, c, d\}$
                                                        $upd \not\subseteq hist[id]$
                                                        $hist[id] = [S(2), S(3), P()]$
                                                        $1:S \leftarrow 2:R \in br_{21}$ ———— R ⟶ $qp = \emptyset$
                                                                                                $upd \subseteq hist[id]$
                                                        $upd[S] = \{2, 3\}$ ⟵ $\{a, c, d\}$ —— $ans = \{a, c, d\}$
                                                        $upd \subseteq hist[id]$
$upd = [R(b), R(c)]$ ⟵ $\{2, 3\}$ —— $ans = \{2, 3\}$
$upd \not\subseteq hist[id]$
$hist[id] = [R(b), R(c)]$
$2:R \leftarrow 1:S \wedge P \in br_{12}$ —— $S \wedge P$ ⟶ $qp = \{S, P\}$
                                                        $hist[id] = [S(), P()]$
                                                        $upd = [S(), P()]$
                                                        $1:S \leftarrow 2:R \in br_{21}$ ———— R ⟶ $qp = \emptyset$
                                                                                                $upd \subseteq hist[id]$
                                                        $upd = [S(2), S(3), P()]$ ⟵ $\{a, b, c, d\}$ —— $ans = \{a, b, c, d\}$
                                                        $upd \not\subseteq hist[id]$
                                                        $hist[id] = [S(2), S(3), P()]$
                                                        $1:S \leftarrow 2:R \in br_{21}$ ———— R ⟶ $qp = \emptyset$
                                                                                                $upd \subseteq hist[id]$
                                                        $upd = [S(2), S(3), P()]$ ⟵ $\{a, b, c, d\}$ —— $ans = \{a, b, c, d\}$
                                                        $upd \subseteq hist[id]$
$upd = [R(b), R(c)]$ ⟵ $\{2, 3\}$ —— $ans = \{2, 3\}$
$upd \subseteq hist[id]$
$ans = \{a, b, c, d\}$

Fig. 3. Answering R

An important aspect of our work is that centralised and decentralised approach can coexist allowing more flexibility in the system design and modifications.

In Information Manifold (IM) [8], a user is presented with a *world-view*, i.e., a set of relations modelling the kind of information that can be obtained from the available sources. Queries to the system are formulated in terms of the world-view relation. On the one hand this frees the user from having to know the specific vocabulary used in every information sources, on the other hand it might force the user to abandon local applications, as usually the local applications strictly depend on the structure of the data stored in the information source and on their data semantics. Unlike IM, our approach doesn't force the user to abandon the local applications. Given a query, IM uses the description of the information sources to generate a query-answering plan. The focus here is to use the world-views in order to decide which are the relevant information sources and generate a plan that prunes as many information sources a possible. In our architecture, the wrapper W is able to generate a query-plan qp and establish (by itself or by querying some other broker agent B) whether there is an information source S' containing information relevant in order to answer Q. We have not developed yet a query planer, because our main purpose was the foundation of the algorithm. Future work will study the applicability (and possible extensions) of existing query planners to our algorithm.

In TSIMMIS [6] a common model, called OEM, and a specific query language, called LOREL, are used for interrogating the system. Mediators perform query propagation, query interpretation, result propagation and result interpretation. Differently from IM, the common model is very simple. The main idea is that all objects have *labels* describing their meaning. Therefore no tight integrated view is provided and the rules contained in the mediator for retrieving information from the different sources are similar to closed (without free variables) deductive bridge rules. Unlike TSIMMIS, our approach does not provide fixed rules for query (result) propagation and query (result) interpretation, and enables to represent more flexible communication protocols among agents.

HERMES [2] provides a declarative language for defining mediators. In HERMES a set of knowledge bases (modelled as deductive databases) may be integrated in a unique amalgamated database, called amalgam. On the one hand, HERMES takes a more general approach than ourselves. Indeed the algorithm is aimed to handle possibly mutually incompatible answers to the same query, interruptible query answering, and formulae with complex sets of truth values and time intervals. On the other hand, adopting a global amalgamated database is the reason of the main drawbacks of this system. First, the algorithm is executed on a unique amalgamated database. A parallel execution of different parts of the algorithm is therefore difficult to achieve. Second, using a unique amalgamated database forces the databases to agree on a common domain. Indeed, if the query $\leftarrow p(X) : [1, 2, t]$ (asking for the joint values that $p(X)$ assumes in databases 1 and 2, with truth value t) is submitted to the amalgam and the fact $p(a) : [1, \top]$ (saying that $p(a)$ is in database 1 with truth value $\top$) is in the amalgam, then the most general unifier of $p(X)$ and $p(a)$ is $\theta = \{X = a\}$. The hypothesis underlying the answer to this query is that a denotes the same individual in both databases. Indeed if there are different objects in different databases denoting the same "real world" object, the algorithm is not able to relate them and provide the correct answer. Differently, the use of domain relations enables the retrieval of joint information in different databases about without imposing a common domain. Third, all the internal clauses are duplicated in the amalgam.

A significant formalism based on description logics for information integration is described in [4]. This approach is mainly focused on defining algorithms for query subsumption. As we don't have any integrated global view, we cannot easily provide the same reasoning service.

In conclusion, we have presented here an architecture for an agent-based integration of information that enables system designers to adopt both the centralised and decentralised approach. We have briefly described a formal model for such architecture, in the case of integrating deductive databases, and a well founded algorithm for query answering. We have considered here deductive databases, but the same ideas can be applied to other logical models of data.

References

1. S. Abitebul, R. Hull, V. Vianu. *Foundation of Databases*. Addison-Wesley, 1995.
2. S. Adali and V.S. Subrahmanian. Amalgamating knowledge bases, III: algorithms, data structures, and query processing. *Journal of Logic Programming*, 28(1), 1996.
3. C. Casotto. Un algoritmo distribuito per l'interrogazione di basi di dati federate. Technical Report Thesis 9801-05, IRST, Trento, Italy, 1998.
4. T. Catarci and M. Lenzerini. Representing and using interschema knowledge in cooperative information systems. *International Journal of Intelligent and Cooperative Information Systems*, 2(4):375–398, 1993.
5. Fipa '97 draft specification. Revision 2.0 available at `drogo.cselt.stet.it/fipa/`.
6. H. Garcia-Molina, Y. Papakonstantinou, D. Quass, A. Rajaraman, Y. Sagiv, V. Vassalos, and J. Widom. The TSIMMIS Approach to Mediation: Data Models and Languages. *Journal of Intelligent Information Systems*, 8(2):117–132, 1997.
7. C. Ghidini and L. Serafini. Distributed First Order Logics. In *Frontiers Of Combining Systems 2*, Studies in Logic and Computation. Research Studies Press, 1998.
8. A. Levy, A. Rajaraman, J.Ordille. Query answering algorithms for information agents. In *Proc. of 13th National Conf. on Artificial Intelligence (AAAI)*, 1996.
9. C. Türker, G. Saake, and S. Conrad. Modeling Database Federations in Terms of Evolving Agents. In *Poster Proc. of the 10th Int. Symp. on Methodologies for Intelligent Systems (ISMIS'97)*, 1997.
10. G. Wiederhold. Mediators in the architecture of future information systems. *IEEE Computer*, 25(3):38–49, 1992.

An Interactive Cryptanalysis Algorithm for the Vigenere Cipher

Mehmet E. Dalkilic and Cengiz Gungor

Ege University, International Computer Institute
Bornova 35100 Izmir, TURKEY
{dalkilic,cgungor}@bornova.ege.edu.tr

Abstract. Though it dates back centuries, Vigenere Cipher is still a practical encryption method that can be efficiently used for many applications. We have designed and implemented an interactive cryptanalysis software based on the Kasiski test, and a novel use of the Index of Coincidence (IC) concept. Our results show that cryptanalysis is possible for very short text lengths where classical cryptanalysis methods fail. Furthermore, we have observed that our software which is designed to work on English based ciphertexts, can be successfully executed on ciphertexts based on other languages. Along the way, we also compute and report the IC values for Turkish and some other languages under a different number of enciphering alphabets.

1 Introduction

The use of cryptosystems for sensitive communications for military and diplomatic purposes dates back to the time of ancient Egyptians as the story told by Kahn [1], the best known reference to the history of the secret writing, or cryptography. Traditionally cryptography has been exclusively used by governments and large corporations. However, with the widespread use of computers and the high availability of sophisticated communication networks, now most individuals and corporations need special protection of their private or business data[1] by cryptosystems.

The conventional cryptosystems are formulated by C.E. Shannon in a landmark paper at 1949 [3]. A cryptosystem is a five-tuple [4] $(\mathcal{P}, \mathcal{K}, \mathcal{C}, \mathcal{E}, \mathcal{D})$ where $\mathcal{P}$ is a finite set of plaintext, $\mathcal{K}$ is a finite set of keys, $\mathcal{C}$ is a finite set of ciphertexts, and for each $k \in \mathcal{K}$, there is an encryption rule $E_k \in \mathcal{E}$ and its inverse (decryption) rule $D_k = E_k^{-1} \in \mathcal{D}$.

An encryption rule is a one-to-one function that takes any plaintext message, $p \in \mathcal{P}$, and transforms it to an unintelligible form called a ciphertext message with the intend to conceal meaning. The decryption rule recovers the original plaintext from a given ciphertext. A simple crytosystem model is illustrated in Figure 1.

---

[1] Storing data can be seen as transmission of the data in the time domain [2]. Therefore, the term transmission (or communication) refers to any situation where data are stored and/or transmitted.

T. Yakhno (Ed.): ADVIS 2000, LNCS 1909, pp. 341–351, 2000.

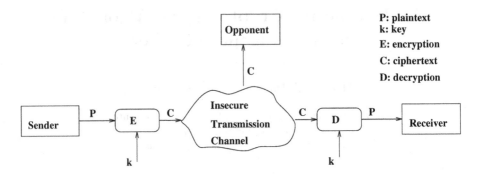

P: plaintext
k: key
E: encryption
C: ciphertext
D: decryption

Fig. 1. A simple model of a conventional cryptosystem

In this paper, we have studied a well-known classical cipher, Vigenere. Dating back to 16th century [5], Vigenere cipher is a polyalphabetic substitution cipher. With a random key equal to the size of the plaintext, Vigenere cipher becomes a *one-time pad* which is the only proven unbreakable cipher in the history. We have developed an interactive computer program to cryptanalyze the Vigenere cipher. Our program brings together classical methods such as Kasiski Test and the Index of Coincidence with a new concept, Mutual Index of Coincidence [6].

In the next section, we briefly describe the Vigenere cipher, and explain our approach to determine the cipher period, and the keyword. Section 3 outlines our interactive cryptanalysis algorithm followed by the experimental results in section 4. Conclusions and further research are outlined in the last section.

2   Cryptanalysis of the Vigenere Cipher

Suppose we have an n-character alphabet $A = (a_1, a_2, ..., a_n)$, an m-character key, $K = (k_1, k_2, ..., k_m)$ and a t-character plaintext, $M = (m_1, m_2, ..., m_t)$. Then, we define a Vigenere cipher $E_K(a_i) = (m_i + k_i \bmod n)$ and $D_K(c_i) = (c_i - k_i \bmod n)$.

The Vigenere cipher uses m shift ciphers where each k_i determines which of the n mono-alphabetic substitutions to be used.

The strength of the Vigenere cipher is that a plaintext letter can be represented by up to m different ciphertext letters. Nevertheless, the Vigenere cipher can not obscure the structure of the plaintext completely.

Suppose we have the following ciphertext that we believe it was encrypted using a Vigenere cipher.

```
WERXEHJVYSOSPKMUVCOGSIXFUFLTHTVYCBTWPTMCLHTRGCMGQEAGRDVFEGTDJPFP
WPGVLIASCSGABHAFDIASEFBTVZGIIIHDGIDDKAVYCCXQGJQPKMVIYCLTQIKPMWQE
QDYHGEMCTPCKRAXTKVJSPWVYJXMHNVCFNWRDCCMVQNCKXFVYCSTBIVPDYOEFBTVZ
GIIQXWPXAPIHWICSUMVYCTGSOPFPLACUCXMSUJCCMWCCRDUSCSJTMCEYYCZSVYCR
KMRKMVKOJZABXHJFBXGGVRLIEMKWLTXRDV
```

2.1   Determining the Cipher Period, m

Kasiski and the Index of Coincidence tests are commonly used to determine the key size. **Kasiski test** is based on the observation that if two identical fragments of length three or more appears in the ciphertext, it is likely[2] that they correspond to the same plaintext fragment. The distance between the occurrences has to be a multiple of the period.

In our example, the segment VYC appears four times beginning at position 31, 103, 175, 211 and 253.

Starting Position	Distance from Previous	Factors
103	72	2,3,4,6,8,9,12,18,24,36,72
175	72	2,3,4,6,8,9,12,18,24,36,72
211	36	2,3,4,6,9,12,18,36
253	42	2,3,6,7,14,21,42

Common factors 2, 3, and 6 are the key length candidates to be tested by the Index of Coincidence (IC) defined below.

Def: Let $x = \{x_1 x_2 ... x_n\}$ be a string of n alphabetic characters. The index of coincidence of x, IC(x), is the probability that randomly chosen two elements of string x are the same [4,5].

When applied to a natural language, the index of coincidence is like a fingerprint of the standard distribution. Therefore, the IC can be used to determine if ciphertext substrings' frequency distribution resemble standard distribution of the source language[3]. For a small sample the match would not be exact, but a close correspondence is usually sufficient. We call this test **IC-substring test**.

To try $m = 6$ we divide the ciphertext into 6 substrings so that each substring is encrypted by the same key letter i.e., $S_1 = \{c_1, c_7, c_{13}, ...\}, S_2 = \{c_2, c_8, c_{14}, ...\}, ..., S_6 = \{c_6, c_{12}, c_{18}, ...\}$. If our guess about m is correct, each of these substrings will have an IC value close to IC of the source language, in our example English. For instance, $IC(S_1) = 0,0714, IC(S_2) = 0,0672, IC(S_3) = 0,0771, IC(S_4) = 0,0745, IC(S_5) = 0,0585$ and $IC(S_6) = 0,0674$. Considering, the IC of standard English is around 0,068, $m = 6$ is correct with very high probability. If our guess is not correct, then the substrings will be more like random strings with much smaller IC values. For instance, if we have tried $m = 5$ that is $S_1 = \{c_1, c_6, c_{11}, ...\}, S_2 = \{c_2, c_7, c_{12}, ...\}, ..., S_5 = \{c_5, c_{10}, c_{15}, ...\}$. Then we would obtain $IC(S_1) = 0,0448, IC(S_2) = 0,0369, IC(S_3) = 0,0484, IC(S_4) = 0,0375$ and $IC(S_5) = 0,0478$.

Another use of the IC is that it can directly predict the period given that the amount of ciphertext is sufficiently large and the original plaintext has a normal

---

[2] The likelihood of two three-letter sequences not being from the same plaintext fragment is $1/n^3 = 0,0000569$ for n=26 [7].

[3] The IC values for Turkish are computed for three different letter frequency studies and it is shown in Table 1. First ten letters with highest frequency are also listed and almost perfectly matches in all three study.

Table 1. Index of coincidence (IC) values for Turkish

	IC	IC†	First ten	First ten†	Sample size
Koltuksuz [8]	0,0582	0,073083	{a,e,i,n,r,l,ı,d,k,m }	{i,a,e,n,r,l,u,s,d,k }	5,321,885
Goksu [9]	0,0608	0,072213	{a,e,i,n,l,r,ı,k,d,t }	{i,a,e,n,l,r,u,s,k,d }	574,004
Dalkilic [10]	0,0597	0,071455	{a,e,i,n,r,l,ı,k,d,m }	{i,a,e,n,r,l,u,s,k,d }	1,115,919

†: Turkish text written in English Alphabet

distribution. It is possible to predict the expected IC value, IC_E, for a cipher of period m, and tabulate the results in a table such as Table 2 for different m values using the formula [5,11]

$$ IC_E = \frac{S - m}{m(S - 1)}(IC(SourceLang)) + \frac{(m - 1)S}{m(S - 1)}(IC(RandText)) \quad (1) $$

where S is the length of the ciphertext, $IC(SourceLang)$ is the IC of the plaintext source language of the ciphertext and $IC(RandText)$ is the IC of random characters which is $1/n$, n alphabet size. By comparing IC(C) obtained from the ciphertext at hand to the values given in Table 2, we can get an estimate of m which may support the predictions of Kasiski test. For instance, in the ongoing example $IC(C) = 0,0429$ and from Table 2 (using the IC(English) row) we obtain the closest IC value is under $m = 6$ that matches the prediction of the Kasiski test. Nevertheless, IC is a good estimator of the cipher period only for small m, but its predictions are less accurate for larger m values. We call this test **IC-predict-m test**.

Table 2. Cipher period vs. index of coincidence

	$m = 1$	$m = 2$	$m = 3$	$m = 4$	$m = 5$	$m = 6$	$m = 7$	$m = 10$	$m = \infty$
IC(Turkish)	0,0597	0,0470	0,0428	0,0407	0,0395	0,0386	0,0380	0,0370	0,0344
IC(French)	0,0778	0,0589	0,0526	0,0494	0,0475	0,0463	0,0454	0,0437	0,0400
IC(German)	0,0762	0,0573	0,0510	0,0478	0,0460	0,0447	0,0438	0,0422	0,0384
IC(English)	0,0667	0,0525	0,0478	0,0455	0,0441	0,0431	0,0424	0,0412	0,0384
IC(Russian)	0,0529	0,0431	0,0398	0,0382	0,0372	0,0365	0,0361	0,0352	0,0333
IC(Tr-Eng)	0,0715	0,0549	0,0494	0,0467	0,0450	0,0439	0,0431	0,0417	0,0384
IC(Tr-Eng‡)	0,0720	0,0533	0,0475	0,0473	0,0450	0,0438	0,0418	–	–

Tr-Eng: Turkish text written in English Alphabet
‡: Empirical results

2.2 Determining the Keyword

We employ a new technique suggested by Dan Velleman [6] which uses mutual index of coincidence (MIC) in a very smart fashion.

Shift	String 1	String 2	String 3	String 4	String 5	String 6	Key
26	0,0314	0,0382	0,0344	0,0433	0,0356	0,0422	A
25	0,0358	0,0434	0,0315	0,044	0,0309	0,0385	B
24	0,0698	0,0429	0,0411	0,0461	0,0353	0,0317	C
23	0,0419	0,0331	0,0266	0,0262	0,0344	0,0499	D
22	0,038	0,0503	0,0364	0,0439	0,0498	0,0418	E
21	0,026	0,0463	0,0331	0,0335	0,0393	0,0409	F
20	0,0375	0,0389	0,0278	0,0336	0,0487	0,026	G
19	0,034	0,0358	0,0338	0,0285	0,04	0,0406	H
18	0,0409	0,0321	0,0416	0,0388	0,048	0,0394	I
17	0,0401	0,0265	0,0518	0,0377	0,0363	0,0307	J
16	0,036	0,0402	0,0422	0,0431	0,0388	0,0369	K
15	0,0329	0,0392	0,0412	0,0467	0,0332	0,0415	L
14	0,0315	0,0344	0,0332	0,0349	0,04	0,0361	M
13	0,0491	0,043	0,039	0,0302	0,0303	0,0293	N
12	0,0413	0,0337	0,0412	0,0444	0,0257	0,0645	O
11	0,0399	0,0342	0,0399	0,0686	0,0394	0,0469	P
10	0,0381	0,0449	0,0351	0,0364	0,0357	0,0303	Q
9	0,053	0,0645	0,0373	0,0334	0,0376	0,0239	R
8	0,0376	0,0353	0,0336	0,0263	0,0375	0,0504	S
7	0,0285	0,0292	0,0319	0,0368	0,0663	0,0402	T
6	0,0274	0,0444	0,048	0,0331	0,04	0,0336	U
5	0,0457	0,04	0,0374	0,0367	0,0364	0,0343	V
4	0,0368	0,0322	0,0322	0,0318	0,0313	0,0381	W
3	0,0319	0,0358	0,0416	0,039	0,041	0,0241	X
2	0,0435	0,0369	0,0704	0,0443	0,0336	0,0347	Y
1	0,0315	0,0248	0,0376	0,0387	0,035	0,0543	Z
	C	R	Y	P	T	O	

Fig. 2. Mutual indices of coincidences for the example ciphertext

The mutual index of coincidence of $\mathbf{x}$ and $\mathbf{s}$, denoted MIC(x,s), is defined to be the probability that two random elements, one from each, are identical. Let $\mathbf{x}$ is a string of standard source language e.g. English. Let string $\mathbf{x}$ has length l and frequency distribution $r_1, r_2, ..., r_n$ where n is the size of the alphabet e.g. 26. Clearly, for large l, probability distributions of n letters r_i/l will be similar to the standard probability distributions p_i of the source language. Now, let string $\mathbf{y}$ has length l' and frequency distribution $r'_1, r'_2, ..., r'_n$. Then,

$$MIC(\mathbf{x}, \mathbf{s}) = \frac{\sum_{i=1}^{n} r_i r'_i}{l l'} \simeq \frac{\sum_{i=1}^{n} p_i r'_i}{l'} \tag{2}$$

Suppose $\mathbf{s_1}, \mathbf{s_2}, .., \mathbf{s_m}$ be the m substrings of the ciphertext $\mathbf{s}$. Each of these substrings are obtained by shift encryption of the unknown plaintext. Though plaintext is unknown at the moment, its probability distribution and therefore its IC is known. Consider, substring s_j $(1 \leq j \leq m)$ is encrypted by unknown key k_j. Suppose we shift s_j by b $(1 \leq b \leq n)$ and obtain n different s_j^b each of which corresponds to a decryption with a different key value. If s_j has frequency distribution $r_1'', r_2'', ..., r_n''$ and length l'', then

$$MIC(x, s_j^b) \simeq \frac{\sum_{i=1}^{n} p_i r_{i-b}''}{l''} \qquad (3)$$

It is expected that $MIC(x, s_j^b) \simeq IC(\text{source language})$ if b is the additive inverse of the correct key modulo n; otherwise we obtain a much smaller MIC value. For instance if $n = 26$, and maximum MIC value for a substring is obtained when $b = 5$, then the probable key for that substring is $-5 \bmod 26 = 21$. For the example ciphertext our prediction is $m = 6$. So, we divide the ciphertext into 6 substrings, and we shift each substring by one, two, up to 26 and compute the MIC values shown in Figure 2. By simply selecting the maximum MIC value for each substring we get a probable keyword, in our example it is CRYPTO. To see whether the keyword we have obtained is correct we decrypt the ciphertext discovering[4] a passage from Tanenbaum [12].

Until the advent of computers one of the main constraints on cryptography had been the ability of the code clerk to perform the necessary transform ations often on a battlefield with little equipment However the danger of a code clerk being captured by the enemy has made it essential to be able to change the cryptographic method instantly if need be

3   Interactive Cryptanalysis Algorithm

Figure 3 outlines the basic algorithm used in our interactive cryptanalysis program. For a given ciphertext, first apply the Kasiski test, and then run the IC-predict-m test. Then, using the predictions about the key length, m, select a likely m value. Note that, the results of the Kasiski test are sorted giving priority to those obtained from the occurrence of the largest segment size. For instance, suppose possible m values 6, 9 obtained from a segment of size three and 5, 11 obtained from a segment size of four. In that case, we give priority to 5, 11 values. Next, apply IC-substring test; If the results are promising, IC values of substrings are close to the IC of standard plaintext distribution, then we continue with the Mutual Index of Coincidence (MIC) test; otherwise, go back and select another m value.

After the MIC test, one gets a key string. If the key is correct then we have a **perfect run**; that is, the cryptanalysis program found the key in a single shot. If the key is not correct (i.e., the plaintext does not seem right) but there seems to be off by several positions (e.g., small segments of text are recognizable) then interactively modify the key with the help of the substring key alternatives provided by the cryptanalysis program. For instance, in Figure 2 if the key "C" were not the right one for substring S_1, then the second choice would be the key "N" which has the second highest MIC value at 0,0491. Our program displays on the screen the first six choices for each substring. If you reach the solution

---

[4] Spaces are added to ease the reading.

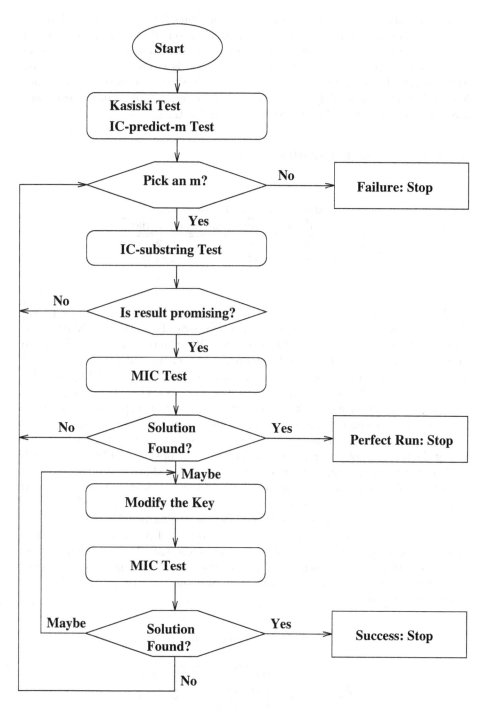

Fig. 3. The Basic Flowchart of the Program

either after trying multiple m values or test multiple key alternatives, that run is marked as **success** but not a perfect run. However, if you think you are not getting any closer to a solution, then go back to select another m. This process of trial and error continues until either a solution is found or all m values are exhausted. If the ciphertext could not be broken within a preset time limit, mark it as a **failure**. Fortunately, as the following section demonstrates, most trials are either perfect runs or leads to success in a few trial.

4   Experimental Results

Table 3. The Kasiski test results

No Prediction	Correct Prediction	Wrong Prediction
10	40	2
19.23%	76.92%	3.84 %

To test our interactive cryptanalysis program we used the following procedure. Each author prepared thirteen plaintext files of length 50, 75, 100, 125, 150, 200, 250, 300, 400, 500, 600, 800, 1000. We allowed up to 5% deviation on the size of the files. Then, each plaintext file is encrypted with two different keys. Key lengths vary from 2 to 11 characters. Thus, a total of fifty-two different ciphertext files are attained. Then, the authors exchanged the ciphertext files but no plaintext files or key information. Next, the cryptanalysis program run on each of the ciphertext files. While runs on larger ciphertexts were almost always ended in perfect runs, the smallest ciphertext files were harder to break; five of them could not be broken within half an hour of time and left alone. Both authors have acquired similar success rates with their share of the test cases.

As shown in Table 3, for about one fifth of the test cases Kasiski test could not produced any key length (m) predictions. These cases happened to be the shortest ciphertext files where there were no repeated patterns in the ciphertext. Interestingly, there were also two cases where repeated patterns are accidental, thereby causing Kasiski test to output wrong predictions. Nevertheless, the Kasiski test demonstrated its strength by generating the correct key length for over 76% of the cases where among more than half the time repeated patterns of length five discovered.

Table 4. The IC-predict-m test results

No Prediction	Correct Prediction	Wrong Prediction under	over
0	16	15	21
0%	30.76%	28.84%	40.38%

As it is shown in Table 4, the IC-predict-m test was on target less than one third of the time. For the remaining cases, overestimates of the key length (21 cases) were higher than the underestimates (16 cases). As expected, the IC-predict-m test is less reliable than the Kasiski test. However, when the Kasiski test has no output the IC-predict-m test is useful to guide the key search.

Table 5. Overall test results for the program

No Result	Perfect Runs	Success		
		cases	m-trials	avg. difference
5	26	21	3.01	0.6745
9.62%	50%	40.38%	–	–

The overall results of our tests given in Table 5 show that half of the time our cryptanalysis program had perfect runs while less than one tenth of the time it failed. For the remaining two fifth of the cases, the key attained by interactively trying on the average 3.01 different m values. It usually takes few minutes of wall clock time to get a result on a test case. For the multiple-m trial cases when the correct key length is provided, the MIC test acquired the exact key at 38% of the time. For the remaining multiple-m trial cases, the average difference between the key generated by MIC test and the correct key was 0,6745 i.e., less than a single character. If we add perfect runs (50%) to those multiple-m trial cases where MIC test attains the exact key (38% of 21 cases out of a total 54 cases i.e., 15.34%), we see that for about two third of the time MIC test has produced perfect results.

In another test case we have fed plaintext files as ciphertext to the program to see if we can fool it. However, IC-predict-m test has immediately detected that the IC of the input is close to normal plaintext distribution and concluded that m=1; that is, either we have a plaintext or a shift of a plaintext. The IC-substring test has determined that shift is 0 and returned our artificial ciphertext as plaintext.

Our Vigenere cryptanalysis program assumes that the underlying plaintext is in English. We wanted to see if the program will work if we feed it with ciphertexts where the source plaintexts are in another language e.g., Turkish. The success rates achieved were close to that of English based ciphertexts. Clearly, the Kasiski test finds repeated patterns and it is not in any way affected by a source language change. The index of coincidence tests (IC-predict-m and MIC test) are closely coupled with the letter distribution of the source language, and therefore they will be effected. Interestingly, the index of coincidence value of a Turkish text written in English alphabet, 0.0715 (see Table 2), is quite close to the IC value of standard English texts, 0,0667. Therefore, the IC based tests in our program still were able to generate results almost as good as those where the underlying text is English. Note that, from table 2, the IC value of Turkish (written in Turkish alphabet) is 0,0597 highly distanced to the IC of English at 0,0667. Therefore, it is plausible to assume that most natural languages written in English alphabet will have their index of coincidence values to move closer to that of English. As a consequence, our cryptanalysis program designed to work with English as its source language will work, with small degradation, on ciphertexts based on other languages.

Classical textbooks [e.g., Pfleeger [7]] state that the Kasiski test and IC-predict-m test works if there is a large body of ciphertext. Nevertheless, our experience shows that even for cases where under 100 characters available, satisfactory results can be attained with our approach. The shortest ciphertext instance that the program cryptanalyzed successfully is consisted of only 50 characters, and with a key length of 7 each substring had less than 8 characters to work with.

5   Conclusions and Future Work

We have reported an interactive cryptanalysis program for the Vigenere cipher. In addition to the classic cryptanalysis techniques such as Kasiski test and the IC-predict-m test, we have explored a recently proposed test which is based on the tabulation of the IC values of the shifted substrings for a possible key length. We have evaluated the performance of the individual tests and the overall performance of our cryptanalysis program. The new use of the IC concept proved to be exceptionally good leading to over 90% success rate for the program. In addition, our results show that cryptanalysis is possible for short text lengths where classical cryptanalysis approaches fail. We have also reported the index of coincidence values for Turkish and few other languages under different number of enciphering alphabets.

Using the facts that (i) multi-round encryption greatly increases the strength of a cipher, and (ii) a ciphertext running key is extremely difficult to cryptanalyze, we work on the design and implementation of a multi-round auto-cipher-key Vigenere cryptosystem. A formal strength analysis of our multi-round cryptosystems is due.

References

1. D. Kahn, *The Codebreakers: The story of Secret Writing*. NewYork: Macmillan, 1967. (abriged edition, NewYork: New American Library, 1974).
2. H. C. van Tilborg, *An Introduction to Cryptology*. Kluwer Academic Publishers, 1988.
3. C. Shannon, "Communication Theory and Secrecy Systems," *Bell System Technical Journal*, vol. 28, pp. 656–715, Oct 1949.
4. D. R. Stinson, *Cryptography: Theory and Practice*. CRC Press, 1995.
5. A. Menezes, P. van Oorschot, and S. Vanstone, *Handbook of Applied Cryptography*. CRC Press, 1997.
6. D. R. Stinson, "A More Efficient Method of Breaking a Vigenere Cipher," 1997. unpublished manuscript.

7. C. P. Pfleeger, *Security in Computing*. Prentice-Hall, 1989.
8. A. H. Koltuksuz, *Simetrik Kriptosistemler icin Turkiye Turkcesinin Kriptanalitik Olcutleri*. PhD thesis, Ege University, 1996. (in Turkish).
9. T. Goksu and L. Ertaul, "Yer Degistirmeli, Aktarmali ve Dizi Sifreleyiciler Icin Turkce'nin Yapisal Ozelliklerini Kullanan Bir Kriptoanaliz," in *Proc. of the 3rd Symposium on Computer Networks*, pp. 184–194, June 1998. (in Turkish).
10. M. E. Dalkilic and G. Dalkilic, "Language Statistics of Present-Day Turkish with Computer Applications," July 2000. (working paper).
11. J. Seberry and J. Pieprzyk, *Cryptography: An Introduction to Computer Security*. Prentice Hall, 1989.
12. A. S. Tanenbaum, *Computer Networks, 3rd Ed.* Prentice Hall, 1996.

Mappings and Interoperability: A Meta–modelling Approach[*]

E. Domínguez and M. A. Zapata

Dpt. de Informática e Ingeniería de Sistemas. Facultad de Ciencias.
Universidad de Zaragoza. E-50009 – Zaragoza. Spain.
e-mail: ccia@posta.unizar.es

Abstract. In this paper we propose a meta–modelling technique for helping in the construction of mappings which make feasible the interoperability demanded between different modelling techniques. A concrete method for conducting the construction of a mapping is presented. In order to facilitate the understanding of this method, we explain, as an example, the construction of a mapping from the Cooperative Objects Meta–Model to the Classes with Objectcharts Meta–Model.

1 Introduction

The existence of many different modelling techniques and the necessity of translating models, which arise from different methods, has raised a growing interest in developing mapping techniques between models. A well–known issue has been the translation of an entity–relationship schema into a relational database schema (see, for example, [3,16]). In turn, the reverse mapping, from the relational model into the E/R model, is also a subject of research due to the need of reverse engineering of database (see, for example, [13,6]). In the area of interoperable information systems, translation techniques have been used to interoperate between separately developed systems ([17,14]), and model mappings have also been proposed as tools for integrating different notations in the development of a software system [10].

This diversity of contexts has given rise to the existence in the literature of different approaches to achieving model translations. In the case of a translation only being required between two specific modelling techniques, the authors generally propose a specific method or procedural algorithm. This is the case of the proposals whose goal is to map an E/R schema into a relational schema [3, 16] or, the other way round, to map a relational schema into an E/R schema [6]. The advantage of giving a specific method is that it ensures great realibility of conversion. However, since they are particular solutions, they cannot be reused for defining translations between other modelling techniques.

With the aim of providing more general mapping techniques, another approach consists in defining only one meta–model which, like an ontology, must generically represent the building blocks of data models ([13,14]). Instead of

---

[*] This work has been partially supported by the project DGES PB96-0098C04-01.

using only one meta–model, other authors propose the construction of a meta–model for each methodology representing the concepts and rules by which the models are constructed within a method ([10,11,17]). In these cases, each author proposes a different metamodelling notation and also a different way of carrying out the interoperability. For instance, in [10], the interoperability between each pair of meta–models is achieved by means of a mapping which is based on the construction of a meta–model integrating the concepts of the initial ones. In [17] a fixed semantic data model is used, in which the rest of the meta–models are mapped. On the other hand, within the COMMA project [11], an object oriented core meta–model is proposed so that the COMMA–compliant methodologies can be derived from it, allowing a smooth translation of the core concepts to be done. Let us point out that this last approach considers not only the static aspects of methods (as the others do) but also the dynamic ones.

The approach we present in this paper is closer to these three aforementioned papers than to the other previous ones. It is similar to these approaches since we also propose a metamodelling notation, taking into account, as in [11], both the static and dynamic aspects. However, unlike them, we try to achieve a meta–modelling technique which contains, in particular, generic mechanisms for constructing mappings between meta–models, in which a concrete type of intermediate meta–model does not have to be defined. The use of a type of intermediate meta–model can somehow be seen as a way of using our proposal.

In [8] we proposed a concept–based technique for constructing meta–models aimed at being flexible enough to allow modelling knowledge of methods, taken from different application fields, to be expressed. As the main goal of this paper, we extend our technique by adding mechanisms for constructing mappings between meta–models.

In order to illustrate our proposal we present, as a practical application, a mapping from the Cooperative Objects [1] Meta–Model to the Classes with Objectcharts [5] Meta–Model. We have chosen these formalisms as an example mainly because we consider them to be complex enough to allow us to provide evidence of the potential feasibility of our approach. On the one hand, they take into account both structural and behavioural aspects and, on the other hand, they use different approaches to express the dynamics of a class: the Objectcharts are a state–based formalism and the PNOs are a Petri Net–based formalism [15]. The full formal presentation of the example is very extensive and it is beyond the scope of this paper. For this reason, we have only presented formally the supports of the meta–models and the specification of the mapping.

This paper is organised as follows. The next section describes our meta–modelling technique. The notion of mapping is introduced in section 3, explaining in section 4 a method for mapping construction. Finally, conclusions are given.

2   Conceptual Meta–models

This section reviews the main characteristics of the conceptual meta–models constructed by means of our meta–modelling technique [8]. Since in [8] some

Table 1. Notation for the supports

NAME	REPRESENTATION	NAME	REPRESENTATION
basic concept	▭	univalued attribution	———
simple attribute	⬭	set attribution	—(,)—
structured attribute	⬭	ordered set attribution	—(,)—¹²³
specialization	↑	tuple attribution	—(,)—
		multi-set attribution	—m(,)—

elements were not explained in enough depth, in this paper we explain in more detail those elements which are necessary for understanding the practical examples we present.

A conceptual meta–model contains a perspective, an anchoring system[1] (called system of concepts in [8]) and a family of supports. A *perspective* is a description of the meta–model with the aim of presenting its purpose and centering the context. An *anchoring system*, like a dictionary, consists in accurately establishing the meaning of the elements the meta–model provides, as well as the principles the models must fulfill. For the sake of brevity, we will not include any examples of perspective and anchoring system in this paper. A *support* is a family of concepts, which are contained in the anchoring system, related to each other by specializations and attributions (table 1 and figure 1). Some concepts of a support, called *basic concepts*, represent the components which can be independently perceived in a model. All the other concepts, called *attributes*, represent the aspects we can or must use to specify these components. The basic concepts are related to each other by specializations (with the usual inheritance property) constituting an IS–A hierarchy of concepts with only one sink. This hierarchy holds the following principle: each instance of a non–source basic concept C is an instance of some specialization of C. Specialization relationships can also be established between the attributes.

In order to construct a support, five types of attribution relationships can be used (table 1), from which the last four types are distinguished in terms of the underlying mathematical structure of the instances' multiplicity. As is usual, we consider optionality constraints (represented by means of a white circle) and obligatory constraints (black circle) with their habitual meaning.

As for the types of attributes, they can be *simple* when their instances are values of an atomic domain, or *structured*, on the contrary. Among the structured ones we highlight those, called *aggregation–based attributes*, whose instances are the aggregation of instances of their attributes. We also point out, due to their importance, those called *support–based attributes* whose instances are models of another support, which must also be defined. In this way, the support–based attributes allow the construction of supports to be modularized, making their specification easier. In order to graphically differentiate the support–based attri-

---

[1] This name has been taken from [4] but we use it here with a slightly different meaning.

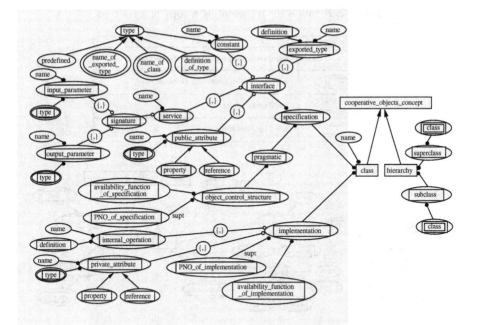

Fig. 1. Cooperative Objects Support

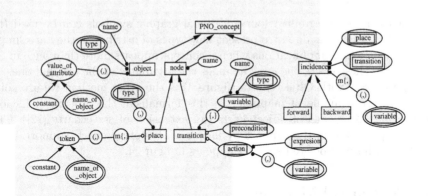

Fig. 2. Petri Nets with Objects Support

butes from the other structured attributes they are labelled with the expression 'supt'. For example, the Petri Nets with Objects (PNO) are used in the Cooperative Objects formalism to represent the specification and implementation of the objects' behaviour. This fact has been captured in the Cooperative Objects Support (figure 1) by means of the support–based attributes 'PNO_of_specification' and 'PNO_of_implementation' which refer in a modular way to the PNO Support (figure 2).

Table 2. Symbols for representing a copy of a concept

GRAPHIC REPRESENTATION	MEANING
	copy of a basic concept in the actual hierarchy of the support
	copy of a basic concept with the role of structured attribute
	copy of a simple attribute
	copy of a structured attribute

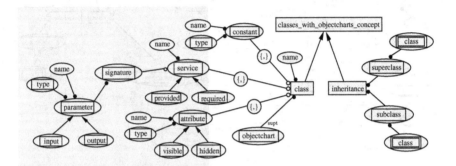

Fig. 3. Classes with Objectcharts Support

For the sake of legibility, four additional graphic symbols can be used (table 2). These symbols do not represent new types of attributes, they are simply elements allowing simplifications when a concept appears more than once in the support. In these cases, the copy assumes the attributes of the original one.

The Classes with Objectcharts (figure 3) is the other method we are going to use in the example of mapping. In this formalism, the objects' behaviour is described by means of objectcharts (an extension of statecharts [12]). This fact is represented in the support by means of the 'objectchart' support–based attribute whose associated support appears in figure 4.

3   Conceptual Mappings

A conceptual mapping from one modelling technique T to another T' refers to a method which allows a model of T' to be determined starting from a model of T. In this way, mappings allow model translations between methodologies providing a way of achieving interoperability between semantically heterogeneous information systems. The problem lies in the fact that, in general, finding a well defined mapping between two given modelling techniques is a complex task. In order to construct such a mapping, it is important to analyse and understand similarities and differences between the specification elements each modelling technique provides. This analysis will allow the mapping to be defined more easily since it will help to establish the way in which elements of the source technique can be

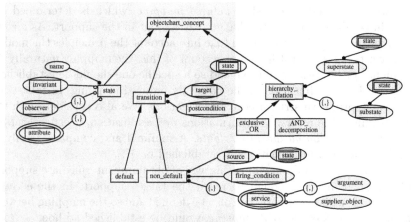

Fig. 4. Objectcharts Support

Table 3. Some primitive structured transformations

Name	Meaning
specialization	addition of a series of concepts as the refinement of a concept
generalization	addition of a concept as a generalization among previously defined concepts
aggregation	addition of an aggregation–based attribute as an aggregation of a sub-set of attributes of a predefined concept so that the added attribute becomes an attribute of the concept
addition of attribute	addition of a new attribute to a concept

translated into elements of the target technique. These similarities and differen-ces, in our opinion, will be identified more easily and more formally by means of a meta–modelling technique since meta–models reveal aspects of methodologies which can go unnoticed in informal comparisons [9].

In our technique, supports formally establish and classify the specification elements that can be used to construct models. In this way, they are a suitable means for comparing two methodologies and, therefore, for constructing a map-ping between them. For this reason, we propose an approach to facilitate the definition of mappings which tries to take maximum advantage of the supports. Essentially, within our approach, the construction of a mapping between a mo-delling technique T, with support S, and a modelling technique T', with support S', consists of considering a finite chain of supports $S=S_1, S_2,..., S_n=S'$ so that: 1) on the one hand, two consecutive supports S_i, S_{i+1} must differ slightly in their semantics in order to make the development of a mapping between their models easier; 2) on the other hand, the composition of these easier mappings provides a complete mapping, which must correspond to our interests.

One way of obtaining a support S_{i+1} starting from S_i, rests on making struc-tured transformations in S_i, for instance, specializing a concept or removing an existing one. Every structured transformation defines, in a natural way, a mapping between the models, called *structured mapping*, which is determined expressing, in terms of models, the changes carried out in the support. As for the definition of this mapping, we should take into account the principles the models of the target technique must hold. On account of this, the mapping naturally induced might not be entirely suitable and so a specific one should be established, which would be a modification of the naturally induced mapping. In table 3 we indicate some primitive structured transformations. We also consider, with the natural meaning, the inverse transformations *despecialization*, *degeneralization*, *deaggregation* and *elimination of attribute*. A minimal and complete family of structured support transformations is established in [7].

There will be situations in which this way of obtaining intermediate supports of the chain will not be enough to reach the target support. In these cases, particular intermediate supports can also be defined and so the mapping between the models, called *unstructured mapping*, would be established ad hoc.

4   Mappings Construction Process

In this section we are going to describe how to conduct the construction of a mapping between two meta–models. The process we propose consists of five steps, expressed in a linear manner for purposes of exposition, but which in practice must be used iteratively. As we explain each step, we will use the example of mapping from the Cooperative Objects Meta–Model to Classes with Objectcharts Meta–Model in order to facilitate the understanding of the process.

Step 1. Setting the mapping context. The goal of this step is to determine in which models of the source meta–model the mapping into a model of the target meta–model makes sense; that is, within what context, the interoperability translation is meaningful. In order to do this the ranges of application of the involved meta–models must be compared, determining if every system, which can be represented by means of a meta–model, can also be represented by the other meta–model. For instance, the Classes with Objectcharts Meta–Model has the restriction of representing only static systems of objects and avoiding dynamic object creation and deletion; furthermore, this formalism imposes the following technical restrictions: 1) multiple inheritance cannot be used and; 2) the execution of a service cannot involve a nested request for another service. Therefore, the mapping we are going to define must only be applied to Cooperative Objects holding these restrictions.

Step 2. Understanding the similarities and differences. The basic idea is to get informally a general perspective of the mapping that will be carried out. To do this, the supports must be compared bringing out their similarities and differences. For instance, the basic concepts of the Cooperative Objects Support are similar to the ones of the Classes with Objectcharts Support, the differences mainly lying in the description elements of the classes. For example, in the Cooperative Objects (figure 1), specification and implementation of classes are clearly separated, whereas in the Classes with Objectcharts (figure 3) they are mixed up. Another important difference between them is that the dynamics

Table 4. Structured transformations in the Cooperative Objects Support

1. Deagregation of the attributes 'interface', 'object_control_structure', 'pragmatic', 'specification' and 'implementation'.
2. Despecialization of the concepts 'public_attribute' and 'private_attribute' removing the sub-concepts 'property' and 'reference'.
3. Generalization of the concepts 'public_attribute' and 'private_attribute' into the concept 'attribute'.
4. Elimination of the attribute 'exported_type'. The associated mapping must remove the instances of the attribute 'exported_type' but, as the names of the removed types can appear as instances of the concept type, the mapping must also replace these names by their definitions.
5. Despecialization of the attribute 'type'.
6. Despecialization of the attributes 'availability_function_of_specification' and 'availability_function_of_implementation'.
7. Elimination of the attribute 'internal_operation'. The associated mapping must remove the instances of this attribute and, as the name of an 'internal_operation' can appear describing a transition of a Petri Net with Objects, the mapping must also replace, in these cases, the name of the 'internal_operation' by its effects.

Table 5. Structured transformations in the Classes with Objectcharts Support

1. Degeneralization of the concept 'parameter'.
2. Degeneralization of the attribute 'service'.
3. Elimination of the concept 'required_service'. In this case we must define the inverse mapping associated to this transformation. This mapping, starting from a model without the specification of the 'required_service', retrieves this information from the transition specification of the Objectchart.

of classes is expressed by means of two semantically different approaches (statecharts and Petri Nets). In this step, published comparisons about the involved methods can be particularly useful [9], as well as core methodologies [11].

Step 3. Approaching semantically the initial supports. In this step the source and target supports are semantically approached by means of increasingly close intermediate supports. These supports are obtained by making structural transformations in the initial ones with the aim of smoothing over their differences. In order to carry out semantically correct transformations, the meanings of the concepts involved (which are specified in the anchoring system) must be considered. For each transformation, a mapping between the models of the supports must be established (the naturally induced mapping or a modification of it) so that the mapping preserves the principles the models must hold. The two sequences of structured transformations we have applied to our examples appear in tables 4 and 5. In these tables we have also indicated the associated mapping in those cases in which the mapping is not the one naturally induced by the transformation. The two supports obtained as a result of this step are depicted in figures 5 and 6.

Step 4. Determining unstructured mappings. When we consider that the supports cannot be approached any further by means of structured transformations, then particular intermediate supports must be designed, as well as ad hoc mappings between the models. In our example we can establish pairs of synonym concepts–(name, name), (public_attribute, visible_attribute), etc.– between the supports achieved in the previous step so that the instances of one concept

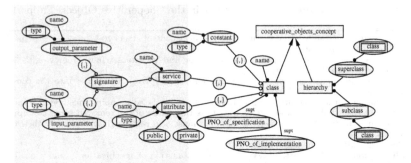

Fig. 5. Modified Cooperative Objects Support

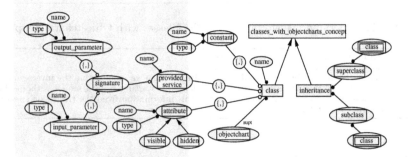

Fig. 6. Modified Classes with Objectcharts Support

can be considered as instances of the synonymous concept. This simple mapping can be defined for all the concepts of the supports except for those representing the dynamics (that is, 'PNO_of_specification', 'PNO_of_implementation' and 'objectchart'). As for these concepts we have had to use an intermediate support owing to the important differences regarding the description of the dynamics. To be precise, an Augmented Transition Networks (ATN) [18] support is used (figure 7), defining an unstructured mapping from the PNOs into the ATNs (which would be applied either to the 'PNO_of_specification' or to the 'PNO_of_implementation', depending on the analyst's interests) and a mapping from the ATNs into the objectcharts. The idea of using, as an intermediate support, that of the Augmented Transition Networks (ATN) [18] arose from the algorithm described in [2], in which, starting from a PNO, an ATN is calculated by means of the construction of the PNO's covering graph. However, we consider it necessary to modify this algorithm adding a register to the ATN for each place of the net. These registers store the tokens of the PNO's places in each reachable marking. Moreover, the preconditions and actions of each arc must be rewritten so that they may manipulate these registers instead of the variables of the net. Let us point out that we do not consider this algorithm as a definitive proposal since it could be improved. However, we think it is enough for illustrating our

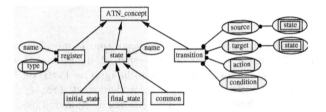

Fig. 7. Augmented Transition Networks Support

Table 6. Mapping from an ATN into an objectchart

1. The states and arcs of the ATN are considered as states and transitions of the Objectchart.
2. A state is added for each connected component of the ATN so that there is an exclusive_OR relation for each one of these states (superstates) with the rest of states of the corresponding connected component (substates).
3. If the ATN has more than one connected component then a state is added so that there is an AND_decomposition relation between this state (superstate) and the previously added states (substates).
4. The visible and hidden attributes of the corresponding class are declared as attributes of the root state.
5. The registers of the ATN are also declared as attributes of the root state. Additionally, these registers must be declared as hidden attributes of the class.
6. Default transitions are added whose targets are the states which correspond to the initial states of the connected components of the ATN. The postcondition of these transitions set the initial values of the registers.
7. The firing condition of each transition is the precondition of the corresponding arc in the ATN.
8. The services of each transition are deduced from the name and the action of the corresponding arc of the ATN.
9. Lastly, the postcondition of each transition is determined from the action of the corresponding arc of the ATN.

technique. Finally, a mapping from the ATNs into the objectcharts is sketched in table 6.

Step 5. Determining the complete mapping. As the last step, the global mapping from the source support to the target one is defined as a suitable composition of those mappings obtained in the two previous steps or their reverse mappings.

5   Conclusions

Throughout this paper we have given notions and mechanisms for defining mappings between meta–models. A method for establishing mappings is proposed in section 4. Starting with two given supports, the method allows to obtain two semantically closer ones, that is, two new supports such that it is possible to define a mapping between their respective models.

Let us point out that these notions and mechanisms can be used within other methods, for example, following the ideas proposed in [10], [11] and [17], the method can be based on a support integrating the initial supports, on a support of a fixed meta–model or on a core support.

References

1. R. Bastide, *Objects Coopératifs: un formalisme pour la modélisation des systemes concurrentes*, Théses de doctorat de l'Université Toulouse I, France, 1992.
2. R. Bastide, P. Palanque, Implementation techniques for Petri Net based specifications of Human-Computer dialogues, in J. Vanderdonckt, *Proc. of Computer-Aided Design of User Interfaces*, Presses Universitaires de Namur, 1996, 285-301.
3. C. Batini, S. Ceri, S.B. Navathe, *Conceptual Database Design: An Entity-Relationship Approach*, The Benjamin/Cummings Publishing Company Inc, 1992.
4. S. Brinkkemper, M. Saeki, F. Harmsen, Meta–Modelling Based Assembly Techiques for Situational Method Engineering, *Information Systems*, 24, 3, 1999, 209–228.
5. D. Coleman, F. Hayes, S. Bear, Introducing Objectcharts or how to use statecharts in object–oriented design, *IEEE Trans. on Soft. Eng.*, 18, 1, January 1992, 9-18.
6. I. Comyn-Wattiau, J. Akoka, Reverse Engineering of Relational Database Physical Schemas, in B. Thalheim (Ed.), *Conceptual Modelling – ER'96*, LNCS 1157, Springer, 372-391.
7. E. Domínguez, M. A. Zapata, Noesis: Towards a Situational Method Engineering Technique, Preprint.
8. E. Domínguez, M.A. Zapata, J.J. Rubio, A Conceptual approach to meta–modelling, in A. Olivé, J.A. Pastor (Eds.), *Adv. Info. Syst. Eng., CAISE'97*, LNCS 1250, Springer,1997, 319-332.
9. G. Eckert, P. Golder, Improving object–oriented analysis, *Information and Software Technology*, 36 (2), 1994, 67–86.
10. J.C. Grundy, J.R. Venable, Providing Integrated Support for Multiple Development Notations, in J. Iivari, K. Lyytinen (Eds.), *Adv. Info. Syst. Eng., CAISE'95*, LNCS 932, Springer, 255-268.
11. B. Henderson–Sellers, A. Bulthuis, *Object–Oriented Metamethods*, Springer, 1997.
12. D. Harel, Statecharts: A visual formalism for complex systems, *Science of Computer Programming*, vol 8, North–Holland, 1987, 231–274.
13. M.A. Jeusfeld, U.A. Johnen, An Executable Meta Model for Re–Engineering of Database Schemas, in P. Loucopoulos (Ed.), *Entity–Relationship Approach – ER'94*, LNCS 881, Springer–Verlag, 1994, 533-547.
14. C. Nicolle, D. Benslimane, K. Yetongnon, Multi–Data models translations in Interoperable Information Systems, in J. Mylopoulos, Y. Vassiliou (Eds.), *Adv. Info. Syst. Eng., CAISE'96*, LNCS 1080, Springer, 1996, 176-192.
15. P. Palanque, R. Bastide, L. Dourte, C. Sibertin-Blanc, Design of User-Driven Interfaces using Petri Nets and Objects, in C. Rolland, F. Bodart, C. Cauvet (Eds.) *Adv. Info. Syst. Eng., CAISE'93*, LNCS 685, Springer, 1993, 569-585.
16. T.J. Teorey, D. Yang, J.P. Fry, A logical design methodology for relational database using the extended entity-relationship model, *ACM Computing Surveys*, 18, 2, June 1986, 197-222.
17. S.D. Urban, A semantic framework for Heterogeneous Database Environments, *First Int. Workshop on Interoperability in Multidatabase Systems*, 1991, 156-163.
18. W.A. Wood, Transition network grammars for natural language analysis, *Communications of the ACM*, 13, 10, October 1970, 591-606.

Reliability Index Evaluations of Integrated Software Systems (Internet) for Insufficient Software Failure and Recovery Data

Mehmet Sahinoglu

Eminent Scholar & Chair, Comp.& Info. Sci.,TSUM, Montgomery, AL
mesa@tsum.edu; http://www.cis.tsum.edu/~mesa

Abstract. If the recovery or remedial time is not incorporated in the reliability of a software module in safety and time-critical integrated system operation, such as in Internet environment, then a mere reliability index based on software failure characteristics is simply not adequate and realistic. Recovery data need also to be collected. In deriving the Sahinoglu-Libby probability density function (pdf) of the software component availability, empirical Bayesian procedures will be used to employ expert engineering judgment through appropriate non-informative and informative prior distribution functions by employing various definitions of risk functions. It is emphasized that the uncontested usage of maximum likelihood estimators (mle) regardless of the insufficiency of historical data is erroneous and misleading. Case studies show this phenomenon of overestimation of reliability in safety and time critical components as well as systems.

1 Introduction

In the case of safety and time-critical systems where rare events may be involved, conventional software reliability models may not be adequate [2]. The omission of recovery or debugging capacity in safety or time-critical phases, such as in a space mission or medical emergency and security software operations can be altogether misleading [14]. Three alternative empirical Bayesian estimators are proposed for individual and system availability indices over the widely used maximum likelihood estimators. This research paper proposes to derive the probability density function (pdf) of the "steady state unavailability of a software module" as a random variable in the form of FOR (Forced Outage Rate) as in hardware terminology. The ratio of unequal-parameter $\text{Gamma}(\alpha_1, \beta_1)$ to the sum of unequal $\text{Gamma}(\alpha_1, \beta_1)$ and $\text{Gamma}(\alpha_2, \beta_2)$ was originally derived [1], [5] in the statistical literature. Subsequent treatment of the same problem can be found in [6] and [4] in the form of G3B(α, β, λ). Therefore the said pdf is henceforth called Sahinoglu-Libby pdf, named after its independent originators [1], [5], [6]. The general problem is also given extensively by a classical textbook on Statistical Distributions [3].

Empirical Bayesian procedures will be used to employ prior engineering judgment by using non-informative and informative prior distributions, with respect to two different loss function definitions, squared error and weighted squared error loss as in

T. Yakhno (Ed.): ADVIS 2000, LNCS 1909, pp. 363-373, 2000.
© Springer-Verlag Berlin Heidelberg 2000

[5]. The empirical updating of the parameters will enable the model to possess a similar effect of a non-homogeneous Poisson process, also used in determining stopping rules for testing [7], [8], [10] with varying failure (λ) and recovery rates (μ). See also [13] for an annotated bibliography. Numerical examples on systems will follow for illustration. It will be shown that often used maximum likelihood estimators of system availability, as default information in case of inadequate data is incorrect and should be replaced by its small sample empirical Bayesian estimators using alternative definitions of loss functions. This is imperative to avoid using optimistic or pessimistic estimators of (un)availability. Further analysis is carried out to implement the component study into the system configurations. Since the probability distribution function for a system index is simply infeasible in closed-form solution to attain, based on independence assumption of components, a product of expectations is utilized to estimate the desired availability indices. Numerical examples on components and various system configurations are given. The goal is to calculate the system availability in a time-critical software network, despite lack of long-term historical data, by empirical Bayes, using small sample estimators.

1.1   Nomenclature

The prior and field parameters are defined in the derivations of un(availability) indices below.

a = number of occurrences of operative (up) times sampled,

x_T = total sampled up time for "a" number of occurrences.

b= number of occurrences of debugging(down) times sampled

y_T = total sampled debugging (down) times for "b" number of occurrences of debugging activity.

c= shape parameter of Gamma prior for failure rate λ,

ξ =inverse scale parameter of Gamma prior for failure rate λ,

d = shape parameter of Gamma prior for recovery rate μ,

η = inverse scale parameter of Gamma prior for recovery rate μ.

q* = Unavailability Index(FOR) estimator using informative prior w.r.t. weighted square loss

q**= Unavailability Index(FOR) estimator using non-informative prior w.r.t. weighted square loss

q_{mle} =Unavailability Index(FOR) maximum likelihood estimator for long run field data: a, b$\rightarrow \infty$

q^=E(q)=Unavailability Index(FOR) Expected Value Estimator using informative prior w.r.t. square loss

r* = Availability Index(1-FOR) estimator using informative prior w.r.t. weighted square loss

r**= Availability Index(1-FOR) estimator using non-informative prior w.r.t. weighted square loss

r_{mle} =Availability Index(1-FOR) maximum likelihood estimator for long run field data: a, b$\rightarrow \infty$

r^=E(r)=Availability Index(1-FOR) Expected Value using informative prior w.r.t. square loss

2 Mathematical Formulation

After some algebra [5], one by using distribution function technique derives the pdf of

FOR,   (Unavailability)$= \dfrac{\lambda}{(\lambda + \mu)}$,   first by deriving $G_Q(q) = P (Q \leq q) = P$

$(\dfrac{\lambda}{(\lambda + \mu)} \leq q)$ and then taking its derivative to find $g_Q(q)$, as follows,

$$g_Q(q) = \frac{\Gamma(a+b+c+d)}{\Gamma(a+c)\Gamma(b+d)} \frac{(\xi + x_T)^{a+c}(\eta + y_T)^{b+d}(1-q)^{b+d-1}q^{a+c-1}}{[\eta + y_T + q(\xi + x_T - \eta - y_T)]^{a+b+c+d}} \tag{1}$$

$$= \frac{\Gamma(a+b+c+d)}{\Gamma(a+c)\Gamma(b+d)} \{ \frac{(\xi + x_T)q}{[\eta + y_T + q(\xi + x_T - \eta - y_T)]} \}^{a+c}$$

$$\{ \frac{(\eta + y_T)(1-q)}{\eta + y_T + q(\xi + x_T - \eta - y_T)} \}^{b+d} \frac{1}{q(1-q)}$$

is the p.d.f. of the r.v. Q=FOR where $0 \leq q \leq 1$. This expression can be reformulated [1], [3], [5], [6] as follows:

$$g_Q(q) = \frac{\lambda^{a+c}q^{a+c-1}(1-q)^{b+d-1}}{B(a+c, b+d)[1-(1-L')q]^{a+b+c+d}}$$

where

$$B(a+c, b+d) = \frac{\Gamma(a+c)\Gamma(b+d)}{\Gamma(a+b+c+d)}, \quad \text{and} \quad L' = \frac{(\eta + y_T)}{(\xi + x_T)} \tag{2}$$

Similarly one obtains the p.d.f. of the r.v. R(Availability) $=1-Q$ as follows,

$$g_R(r) = \frac{\lambda^{b+d}(1-r)^{b+d-1}r^{a+c-1}}{B(b+d, a+c)[1-(1-L')(1-r]^{a+b+c+d}}$$

where

$$B(b+d, a+c) = \frac{\Gamma(a+c)\Gamma(b+d)}{\Gamma(a+b+c+d)}, \quad \text{and} \quad L' = \frac{(\xi + x_T)}{(\eta + y_T)} \tag{3}$$

2.1   Empirical Bayes Estimators

Note that upper case denotes system. System availability is R_{sys} and system unavailability is Q_{sys}. Now, let q^ denote an estimate of the r.v. denoted to be q≡FOR The loss incurred, L(q, q^), in estimating the true but unknown q can be defined at will. Usually, the loss penalty increases as the difference between q and q^ increases. Hence, the squared error loss function, $L(q,q^\wedge)=(q-q^\wedge)^2$ has found favor where the risk $R(q,q^\wedge)=E\{L(q,q^\wedge)\}=E(q-q^\wedge)^2$ would then be the variance of the estimator Q=q to penalize larger differences more in classical least squares theory [5]. Bayes estimator q^ in our problem with respect to squared error loss function is the first moment or expected value of the r.v. Q=q using its p.d.f. in equation (2).

$$q^\wedge=E_Q[q|\underline{X=x,Y=y}]= \int_0^1 qg_Q(q)dq \tag{4}$$

Similarly, the Bayes estimator r^ with respect to squared error loss function is the first moment or expected value of the r.v. R=r using its p.d.f. in equation (3). That is,

$$r^\wedge=E_R[r|\underline{X=x,Y=y}]= \int_0^1 rg_R(r)dr \tag{5}$$

Weighted squared error loss is of considerable interest to engineers as practiced, and has the attractive feature of allowing the squared error $(q-q^\wedge)^2$ to be weighed by a function of q. This will reflect that a given error of estimation often varies in penalty according to the value of q. In software reliability studies, however, the importance of error magnitude e=|q-q^| may be more pronounced in the initiation or termination stages, like at q=0.01 or q=0.99, than at central stages like q=0.5. Then, the weighted squared error loss function used in such cases is as follows,

$$L(q,q^\wedge)^2=\frac{(q-q^\wedge)^2}{q(1-q)}=w(q)(q-q^\wedge)^2 \tag{6}$$

With this loss function, the Bayes estimator is given by the ratio of the two Stieltjes integrals:

$$q^*=\frac{\int_q^{} qw(q)dh(q|X=x,Y=y)}{\int_Q^{} w(q)dh(q|X=x,Y=Y)}= \tag{7}$$

$$\frac{1+\dfrac{(a+c)(\eta+y_T)}{(\xi+x^T)(b+d)}}{2+\dfrac{(b+d)(\xi+x_T)}{(\eta+y_T)(a+c-1)}+\dfrac{(a+c)(\eta+y_T)}{(\xi+x_T)(b+d-1)}}$$

is the Bayes estimator with respect to weighted squared loss, suggested for use in the conventional studies to stress more for tail values in the event of a small sample situation, as opposed to maximum likelihood estimator(mle) that requires long-term data, reflecting insufficient unit history. Here, w(q) was conveniently taken to be $[q(1-q)]^{-1}$. For the special case when placing $\xi = \eta = 0$, c=d=1 in equation (8) for noninformative (flat) priors, q^* becomes q**.

$$q^{**} = \frac{1 + \dfrac{(a+1)(y_T)}{(x_T)(b+1)}}{2 + \dfrac{(b+1)(x_T)}{(y_T)(a)} + \dfrac{(a+1)(y_T)}{(x_T)(b)}} \tag{8}$$

Finally, q^{**} approaches the q_{mle} as a,b $\rightarrow \infty$, as follow in subsequent equations (9) and (10).

If the sample sizes of up and down times , "a" and "b", usually equal, get large such that $\dfrac{a}{b} \rightarrow 1$,

$$q_{mle} = \frac{1 + \dfrac{y_T}{x_T}}{2 + \dfrac{x_T}{y_T} + \dfrac{y_T}{x_T}} = \frac{x_T + y_T}{2(y_T x_T) + (y_T)^2 + (x_T)^2} = \frac{x_T y_T}{x_T} = \frac{y_T}{x_T + y_T} \tag{9}$$

is the forced down (inoperative or debugging) time divided by the total period of study (operative plus debugging). This equation (9) is also the maximum likelihood estimator for the random variable FOR$\equiv$ q $= \dfrac{\lambda}{\lambda + \mu}$ for large sample case and is ordinarily used in calculations assuming large sample all-field-no-prior information data case. Similarly, the expected value of the FOR as in equation (4) for the squared error loss penalty-function will likewise approach the value of mle in (9) as sample sizes a and b get to be large. Similarly, as in equation (10) below.

$$r^* = \frac{1 + \left\{ \dfrac{(a+c)(\eta + y_T)}{(\xi + x^T)(b+d)} \right\}^{-1}}{2 + \dfrac{(b+d)(\xi + x_T)}{(\eta + y_T)(a+c-1)} + \dfrac{(a+c)(\eta + y_T)}{(\xi + x_T)(b+d-1)}} \tag{10}$$

is the Bayes estimator of the posterior r with respect to weighted squared loss, suggested for use in the conventional studies to stress more for tail values in the event of a small sample situation reflecting insufficient unit history as opposed to maximum likelihood estimator(mle) that requires long-term data. Here, w(r) was conveniently taken for equation (7) to be $[r(1-r)]^{-1}$. For the special case when $\xi = \eta = 0$, c=d=1, i.e. for non-informative (flat) priors, r^* becomes r** as in equation (12). Finally, r^{**} approaches the r_{mle} as a,b$\rightarrow \infty$, as follows .

$$r^{**} = \frac{1 + \left\{ \dfrac{(a+1)(y_T)}{(x_T)(b+1)} \right\}^{-1}}{2 + \dfrac{(b+1)(x_T)}{(y_T)(a)} + \dfrac{(a+1)(y_T)}{(x_T)(b)}} \tag{11}$$

If the sample sizes of up & down times , "a" & "b" , usually equal, get that $\dfrac{a}{b} \to 1$, then

$$r_{mle} = \frac{1 + \left\{ \dfrac{y_T}{x_T} \right\}^{-1}}{2 + \dfrac{x_T}{y_T} + \dfrac{y_T}{x_T}} = \frac{x_T + y_T}{2(y_T x_T) + (y_T)^2 + (x_T)^2} \frac{x_T y_T}{y_T} = \frac{x_T}{x_T + y_T} \tag{12}$$

Note that $q^* = 1 - r^*$ if and only if $a+c \approx a+c+1$ and $b+d \approx b+d+1$, i.e. $a+c$ and $b+d$ both very large. This is why for small sample (other than mle) cases, $r=1-q$ no longer holds. Therefore, in evaluating various system availability or unavailability, exact values are used such as πq_i and πr_i. However, $\pi(1-q_i)$ or $\pi(1-r_i)$, can not be used except for mle, as follows:

I)   Series Systems: $R_{sys} = \pi r_i$ and $Q_{sys} = 1 - \pi r_i$; note, π denotes " the product of "
II) Parallel Systems: $Q_{sys} = \pi q_i$ and $R_{sys} = 1 - \pi q_i$
III) Parallel-Series Systems: $R_{sys} = 1 - (1 - \pi r_i)^n$ and $Q_{sys} = (1 - \pi r_i)^n$; note, n= # parallel paths
IV) Series -Parallel Systems: $R_{sys} = (1 - \pi q_i)^m$ and $Q_{sys} = 1 - (1 - \pi q_i)^m$; note, m= # series subsytems

2.2   An Application to Integrated Software Systems

The system structures or configurations used in hardware designs can also be approximately valid in software networks, if not exactly. Computer (software) systems are often used to monitor or control a critical system where failures can have life-threatening or other severe consequences. Internet is a recent sound example of such a network, where the architecture is a collection of series or parallel or mixed series-parallel or mixed parallel-series. Moreover this could be an initial step toward the general problem of estimation of internet reliability.

According to [12], a safety analysis of a critical system begins by identifying and classifying a set of hazards within the context of the operational environment A hazard can be critical such as to shut down a nuclear plant causing severe injuries or occupational illness. This can be catastrophic to lead to massive losses, or breakdown of a stable system. Therefore corrective maintenance is required. Three scenarios are tested and illustrated. See Figures 1–4 to observe the differences.

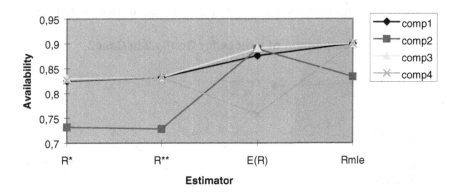

Fig. 1. A Comparison of Bayesian Availability Indices for 4 Different Components

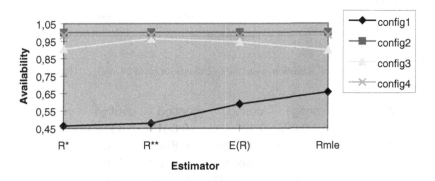

Fig. 2. A Comparison of Bayesian Availability Indices for 4 Different Systems (Identical Components)

The first case involves a comparison of the various Bayes estimators depending on the loss functions for different single components in a non-system case.

The second case depicts the system availability (as in configurations I to IV) where identical components are present.

The third case is the same as the second case except that the components are non-identical and chosen in ascending order while configuring. Find input data and results for Cases 1, 2 and 3 in Table 1.

3 Results and Discussion

For 4 different components above in Table 1, different likelihood functions for the gamma priors (left or right skewed or quasi-symmetrical) are chosen by the analyst due to expert judgment. Empirical Bayes estimators are computed for q* with respect to weighted squared error loss and q^ = E(q) due to squared error loss from the

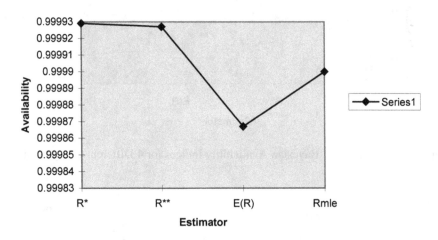

Fig. 3. A Comparison of Bayesian Availability Indices for Case 2 Config.2 (Identical Components)

Index	config1	config2	config3	config4
R*	0,825	0,999921	0,8811	0,9821
R**	0,8318	0,999885	0,8785	0,9781
E(R)	0,5261	0,999808	0,9282	0,9709
Rmle	0,6075	0,999833	0,9525	0,9735

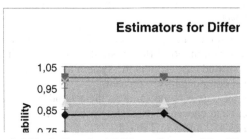

Fig. 4. A Comparison of Bayesian Availability Indices for 4 Different Systems (Non-Identical Components)

respective p.d.f., $g_Q(q)$. Or, q^{**} due to flat priors where scale parameters are ∞ and shape parameters are unity each. Similarly, r^*, r^{**} due to weighted squared error loss and, $r^{\wedge} = E(r)$ due to squared error loss are calculated from its respective p.d.f. $g_R(r)$. Also, the standard deviations are calculated to indicate the dispersion. If for very small (or large) values of q=FOR when the penalty is more pronounced than the rest of the distribution, the weighted squared error loss function is more appropriate and q/(1-q) is to be used. Or by adjusting the constant c in the function cq/(1-q), one may tailor it for the desired loss or penalty function in a specific software reliability evaluation. The same holds for r=1-q, the availability, long-run reliability.

Table 1. Input parameters and Empirical Bayes estimators for cases 1, 2 and 3

Input Parameter	*Component 1*	*Component 2*	*Component 3*	*Component 4*
A	10	5	10	10
B	10	5	10	10
x_T	1000hrs	25hrs	1000hrs	1000hrs
y_T	111.11hrs	5hrs	111.11hrs	111.11hrs
C	0.02	0.2	0.5	1.5
ξ	1	1	1	0.3333
D	0.1	2	2	2
η	1	0.5	4	0.25
Case 1 Estimators	**Component 1**	**Component 2**	**Component 3**	**Component 4**
R*	0.8250	0.7318	0.8295	0.8265
R**	0.8318	0.7282	0.8318	0.8318
R^ =E(r)	0.8759	0.8893	0.8909	0.7580
r_{mle}	0.9	0.8333	0.9	0.9
σ (r)	0.047	0.042	0.328	0.107
Case 2 Estimators	**System Config. I**	**System Config. II**	**SystemConfig. III**	**SystemConfig. IV**
R*	0.4633	0.999929	0.9051	0.9829
R**	0.4787	0.999927	0.9639	0.9801
R^ =E(R)	0.5886	0.999867	0.9458	0.9771
R_{mle}	0.6561	0.999900	0.8980	0.9832
Case 3 Estimators	**System Config. I**	**System Config. II**	**SystemConfig. III**	**SystemConfig. IV**
R*	0.8250	0.999921	0.8811	0.9821
R**	0.8318	0.999885	0.8785	0.9781
R^ =E(R)	0.5261	0.999808	0.9282	0.9709
R_{mle}	0.6075	0.999833	0.9525	0.9735

The main contribution intended in this study is to represent the long run (un)availability of a software module more accurately in time-critical practices, where corrective maintenance, such as recovery or debugging activity has to be recorded. Further, it is of equal importance that in the case of lack of extensive field or historical data which provides maximum likelihood estimator, existence of expert engineering judgment through prior functions of failure and repair rates will enable the analyst to compute the expected values of the unknown component or system availability or unavailability, with respect to various risk definitions. Otherwise, assuming only an mle for all circumstances, be it a large or small sample (historical) data collection is incorrect and misleading. See the legend "comp4" of Case1 in Figure 1. Also observe "config1" and "config2" of Case2 in Figure 2 and 3. Similarly, "configs 1, 2 and 4" of Case3 in Figure 4. These figures are extracted from tabulations in Table 1. Component or system availability may sometimes be pessimistic or other times optimistic using the maximum likelihood estimates due to lack of adequate history. This is why empirical Bayesian estimation through informative and non-informative prior functions is more realistic. The updating of the failure and recovery rates through more sampling history highly resemble that of a non-homogeneous Poisson process, which improves initial assumptions of exponential failure and debugging parameters.

Following a component analysis in the initial section of Table 1, the authors generalize this analysis to system applications where a number of software modules may be operating in series, parallel or mixed configurations. The programming allows for a balanced number of modules in each series or parallel subsystem, or any indefinite number of components in a simple series or active parallel arrangement. Thus, analyst can see the trend from highly informative priors to non-informative (flat-like uniform) priors in the case of two often used loss-function definitions. For a similar work, see [9]. Secondly, extensive field data in the steady state will lead to maximum likelihood estimators. The component data in Table 1(Case 1) is now utilized for such purposes in the case of identical in Table 1(Case 2) or non-identical components in balanced complex system structures in Table 1(Case 3).

Authors show that one does not have to have only maximum likelihood estimator to conduct an availability study. By informative and non-informative prior judgment to respect non-homogeneity of software failure and recovery rates as opposed to hardware-like constant-failure or recovery rates, and obeying a certain loss definition, like squared error loss or weighted squared error loss, one can calculate the availability indices in complex software system structures. As one obtains more in-field data, it is observed that the above outlined (*), (**) and E(.) estimators will approach that of the maximum likelihood (mle) which is conventionally used in all calculations invariably and at times erroneously such as in the case of small samples. In the absence of large field data, an optimistic guess using an mle despite lack of large field data may result in undesirable consequences like in airplane system composed of active parallel software components. See figures for Case 2 that depict this anomaly in two out of four configurations in the Appendix. Therefore, this approach takes into account of the experience and learning curve that an engineer or analyst has had in working with a particular system. With more estimators available based on the degree of availability of historical data, a better course of corrective maintenance can be effectively applied under different scenarios [11].

Acknowledgement

The authors thank Dr. Christopher Butler, from Math. Department of CWRU for his assistance in computing some complex integral equations with MATHEMATICA. Cordial thanks go to Prof. Kishor Trivedi at Duke University and Profs. Joe and Nell Sedransk from Statistics Dept. at CWRU for guidance, and to Prof. C. Bayrak at CIS of TSUM for graphics. TUBITAK-Ankara and Dokuz Eylul University, Izmir, Turkey are acknowledged for their initial financial support towards this research.

References

1. Sahinoglu M., Longnecker M.T., Ringer L.J., Singh C., Ayoub A.K., "Probability Distribution Functions for Generation Reliability Indices- Analytical Approach", IEEE-PAS, Vol. 102, No.6 (1983) 1486–1493
2. Lyu M.R., Handbook of Software Reliability Engineering, ed. by ; Chap.13 : Field Data Analysis by W.D. Jones and M.A. Vouk, IEEE Comp. Soc. Press, McGraw Hill (1995)
3. Johnson N.L. et al., Continuous Univariate Distributions, Vol.2, Ed.2, John Wiley and Sons Inc. (1995)
4. Pham-Gia T., Duong Q.P., "The Generalized Beta and F Distributions in Statistical Modeling", Mathematical and Computer Modelling, Vol.13, (1985) 1613-1625
5. Sahinoglu M., Statistical Inference on the Reliability Performance Index for Electric Power Generation Systems, Ph.D. Dissertation, Institute of Statistics/Electrical Engineering Department, TAMU (1981)
6. Libby D.L., Novick M.R., "Multivariate Generalized Beta-Distributions with Applications to Utility Assessment", J.educ. Statist. 7(4), (1982) 271-294
7. Sahinoglu M., "Compound Poisson Software Reliability Model", IEEE Trans. on Software Engineering, 18(7), (1992) 624-630
8. Sahinoglu M., Can Ü., "Alternative Parameter Estimation Methods for the Compound Poisson Software Reliability Model with Clustered Failure Data", Software Testing Verification and Reliability, 7(1), (1997) 35-57
9. Deely J.J., Sahinoglu M., " Bayesian Measures to Assess Predictive Accuracy of Software Reliability Methods", The Ninth International Symposium on Software Reliability Engineering (ISSRE'98), Germany, (1998) 139-48 (due to be published in IEEE Trans. Reliability in Fall 2000)
10. Sahinoglu M., Mayrhauser A.V., Hajjar A., Chen T., Anderson Ch.., "How Much Testing is Enough? Applying Stopping Rules to Behavioral Model Testing", Proc 4th International Symposium on High Assurance System Eng. , Washington D.C. (1999) 249-256,
11. Sahinoglu M., Chow E., Empirical-Bayesian Availability Index of Safety & Time Critical Software Systems with Corrective Maintenance, Proc. Pacific Rim International Symposium on Dependable Computing (PRDC1999), (1999) Hong Kong
12. Dugan J., Handbook of Software Reliability Engineering (Chap. 15), Ed. by M. Lyu, IEEE Comp. Soc. Press, Mc Graw Hill (1995)
13. Xie M., "Software Reliability Models- A Selected Annotated Bibliography", Software Testing Verification and Reliability, Vol.3, (1993) 3-28
14. Friedman M.A., Voas J.M., Software Assessment-Reliability, Safety, Testability; John Wiley and Sons Inc., New York (1995)

FedeRaL:
A Tool for Federating Reuse Libraries over the Internet

Murat Osman Ünalır[1], Oğuz Dikenelli[1], and Erden Başar[2]

[1]{unaliro, dikenelli}@staff.ege.edu.tr
[2] basar@compenet.emu.edu.tr

Abstract. A reuse library is an important element of formalizing the practice of reuse. It provides the mechanism to properly manage reusable components and make them available to software systems developers. Software developers welcome to the emergence of a robust marketplace of software components, as a highly competitive marketplace works to the advantage of both producers and consumers of software components. However, two essential requirements for a component marketplace have been slow to emerge: standard, interchangeable parts and the consumers' ability to find the right parts for the job at hand. Fortunately, recent advances and web technology are, at last, providing the means for satisfying these requirements. The goal of FedeRaL is to render the whole Internet as a federated reuse repository for reusable components.

1 Introduction

Reuse libraries are an essential element in the strategy to reuse software and related information [1], [4], [8], [9], [10]. Software reuse has taken place in the past through personal and organizational information preserving structures [12]. Formalizing the preservation and location processes will make reuse possible for a wider range of people and activities. The reuse library is the proper center of activity for these formalized processes and the proper access point for sharable knowledge products.

Application development has been evolving in both techniques and participants since its inception. Throughout the evolution period, the emphasis has been on applying structure and technology in an effort to shift application development from a highly skilled craft to a repeatable engineering discipline [2]. An application is no longer a discrete executable but rather a collection of cooperating software components and shared data. A key enabler for this vision involved selecting and provisioning a repository tool that could define, store, browse and search information about software related components, as well as retrieve the files associated with the component.

The evolution of Internet as a mainstream communication medium has been super-accelerated. Vast amount of reusable components are made available at ever increasing pace. In essence, there are vast amount of reusable components but an acute lack of viable infrastructure and federated search mechanisms.

It is widely recognized that infrastructure of reuse libraries differ from each other in this dynamic and open information universe [11]. The reuse libraries are lack of the support of extensible distributed repository management and dynamic interoperability

T. Yakhno (Ed.): ADVIS 2000, LNCS 1909, pp. 374–383, 2000.

among them. Hence, a common problem facing many reuse libraries today is the uniform and scalable access of multiple repositories. Uniform access implies the effectiveness of remote data access and delivery. On the other hand, scalability refers to the ability of distributed object management services to scale the process of delivering information (reusable assets) from a set of data sources to a typically larger set of consumers.

There are many studies related to reuse libraries [1], [3], [15], [16]. Agora is a prototype being developed by Software Engineering Institute at Carnegie Mellon University [18]. It's object is to create an automatically generated, indexed, worldwide database of software products classified by component type, especially JavaBeans or ActiveX control. Therefore, Agora can be seen as a search-engine for reusable software components. A traditional approach has been to develop large-scale software repositories as large central databases containing information about components and the components themselves. Such efforts are historically failed, principally as a result of their conception as centralized systems. The reasons for these failures include limited accessibility and scalability of the repository, exclusive control over cataloged components and poor economy of scale.

There are many examples of large interoperability projects that have had significant impact on technology. Software library programs include Electronic Library Services and Application (ELSA), Asset Source for Software Engineering Technology (ASSET), Comprehensive Approach to Reusable Defense Software (CARDS), Computer Software and Management Information Center (COSMIC), Defense Software Repository System (DSRS), Reusable Ada Avionics Software Packages (RAASP) and NetLib. Missing from these approaches, however, is a standards-based strategy for integrating capabilities to enable the entire software community to share beneficial products and services. FedeRaL provides valuable insight into the issues of heterogeneous library interoperability.

ELSA, COSMIC and CAE Inc. used the X.500 directory services to accommodate the data model defined in the RIG's BIDM [19]. This experiment defined a new X.500 object class to represent the BIDM. Each library independently provided its assets' metadata in its own X.500 Directory Service Agent (DSA).The directory services allowed users to search the BIDM descriptions of assets and identify the locations of assets desired. Options controlled the depth of a search operation to find exact, wildcard or approximate match, to search aliases or direct nodes.

The Multimedia-Oriented Repository Environment (MORE) uses a structured interface to a database maintaining information about assets [20]. The structured information translated on-the-fly into HTML and transmitted via HTTP. This type of integration allows a web browser to link into many different libraries presenting a common user interface.

FedeRaL has two main contributions to reuse library studies. The object of this tool is to create a scalable and federated reuse libraries over the Internet. By "scalable", we mean a reuse library constructed from scratch and then extended using the library extension framewok explained in section 5. By "federated", we mean integrating an existing reuse library into "distributed" reuse library architecture. Moreover, federation is supplied by schema integration. However, scaling is supplied by schema extension.

2 The Architecture of FedeRaL

FedeRaL is a client-server application developed using Microsoft tools as depicted in
Figure-1. The Internet Explorer user connects to a web page designed to access a
reuse library. The end user enters a question in English. Internet Explorer then passes
the question to the Microsoft Internet Information Server along with the URL of the
ASP page that executes VBScript. The script passes the question to English Query for
translation into SQL. English Query uses domain knowledge about the target database
to parse the question and translate it into SQL. The script then retrieves this SQL,
executes it using ASP database control, formats the result as HTML and returns the
result page to the user [13].

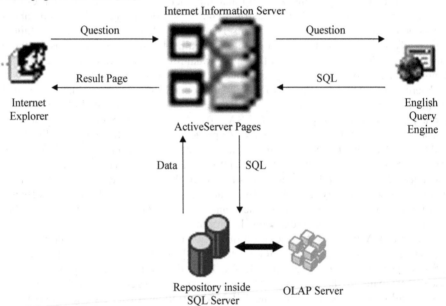

Fig. 1. Architecture of FedeRaL

The Microsoft Repository technology is used as the core of **FedeRaL**. Information
about software related components, metadata, has been organized and catalogued into
the repository. It is realized that for the repository to be a success it needed to be able
to easily locate the software component a user desires. Key factors related to this are
the user interface of the tool, the scalability of the tool and the manner in which the
components in the repository are classified.

Section 3 of this paper introduces the adaptation of IEEE standard 1420.1-1995
into Microsoft Repository environment. Section 4 presents the services offered by the
tool from two perspectives: administrative and user. The extensibility and
interoperability issues are discussed in section 5. Finally, section 6 presents the
current status of the tool and concludes with future considerations.

3 Microsoft Repository: The Core of FedeRaL

One of the most difficult but most important part of defining the repository is to define the schema. The schema defines the metadata of the components that will be stored in the reuse library. Our tool accepts the IEEE standard 1420.1-1995, Data Model for Reuse Library Interoperability: Basic Interoperability Data Model (BIDM) as the basis of schema [14]. The purpose is to use BIDM as the minimal set of information about components that reuse libraries should be able to exchange to support interoperability and by the way, it provides a common framework for the entire federated reuse library.

Microsoft Repository [17] provides a common place to persist information about objects and relationships between objects. In doing so, it provides a standard way to describe object-oriented information used by software tools. Our tool is modeled in Microsoft Repository through an information model. Each information model is stored in the repository database as a repository type library. A repository database is dynamic and extensible. Little of the actual structure is fixed; tables and columns of data are added as new models are defined and as existing models are extended. It enables a user to extend existing models.

Microsoft Repository's ActiveX interfaces are used to define open information model based on BIDM. Repository engine is the underlying storage mechanism for the information model. It sits on top of Microsoft SQL Server. Therefore, Microsoft Repository stores the metadata information about software components.

The features that Microsoft Repository helps us while developing our tool are:

- It already contains useful information. For example, it has its own type information models like Unified Modeling Language (UML) and Repository Type Information Model (RTIM).
- It is object-oriented. Its methods recognize the object as a unit of persistence.
- It enables manipulation of object models using the same techniques used to manipulate your other data.
- It supports COM and Automation programming.

3.1 Implementing BIDM Using MS Repository

BIDM is developed by the RIG. The purpose of the BIDM is to define the minimal set of information about components that reuse libraries should be able to exchange to support interoperability. MS Repository Type Information Model (RTIM) is used to create BIDM's information model. Therefore, the basic building blocks used to create BIDM are classes, interfaces, properties, methods, relationship types and collection types.

In order to be an information model extensible, each class must expose its properties, collections and behaviors through interfaces. To have the instances of a class exhibit certain behaviors or have certain properties or collections, each class implements an appropriate interface. A class in BIDM can be adapted to a class in RTIM by using an interface and a class. For example, for the class "Asset", there must be a class named as "Asset" and the interface "IAsset" implemented by it. However, "Asset" class cannot have properties. It implements "IAsset" interface and by the way it has the properties exposed by that interface. The figure below gives a pictorial overview for the classes, interfaces and properties.

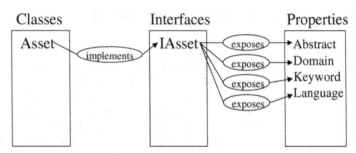

Fig. 2. Classes, interfaces and properties.

A relationship can be read in two directions. For example, consider "WasCreatedBy" relationship. This relationship relates "Asset" and "Organization" classes. It shows which organizations originated or produced an asset. Although awkward, "WasCreatedBy" relationship can be paraphrased by these two sentences:

- An asset is in the set of assets created by an organization.
- An organization is in the set of organizations created an asset.

Therefore, each relationship belongs to two collections. As a consequence, an information model is the mixed of various RTIM constructs such as relationship types, collection types, properties, methods, interfaces and classes. In the figure below, a part of BIDM as developed by using RTIM constructs is presented.

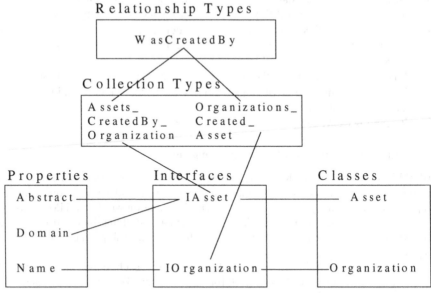

Fig. 3. Relationships and collection types

When the company plans to reuse, first of all, it must articulate the kinds of data that the reuse library will manipulate. The definitions of these kinds of data, called types, are stored in the repository by creating an information model.

The steps involved in developing an information model can be obtained by answering the following questions.

- What kinds of objects will the repository store? That is, what are the classes to which the repository's objects must conform?
- What kinds of relationships will the repository store? That is, what are the relationship types that describe how objects can be related?
- What properties apply to the objects of each class or the relationships of each relationship type?

An information model is a template for data. It is an object model stored in MS Repository. It accommodates the kinds of things that the repository manipulates and thus includes classes, relationship types and properties.

4   Services in FedeRaL: Administrative and User Perspectives

4.1   Administrative Services

- Define a reuse library information model, based on BIDM, articulating fundamental reuse library concepts: This service is handled through an Internet downloadable program written in Visual Basic. Reuse library information models can easily be created with the aid of user friendly user interface.
- Extending reuse library information models using existing reuse libraries: Reuse administrators may adapt an information model to their reuse based software engineering methodology. The base information model can be any information model based on BIDM.
- Define the facets for the classification scheme used by OLAP: Querying large collections of asset metadata causes performance degradations in system response time. If the faceted classification scheme (5, 6) is applied to an information model by this service, performance of the system will get better.
- Defining, viewing and updating asset descriptions based on a reuse library information model: Reuse administrators have the right to define, to view, and to update the instances of the information model. The user interface of these services are implemented by Active Server Pages –on the server side- and by ActiveX Document DLL's -on the client side-.
- Establish and maintain user access to reuse library: Administrator of any reuse library defines the access rights of the ordinary reusers.

4.2   User Services

- Querying asset descriptions using a natural language processor which is constructed from a reuse library information model and its classification scheme.
- Interlibrary asset exchange using XML: EXtended Markup Language is a text-based format for representing structured data. It is the subset of the Standard Generalized Markup Language (SGML), which has achieved a growing

momentum in the Internet world. Our reuse library information models are OIM compliant. XML Interchange Format (XIF) defines a set of rules that govern the encoding metadata objects described by our information model in XML format. XML ensures that structured data will be uniform and independent of applications or vendors. XML is used in **FedeRaL** from these point of views:

- To describe metadata about reusable components.
- To publish and exchange repository database contents.
- As a messaging format for communication between federated reuse libraries.

XML provides interoperability using flexible, open, standards-based format. It separates the data from the presentation and the process, enabling you to display and process the data as you wish by applying different style sheets and applications. With XML, structured data is maintained separately from the business rules and the display.

- Querying library usage metrics.
- Subscribing to different reuse libraries.
- Constructing reuse experience reports.
- Support of different data delivery protocols: In the context of **FedeRaL**, data delivery can be defined as the process of delivering information from reuse libraries as servers, to reusers as clients. There are possible ways that servers and clients communicate for delivering information to clients, such as:
  i.   clients request and servers respond,
  ii. servers publish what are available and clients subscribe to only the information of interest, or
  iii. servers disseminate information by broadcast.

  FedeRaL incorporate different types of information delivery in separate parts of system. "Clients Request and Servers Respond" protocol is used within the asset search mechanisms in **FedeRaL**. The "Publish/Subscribe" protocol is beneficial when delivering new or modified data to clients that subscribe to reuse libraries. The "Servers Broadcast" protocol is useful for delivering information to specific user group in a reuse library.
- Inspecting asset usage history

5   Extensibility and Interoperability in FedeRaL

Successful software information exchange will require the integration and extension of existing standards; library science standards, CASE tool interoperability standards and network protocol standards.

FedeRaL addresses 3 of the standards for successful scalability of reuse libraries. It uses BIDM as the library science standard. **FedeRaL** offers an interoperable reuse library architecture by using open information model which enables reuse administrators to construct the information models on top of their reuse based software engineering methodology. **FedeRaL** uses DCOM for interoperating reuse libraries.

The aim is to start with a reuse library based on **FedeRaL**. Whenever the library information model does not accord with the data produced by the reuse based software engineering methodology, reuse administrator must have to extend the existing information model or create a new one based on the minimum information model (MIM) that satisfies the needs of BIDM. FedeRaL supports the creation of new information models based on MIM. In other words, creating a new information model is extending an existing information model. If we extend from an existing information model, the interoperability degree increases since the information models are derived from the same information model, namely MIM.

For two libraries to be different, their information models must be different. In other words, two reuse libraries differ in their information models and if the base information models are the same they can easily interoperate.

A reuse library information model corresponds to a Type Information Model (TIM) in Microsoft Repository. In one repository database, there can be many TIMs and reuse library information models as well.

FedeRaL proposes framework oriented extensibility. A framework is mainly a library information model used to hold the metadata structure of reuse library. The basic building blocks of an information model in terms of Microsoft Repository TIM are ClassDef, InterfaceDef, RelationshipDef, PropertyDef and MethodDef. Extending an information model means defining a new class, interface, relationship, property or method. In the FedeRaL, reuse library extensibility is supported through use of interfaces. Extensibility can be expressed in terms of these extensions:

- Defining a new property.
- Defining a new interface.
- Defining a new relationship.

As information models are extended from existing information models, a library of information models will begin to emerge.

Beside extending an information model, an information model can be created using two or more information models. In this situation, the newly created information model is constructed by integrating the participating information models.

There is one extra relationship between two information models. Even if the same information model is used for modeling the reuse based software engineering methodology, the assets produced by them may differ in the application domain. This relationship is called **clones** relationship between information models.

The next figure depicts the relationships among information models where each information model is handled as a framework. In this figure, Framework-1 is a MIM. If a corporate that will participate in the federated reuse library architecture, wants to use BIDM as its own reuse information model then it extends Framework-1 which results in Framework-2. For example, if the corporate is using Visual Basic as its development environment and if the products created by the corporate will be reused then the newly created information model must accommodate the information model for Visual Basic development environment. Hence, the reuse library architecture will expand on the Internet. On the other hand, Framework-3 may be created by a corporate that uses Java development environment. Moreover, if any corporate that develops software both using Visual Basic and Java wants to establish a reuse information model, it simply uses existing frameworks, namely Framework-2 and Framework-3, resulting in Framework-4. Framework-5 clones Framework-4 because the components produced by them may differ in the application domain.

Moreover, this implies a bit more powerful structure. If a new property or method is added to the information model, it is directly added to the interface. Hence, all the classes implementing this interface are automatically affected by that change. On the other hand, if a new interface is defined, it is easily be implemented by any class. Users of the information model do not aware of the existence of interfaces. They just see the classes of the information model.

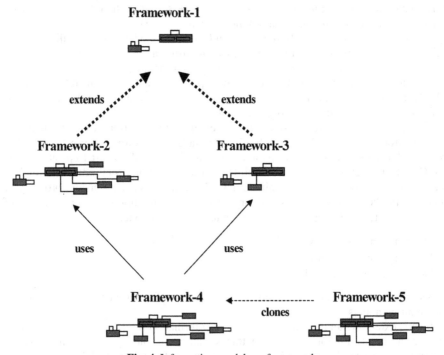

Fig. 4. Information models as frameworks

Each information model contains an interface model and a default class model. All interface models define a set of interfaces through which **FedeRaL** can manipulate repository objects. These interfaces define a standard through which distributed/federated reuse libraries can interoperate. All class models define a default set of classes that may be used to create repository objects. These are not a standard and many tools using the MS Repository will typically define their own class models (and additional interface models). Normally, only the tool that creates an object needs to know its class. All other tools manipulating the object will only be concerned with the interfaces it supports.

To achieve interoperation between tools, tools must be written to manipulate objects through their interfaces. Although default classes are provided, tools should not make assumptions about the classes used by other tools to create objects.

6 Conclusion and Future Considerations

FedeRaL is a tool that enables reuse administrators to construct their reuse library information model based on BIDM. In this sense, a series of reuse libraries will begin to emerge as the reuse administrators constructed their own reuse libraries and FedeRaL acts as a glue among them. It is interoperable because its administrative and user services enable reusers to browse and search different reuse libraries all of which are built on top of BIDM. However, up-to-time reuse library extensibility can be completed, but the user interface of the FedeRaL are not yet implemented. Also, there are some problems while adapting faceted classification to OLAP technology. On the other hand, FedeRaL uses the most recent methodologies that the technology is offering.

References

1. Fugini, M.G., Nierstrasz, O., Pernici, B.: Application Development through Reuse: the Ithaca Tools Environment. ACM SIGOIS. Bulletin (1992). 13-2, 38-47
2. Nierstrasz, O., Gibbs, S., Tsichritzis, D.: Component-Oriented Software Development, Communications of the ACM, (1992) 35-9, 160-165
3. Mahalingam, K., Huhns M.N.: A Tool for Organizing Web, IEEE Computer. (1997) 80-83
4. Solderitsch, J., Bradley, E., Schreyer, T.M.: The Reusability Library Framework – Leveraging Software Reuse. System and Software Engineering Symposium. (1992) 1-19
5. Mili, H., Ah-Ki, E., Godin, R., Mcheick, H.: Another Nail to the Coffin of Faceted Controlled-Vocabulary Component Classification Retrieval. ACM SSR. (1997) 89-98
6. Prieto-Diaz, R., Freeman, P.: Classifying Software for Reusability. IEEE Software. (1987) 6-16
7. Kramer, R.: Databases on the Web: Technologies for Federation Architectures and Case Studies. ACM SIGMOD. (1997) 503-506
8. Eichmann, D.: Software Engineering on the World Wide Web. International Conference on Software Engineering. (1997) 676
9. Ning, J.Q.: ADE – An Architecture Design Environment for Component-Based Software Engineering. International Conference on Software Engineering. (1997) 614-615
10. Kramer, J., Magee, J.: Distributed Software Architectures. International Conference on Software Engineering. (1997) 633-634
11. Lynch, C.A.: Networked Information Resource Discovery: An Overview of Current Issues. IEEE Journal On Selected Areas In Communications. (1995) 13-8, 1505-1521
12. Krueger, C.W.: Software Reuse. ACM Computing Surveys. (1992) 24-2, 131-183
13. Nguyen, T., Srinivasan, V.: Accessing Relational Databases from the World Wide Web. ACM SIGMOD. (1996) 529-540
14. Browne, S.V., Moore, J.W.: Reuse Library Interoperability and the World Wide Web: ACM SSR. (1997) 182-189
15. Henninger, S.: Supporting the Process of Satisfying Information Needs with Reusable Software Libraries: An Empirical Study. ACM SSR. (1995) 267-270
16. Isakowitz, T., Kauffman, R.J.: Supporting Search for Reusable Software Objects. IEEE Transactions on Software Engineering. (1996) 22-6, 407-423
17. Bernstein, A., Philip, et.al.: The Microsoft Repository. Proceedings of the 23rd VLDB Conference. (1997)
18. Seacord, R.C., Hissam, S.A., Wallnau, K.C.: Agora: A Search Engine for Software Components, Technical Report CMU/SEI-TR-98-011. (1998)
19. Tri-lateral Interoperability Phase 2 Technical Report. Advanced Research Projects Agency.
20. Eichmann, D., McGregor, T., Danley, D.: Integrating Structured Databases Into the Web: The MORE System. First International Conference on the World Wide Web. (1994)

Multimedia Information Systems: Towards a Design Methodology Based on Spatio-Temporal Relations, Petri Nets, and ECA Rules

Sabine Boufenara [1], Zizette Boufaïda[1], Mahmoud Boufaïda[1], and
Denis Poiterenaud[2]

[1]Lire Laboratory, Computer Science Department
University of Mentouri, Constantine
25000 Algeria
{boufenara, zboufaida, boufaida}@hotmail.com
[2]LIP6 Laboratory, Pierre & Marie Curie University
Paris 6, France
denis.poiterenaud@lip6.fr

Abstract. We present a methodology for developing multimedia information systems. These latter are considerably different from the traditional ones because they deal with very complex and heterogeneous data such as sound, still- image, text and video. The complexity of data is principally due to their dependency of time. During their presentation, multimedia data must be synchronized on both temporal and spatial scales. We define a conceptual object model and give an outline of a design methodology. The presented model is a combination of object oriented concepts and Allen's relations. In order to make the designer task easier and to help him during the development of his model, we have based this latter on existing formalisms such as those offered by OMT. The defined method is based on the object specification, and a number of models that are designed during the development process according to the water-fall approach.

1 Introduction

Nowadays, electronic information management is in a permanent evolution. Many information types that were traditionally considered as analogic are now processed in a digital way. They consist of complex types such as images and sound. Considering this evolution, users or more precisely societies managers have manifested their needs to manipulate information in its natural and expressive form, i.e. MultiMedia (MM) information.

The process of MM information gives birth to a new type of ISs (for Information Systems) called MMISs (for MultiMedia ISs). When speaking about MM, we refer to data such as video, sound, still-image and text. These data are heterogeneous and voluminous requiring enormous storage means. MMD (for MultiMedia Data) are dependent of time since their presentation on screen takes a certain duration that can be pre-defined for dynamic data such as video and sound data, or free for static data such as text and still-images. The MMD overlapping or sequence needs

T. Yakhno (Ed.): ADVIS 2000, LNCS 1909, pp. 384–397, 2000.

synchronization on temporal scale. MMD are also dependent of space since they occupy a surface on the screen (when the data are different from sound). In order to avoid non-ergonomic presentations and to use them later in the search-based presentation, their organization on the spatial scale must be managed by spatial synchronizations.

In this paper, we present a solution for designing MMISs, by defining a conceptual object model and giving an outline of a MMIS development method. The MOM (for Multimedia Object Model) is a combination of the object-oriented concepts and Allen's relations [5]. It allows the designer to specify temporal relations as well as spatial ones in an object-oriented environment. This latter is adapted for ISs development. The model integrates a set of rules based on ECA (for Event/Condition/Action) rules [8]. These rules are used to manage the unexpected manipulations of the user, over the model. The defined method contains six steps that are ordered according to the water-fall approach [7]. It is based on the specification of the object by means of the conception of a set of models (used to complete the object's specification). We also use a stochastic Petri net [13] in order to model temporal presentations. This one is used to validate temporal presentations in the MOM. A set of mapping rules is defined to transform the valid MOM into a database model. To accomplish this task, we use the logical model of the DBMS (for DataBase Management System) O2 [11].

The paper is structured as follows: In Section 2, we present related works done in the area of MMISs. Then, we introduce the developed model via its concepts in Section 3. In Section 4, we give the proposed method for developing MMISs. Finally, we conclude our work by giving some perspectives for the future.

2 Related Work

The temporal data synchronization is the ordering of data for which the presentation depends of time, on a temporal scale. This synchronization was largely studied ending to many temporal models. These ones are either based on the instant notion or on the interval one. In [17], Weiss defined six operators to express causal relations between multimedia objects: [sequential, parallel, starts, parallel-min, parallel-max, identical, equal]. These operators are not sufficient for modeling all temporal combinations. Whereas Allen's relations introduced in [5] are more global, integrating all binary cases that may exist between two multimedia objects. Allen's relations are the basis of many models such as the temporal logic based model [9], the interval based models [12], or the STORM system [2].

Most of developed models are not concerned with the definition of a conceptual model for the development of MMISs. We can find in [16], a MMIS development approach based on the method O* [14]. This approach deals with the distribution of objects and the federation of data. However, it does not take into account MMD synchronization, neither on the temporal scale nor on the spatial one. This approach does not also specify the multimedia object definition. This latter is considered as a classical one.

In [10], a design method for developing distributed MMISs and a CASE tool integrating the method design, are presented. This method is specific to medical ISs design and the synchronization of MM objects was superficially studied.

In STORM [2], an approach for constructing multimedia DBMSs is given. This one studies the synchronization of objects in time and space. A multimedia presentation is modeled in the object. Unfortunately, this approach creates a certain conceptual redundancy of objects. Also, when a user needs to define a new presentation made of objects belonging to another presentation, he must define a new object containing the new presentation. STORM does not give a global solution for developing MMISs.

Our approach tries to give a solution to some problems we met in other approaches with considering the synchronization of MMD. We also propose a precise methodology for the MMISs development.

3 Definition of the Conceptual Model

In a multimedia environment, constraints to which entities are submitted are the spatio-temporal ones. They are inherent to the complex nature of MMD. Effectively, MMD have got a set of spatio-temporal characteristics that are generally not considered by classical data models. It is essential that a user can be able to identify and organize his objects at the step of design. He should be able to express the object relations in time and space, basing them on a solid platform.

In order to represent the time notion, we have introduced some synchronization concepts inspired from Allen's temporal relations [5]. These concepts will also be used to model the spatial relations between objects.

The developed model is seen as a specification of a high level, making abstraction of all technical aspects such as compression and quality of service. It contains MM objects, MM classes, and temporal and spatial synchronization concepts.

3.1 The Basic Concepts

The MOM model we propose is based on some existing concepts, such as class, associations, object, inheritance link, composition link and abstract classes. Some new notions have been introduced. They concern the MM objects, MM classes and abstract MM classes and link classes used for the MM object synchronization.

The Object in MMISs. We consider two types of objects: the classical object as known in the object oriented approach and the MM object that we define especially for the MMISs development.

The Classical Object : We adopt the concept of object as defined in the O* method [14], where events are defined in an explicit manner and separately from operations. These latter concentrate on 'how' objects are modified without giving the reasons that have triggered this state change. Indeed, events give the object a new dimension and make it more active.

A classical object is defined according to three perspectives: the static one (described by the spatial properties defining *attributes* and the temporal ones defining *events*), the dynamic one (described according to *operations* and *events occurrence rules*) and the behavioral one (described by messages sending or an object inner decision to change when undergoing some changes of its state).

The Multimedia Object : At the moment of their presentation, MMD manifest some time and space dependencies. For example, we can not display the image of the *'Eiffel Tower'* during the whole work session without carrying to erase it. Thus, we assign to each complex object (i.e. an object with a complex type) a specific presentation duration and give the user the possibility to interrupt it. The real world entities dependent of time or space are modeled by multimedia objects [6].

We define a multimedia object as an object that inherits all properties described by the three perspectives. In addition, this object is able to express its temporal and spatial relations. We identify a MM object as Object = (DateP, Content, Duration, δx, δy, dx, dy) where:

- *DateP* is an attribute that describes the creation date of the MM object in the database.
- *Content* is an attribute which identifies by a set of key words known by the designer, the informational content of the sonorous, video, textual or still image sequences.
- *Duration* is an attribute associated to objects from still-image, text, video and sound types. It determines the interval time associated to the presentation of the informational content of the MM object.
- *δx, δy, dx* and *dy* are attributes that determine the spatial position of the multimedia object on the screen. Only objects of Still-Image, Text and Video types are concerned with these four attributes.

Multimedia Classes : We define two types of classes: classes inherent to the domain we are modeling (for instance the classes: Doctor, Echography, ...) and those inherent to the multimedia domain [6]. The creation of the second type of classes is necessary for the MMIS development.

Since we have defined multimedia objects and as an object is a class instance, then multimedia classes are just an abstraction allowing the encapsulation of multimedia objects inherent to the domain application and having the same information type.

We propose four new super classes called *CStill-Image, CText, CSound* and *CVideo.* These are abstract classes. They are instanciated via their derived classes. Each one contains its own abstract operations (*Activate Sound, Activate Video, ...*) and its attributes, symbolizing information about the media type.

Associations : An association describes a group of links, having common structure and semantics. It describes a set of potential links in a same way that a class describes a set of potential objects.

In the model, we define new associations called link classes. These associations represent the synchronization of multimedia objects.

3.2   Temporal Synchronization Concepts

Each multimedia object, perceptible by a user, is in constant interaction with objects of its environment. In this sub-section, we just consider temporal interactions. For instance, an employee photography is displayed for 30 seconds in parallel with a video presentation of the same employee manipulating a factory engine.

In the ISs, data are, in a certain manner, static. With the emergence of MMISs, data became more dynamic due to temporal relations between different entities.

A system must preserve temporal relations between objects via the synchronization process. In order to ensure this synchronization, we have used Allen's relations. These latter are widely used to represent temporal aspects of MMD [1],[3],[10].

Application of Allen's Relations on Multimedia Data. We have defined four abstract classes: *CStill-Image, CText, CSound* and *CVideo*. The instances of the derived classes are time dependent. So, they should be ordered on a time scale. Their interactions are supervised by Allen's relations. In order to ensure the object ordering in time, two types of media have to be connected by one of the seven Allen's relations that are either parallel such as start, overlap, during, finish and equal; or sequential such as meet and before.

For example, the specification [start] ([meet](v1, v2), s1), where v1 and v2 are objects with Video as type and s1 is a sound object, is a temporal composition in which the presentation of the video sequences v1 and v2 is sequential. The relation 'meet' ensures this sequential presentation where v1 precedes directly v2. The beginning of the v1 video presentation triggers the activation of the sonorous object s1. This is ensured by the relation 'start'.

A time duration that coincides with the interval in which the object is perceived by the user is associated to each multimedia object. The duration is noted $T\alpha$ where α represents the multimedia interval.

A total duration Ttr and a duration $T\delta$ (expressing the relative duration between the beginning of the first interval and the beginning of the second one) are associated to each binary relation between two multimedia objects

Integration of Temporal Relations in the Object Model. Our aim is to represent in the same model, all the aspects bound to traditional and MM data. This is done by the combination of Allen's relations and the object model [15] leading to the MOM. We can say that this solution is rational because the Allen's relations can be easily embedded to the object model, leading to a unique model that represents the static aspect (by means of attributes), the dynamic aspect (by means of operations), the behavior aspect (by means of message sending) and finally synchronization aspects (by means of temporal relations).

In the object model, instead of using the association notion to represent relations, we model each temporal relation of Allen by a link class. This choice allows us to make use of the inheritance property. It allows one to avoid the definition of as many associations as relations. It becomes easy to derive all relations between the multimedia objects from a super link class. So, we define seven link classes: MEET, OVERLAP, DURING, EQUAL, FINISH, START and BEFORE corresponding respectively to the Allen's relations.

Constraints of the Temporal Synchronization. In this section, we give some constraints bound to the multimedia objects synchronization. Two objects having the sound type and belonging to the same class can just have a binary sequential relation.

The objects of types still-image and text are considered as static ones, whereas those of type video and sound are dynamic [4]. There is not a specified duration bound to a text or a still-image: they can be presented during 10 minutes or 10 seconds. Whereas video and sound are explicitly bound to time. For instance, a video is played during 15 minutes with the cadence of 30 still-images per second. Thus, the designer must give the duration of video and sound entities at the moment of design.

The duration of a binary relation, noted Ttr can just be computed by Tα and Tβ duration of intervals in relation, and the temporal relation TR (meet, equal, overlap, ...) between them. Temporal parameters must satisfy a temporal relation TR from the seven temporal relations detailed in [6]. A presentation may contain more than two objects. An object may belong to many presentations. These two last properties are ensured by a composition link.

3.3   Spatial Synchronization Concepts

A multimedia object having the text, still-image or video type is perceived by the user. This property is considered via the spatial synchronization relative to the disposition of multimedia entities on screen.

The Notion of Spatial Relations. A spatial relation indicates the relative position of entities on screen. For design uniformity reasons, we have integrated the spatial aspect bound to multimedia entities, in the MOM defined in the previous sub-section. To each multimedia object appearing on the screen, we associate a window containing its visual informational content. The inferior left corner has coordinates noted (δx, δy). The window dimensions, i.e. the height and the width, are specified successively by dy and dx. The repair origin is the inferior left corner of the screen. We suppose that objects, i.e. windows cannot be deformed. The objects' overlapping creates the spatial relation that must be managed.

Modeling the Spatial Aspect. Allen's relations are very powerful in modeling all cases of the binary relations between entities, either in a temporal context or in a spatial one. However, these relations were only defined for a temporal use. We have adapted them to the spatial context and more exactly to the object model defined above. Once the spatial relations are embedded to Allen's relations, their integration to the MOM becomes easy since we have already integrated Allen's relations in the object model. We define, in the MOM, seven new link classes MEETs, BEFOREs, EQUALs, DURINGs, OVERLAPs, STARTs and FINISHs, such as 's' indicates their spatial property. These link classes are derived from the super class DISPLAY, which visualizes the multimedia objects on screen (see Fig. 1.).

Each spatial link class contains methods for computing the object emplacement on screen. These methods must check for each pair of objects, their relative position either on the abscises, on the orders or on both. This is done thanks to the definition of two variables X and Y where $(X, Y) \in \{(0, 1), (1, 0), (1, 1)\}$.

Constraints of the Spatial Synchronization. We notice that objects are disposed on two dimensions. This justifies the choice of two axes rather than only one, as it was the case of the temporal aspect.

The relative disposition of objects on two axes implicates many interpretation possibilities of each spatial relation (SR). This multiple interpretation trains a certain ambiguity when modeling the SR. In order to solve this ambiguity, we have developed a number of constraints that will also be used to check whether spatial attributes satisfy a SR or not. In table 1, we give an example of constraints bound to the application of the relation *Overlap* to the spatial aspect.

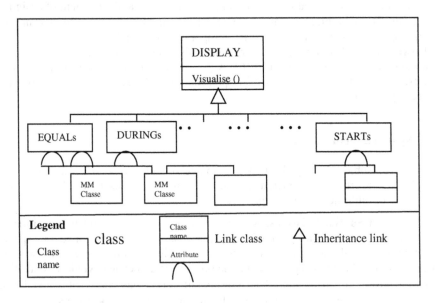

Fig. 1. A part of the spatial class hierarchy

Table 1. Constraints bound to the spatial relation OVERLAPs

Spatial Relation SR	Constraints		
	X=1 and Y=0	X= 0 and Y=1	X=1 and Y=1
OVERLAPs	$\delta y1=\delta y2,$ $\delta x1<>\delta x2$ $\delta x2<\delta x1+dx1$	$\delta x1=\delta x2, \delta y1<>\delta y2$ $\delta y2>\delta y1+dy1$	$\delta x2<\delta x1+dx1$ $\delta y2>\delta y1+dy1$

O1 and O2 are two MM objects where (O1, O2) type $\in$ {still-image, video, text} and O1 ($\delta x1, \delta y1, dx1, dy1$) and O2 ($\delta x2, \delta y2, dx2, dy2$).

3.4   Constraints of the Overlapping of Spatial and Temporal Shadows

The introduction of both spatial and temporal aspects is made in the same object model. So, it may train certain interactions that have to be managed. For this reason, we have defined some constraints given by the following formula:

If $\exists$ TR(O1, O2) / TR $\in$ {BEFORE, MEET} then $\exists$ SR(O1, O2).

One can notice that it is meaningless to construct a spatial relation between two perceptible objects if they are presented at different timing duration.

3.5   Management of Unexpected Interaction Concepts

At the running time, the object model must allow the user to manipulate multimedia presentations in a temporal way. We are speaking about operations such as backward/forward, go to the beginning, go to the end, pause/reprise and even to stop an object presentation among a set of objects being presented. Concerning the last type of manipulations, we do not accept meaningless ones. The operation '*stop*' can only be applied to objects of the 'text' type, since the reading speed varies from a user to another.

The user manipulations may train a time ordering confusion of objects belonging to the same presentation.

In order to manage interactions, we have proposed a set of rules based on ECA ones. As a matter of fact, these rules can be easily modeled in the object, since we defined a particular structure for the objet.

- Events are defined as temporal properties of the object.
- Conditions are defined as an event occurrence rule in the object.
- Actions are the triggering of operations by an object, on other objects belonging to the same presentation.

A counter of seconds is associated to each presentation.

The set of ECA rules is defined on a new class CECA. This class is used so as to abstract all manipulation rules. Presentation classes inherit from this class. In the following, we give two examples of rules corresponding to the allowed manipulations on a presentation.

Rule 1   *On pause*
    If counter < Ttr (temporal presentation not finished yet)
        Then (save MM objects states and stop the presentation) ∧ *(presentation state = pause)*

This rule is triggered when the user stops the MM presentation for a certain duration. It allows him to check if the total temporal duration corresponding to the presentation is not finished yet. In this case, the new state of MM objects is saved in a variable until the user wants to continue the presentation by using the event 'continue'.

Rule 2   *On forward (t1)*
        If t1+counter <= Ttr
        Then (counter = counter + t1) ∧ *(execute the presentation of the link class corresponding to the counter)*

This rule allows the user to put forward his presentation. In this case, we must check if we do not surpass the total duration. In this case, the link class corresponding to the new position (counter+t1) is executed for presentation.

4 Our Methodology for Developing Multimedia Information Systems

The approach we suggest is an object modeling technique based on temporal models and PNs (for Petri Nets). We focus on the temporal and spatial aspects bound to MMD synchronization at a conceptual level.

The proposed method provides two new models, the MOM, which has been presented in the previous section, and the temporal presentation model based on the stochastic PNs. This model is used to validate temporal presentations modeled in the object model. This approach follows the 'water-fall' one for the MMISs development process (see Fig. 2.). It is based on the construction of four independent models: the MOM, the DMM (for Dynamic Multimedia Model), the FMM (for Functional Multimedia Model) and the TPM (for Temporal Presentations Model). The approach generates a unique valid model MOM that will be translated into a database model by a set of mapping rules. This database model is based on the O2 concepts [11].

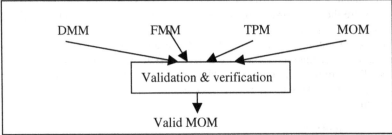

Fig. 2. Development process

Step 1: Construction of the MOM: The construction steps of the MOM can be assumed as follows :

a)   The designer begins by collecting real world entities. For each one, he defines events that may occur along its life. In fact, the resemblance between events can define a new way to gather different entities in a sole entity. Entities are called classes.

b)   Once the classes identified, their properties (operations and attributes) are defined. When the designer finds some MMD, he should process in a different way. Each type of data corresponds to a class, even if some of these MMD are interdependent or dependent on other atomic data. This isolation is due to the fact that MMD require a specific treatment, different from the treatment reserved to atomic data. For example, objects that are of sound type cannot be visualized on the screen. Text and still-image type objects can be deleted from the screen before finishing their appropriate duration without affecting their informational content presentation. Whereas, video and sound type objects affect their presentations (because they are dynamic).

c)   When classes share a common semantic or structure, they make part of an association. This last can have its own attributes.

d)   Collaborating with the user, the designer determines MM objects ordering on spatio-temporal scales. He should not neglect constraints bound to MM objects, such as duration of dynamic objects. Once, objects are ordered on both spatial and temporal scales, their specifications are converted into the pre-defined link-classes. Sometimes, the same ordering of objects is expressed by more than one specification leading to a form of redundancy. To avoid such redundancy, the designer should choose the most complete and optimal specification to be modeled in the MOM. For example, we have the temporal presentation of three objects O1, O2 and O3, where their ordering is the following (see Fig. 3.):

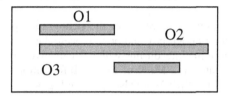

Fig. 3. Example of ordering three objects

This can lead to three specifications
1. Start (Meet (O1, O3), O2)
2. Start (O1, O2) and Meet (O1, O3)
3. Start (O1, O2) and During (O2, O3) and Meet (O1, O3)

The first specification cannot be modeled in the MOM because it generates a composed link-class 'Start'. This latter is composed of an object O2 and a link-class 'Meet'. Whereas, the second and the third specifications are allowed because they are binary and simple. We just note that the third one is more complete than the second one. After having defined the link-classes, the correspondent temporal and spatial attributes are defined on them.
e)   The final step of the MOM construction is the development of the classes hierarchy using inheritance and aggregation links. Let us note that the presentation classes inherit from the super-class CECA.

Step 2: Construction of the DMM: For each class having an important dynamic behavior, a DMM is constructed. It describes the objects life cycle by describing changes in term of events that each object undergoes along its life. The DMM corresponds to the dynamic model defined in the OMT method. It is extended to describe the life cycle of multimedia classes.

Step 3: Construction of the FMM: The FMM describes the system aspects relative to data transformations. It is a description of the data circulation between different processes. For each process, the designer describes input and output data, and gives a preliminary description of each function. He also identifies constraints on data, storage files and actors participation with data flow between them. The FMM is the one described in OMT but dealing too with MMD and then their corresponding functions.

Step 4: Construction of the TPM: A TPM is constructed for each temporal presentation. In order to model the dynamic presentation aspect, we use the stochastic PNs because of their dynamic modeling and their expressive power.
We instantiate the PN to a multimedia modeling by changing some of its descriptions. The new PN is noted SMPN (for Stochastic Multimedia Petri Net), such

as a place P models the informational content of a multimedia object, a transition models the multimedia presentation chaining, a token in a place indicates that informational content of this place is being presented (perceived by the user) and a time duration (corresponding to the guard) associated to each token indicates the time during which the user perceives the multimedia entity.

So, the user describes his wishes concerning the temporal ordering of multimedia entities in presentations; the designer converts these descriptions into SMPNs. These last ones allow the user to understand them easily and thus to check if they correspond to his requirements.

For example, the following narration of the Echography presentation: "I want to perceive the video of the Echography and once I locate the disease traces, I want them to be focused for a certain time during which I can give my comments about the disease" is translated into the SMPN of Fig. 4. Informational content and duration affected to each place of the SMPN are also given in Fig. 4.

After explaining the SMPN to the user, he checks if it really corresponds to his perception. If it is the case, so the SMPN is valid. If not, the designer tries to correct it so as it responds to the user's requirements.

Step 5: Validation of the MOM: The MOM is validated by the three models DMM (by adding missing operations), FMM (by completing the operations by functions and on events occurrence rules by constraints of the FMM) and TPM (by checking if all presentations described in the TPMs are really described in the MOM by link-classes and if presentation specifications modeled are the optimal ones). All models cited above were designed separately. The objective of their construction is to complete the object's specification and to validate presentations without the risk to develop a non-adequate system. The resulting model is the Valid MOM (VMOM) that can now be mapped to a data base model.

Step 6: Mapping the VMOM to a database model: The VMOM uses the concepts of the DBMS O2. This last offers object type description language as well as a programming language. O2 permits the code reuse and thus defines a library of methods associated to the object. It also defines the O2 kit library including some multimedia classes such as Image and Text classes.

In the following, we give an example of mapping rules. We begin by defining our departure hypothesis: all the atomic data are incorporated to the database via a user interface. MMD are supposed existing in the database, each one in a specific file. At the moment of defining MM classes, the user makes reference to storage files described in the FMM, i.e. he assigns a file type to each MM class and a file to each object.

We define as follows the new type VIDEO in O2. It corresponds to the class Cvideo in the VMOM as follows:

```
New type VIDEO is
            {structure tuple of ( DateP : date
                                  Content : list (string)
                    Duration : integer
                    )
            persist as VIDEO
            }
```

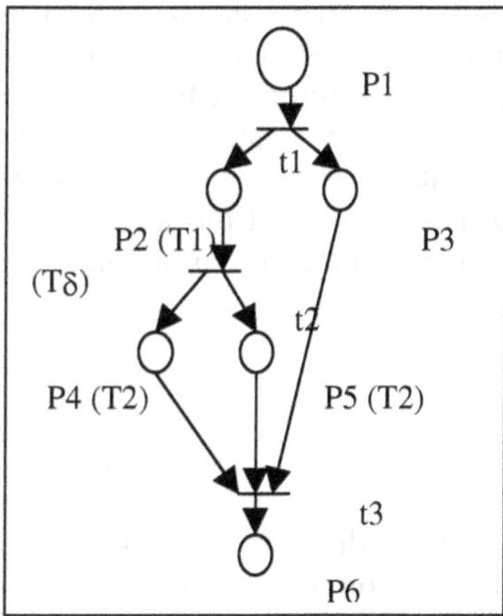

Place	Duration	Informational content
P1	Free	Presentation beginning place
P2	t1	Echography video
P3	tδ	Fictive place without any real content (modeling the waiting time before beginning presentations belonging to P4 and P5)
P4	t2	Image of the disease tracest
P5	T2	Audio comment (the image analysis)
P6	0	The end of the presentation

Fig. 4. SMPN of the Echography presentation

When instanciating this type, we can obtain the values of different videos assigned to different files. For example we can obtain the object Echo1 (from the example given in the last section). This object has the following structure:

```
ECHO1      inherit CECA
           Structure tuple of (Video      12/14/99      "the
echography video taken for P1 after he got all his
medecines" 1,5 minute)
           Methods (Delete: <body of the method>;
                   Modify (content): <body of the method>;
   . . .
                   Visualize (duration=1,5 minute): <body
of the method>)
```

This object inherits all rules defined in the CECA class. This will permit it to manage all user interactions such as: Forward, Backward,...

5 Conclusion

In this paper, we have proposed a MMISs development method. We have focused on the object-based design of the MMD. The management of such data is not easy, especially because of the lack of standard design methods. Furthermore, object oriented methods are not well suited to manage aspects inherent to MM synchronization on both spatial and temporal scales. So, we have introduced some temporal relations that we used to model both temporal and spatial aspects by integrating them in the object model defined in OM. We also used the stochastic PNs to model temporal presentations synchronization. In fact, PNs are the most dynamic models and the most expressive ones. We have used them to validate temporal presentations in the object model. Finally the MOM was transformed into a database model.

We have modeled an approach for MMISs development. However, in the MMISs field, too much work is to be done, notably concerning the modeling of heterogeneous data distribution, the development of request languages permitting a based content database interrogation, the standardization and the normalization of MMISs development methods and the definition of graphical interfaces. For the future work, we aim to represent data distribution, to construct a DBMS based on the proposed model and to construct tools permitting the use of PNs in order to simulate the multimedia presentation at real time.

References

1. Aberer, K., Klas, W.: Supporting Temporal Multimedia Operations in Object Oriented Database Systems. International conference on multimedia Computing and Systems. Boston, (1994)
2. Adiba, M. : STORM : Structural and Temporal Object-Oriented Multimedia Database System. IEEE International workshop on multimedia database systems, Minnowbrook Conference Center, USA   (1995)
3. Adiba, M., Mocelin, F. : STORM : Une approche à objets pour les bases de données multimédias. Techniques et science informatique, Vol. 16. (1997)   897-923
4. Adiba, M. : Serveurs d'objets actifs et présentations multimédias. Networking and Information Systems Journal, Vol. 1. (1998) 39-68
5. Allen, J. F. : Maintaining Knowledge about Temporal Intervals. Common ACM, Vol. 26. (1983) 823-843
6. Boufenara, S., Bboufaïda, Z. : On Designing Multimedia Information Systems: A Model Based on Temporal Relations., Proceeding of the first Mediterranean International Conference on computer Technologies, Tizi-Ouzou, Algeria (1999) 115-128
7. Gaudel, M. C. Marre, B., Schlienger, F., Bernot, G.: Précis de génie logiciel. Masson (eds) Paris 1(996)
8. Hanson, E. , Widom, J.: Rule Processing in Active Database Systems. International journal of expert systems, Vol. 6. (1993) 83-119
9. King, P.: A Logic Based Formalism for Temporal Constraints in Multimedia Document. Lecture Notes in Computer Science, Vol. 1293. Springer-Verlag, Berlin Heidelberg New York (1997) 87-101
10. Lapujade, A., Ravat, F.: Conception des systèmes d'information répartis : Application au milieu hospitalier. Inforsid, Toulouse (1997)
11. Lecluse, C., Richard, P.: The O2 Database Programming Language. Proceedings of the Fifteenth International Conference on Very Large Databases, Amsterdam (1989)
12. Little, T.S.D.C., Ghafoor, A.: Interval Based Conceptual Models for Time-Dependent Multimedia Data. IEEE Transactions on Knowledge and Data Engineering, Vol. 5. (1993)
13. Murata, T.: Petri Nets Properties, Analyses and applications. (1989)
14. Rolland, C. , Cauvet, C.: Modélisation conceptuelle orientée objet. (1990)

15. Rumbaugh, J. , Blaha, M., Premerlani, W., Eddy, F., Lorensen, W.: Object Oriented Modeling and Design. Prentice hall, Englewood Cliffs, New Jersey (1991)
16. Soutou, C., Lapujade, A., Ravat, F., Merlet, J. F., Nadalin, C.: Multimedia Information Systems, From Architecture to a Design Method. Basque International Workshop on Information Technology, Biwit (1994)
17. Weiss, R., Duda, A., Kjfford, D.: Composition and Search with a Video Algebra. IEEE Multimedia, (1995)

Modeling and Information Retrieval on XML-Based Dataweb

Amrane Hocine[1] and Moussa Lo[2]

[1] Département d'Informatique - Université de Pau
B.P. 1155 - 64013 Pau Cedex (France)
Amrane.Hocine@univ-pau.fr
[2] Laboratoire LANI - Université Gaston Berger
B.P. 234 - Saint-Louis (Sénégal)
Lom@ugb.sn

Abstract. The eXtensible Markup Language (XML) is considered as a new standard for data representation and exchange on the web. XML opens opportunities to develop a new generation of Web sites which allows to access and to treat their contents. Indeed disadvantage with HTML Web sites is that mainly HTML language describes how documents are presented to the viewer rather than describing the contents of documents.

In this paper, we propose an approach based on XML to design dataweb offering powerful tools of search and extraction of the information. In this approach the conception of a dataweb is based on three models: a structural model a navigational model and presentational model. The system of search and extraction of information which associated to a dataweb is founded on a concepts base composed of a base of meta-information and a domain thesaurus. It allows an interactive search for relevant documents (or extractions from documents).

1 Introduction

The fast development of the World Wide Web (WWW) has engendered in recent years many Web sites containing a huge amount of information. A large part of this information is stored as static HTML (HyperText Markup Language) pages that are only viewed through a web browser. The disadvantage with HTML Web sites is that mainly HTML language describes how documents are presented to the viewer rather than describing the contents of documents.

Different search engines (Alta Vista, Hot Bot, Lycos, Yahoo!, etc.) provide keyword-based search facilities to help users to retrieve information and the results again come as HTML pages. From one or several words, such engines collect a whole range of information from web documents that use these words, but results are often limited. The search engines do not take account of the semantics of the documents. This approach can be interesting for research on the whole web but becomes obsolete when used on a simple web site. The quality of the results can be improved with integration of meta-information in web sites, that is information on the content, the structure and the organization of the data of the site.

T. Yakhno (Ed.): ADVIS 2000, LNCS 1909, pp. 398–408, 2000.
© Springer-Verlag Berlin Heidelberg 2000

The development of Web sites is a complex task that requires a methodology of adequate conception. Our works concern a new generation of Web sites which offer opportunities of access and of treatment of their contents. Our approach : (i) is based on the notion of dataweb and new emergent standards around XML. The design of a dataweb is founded on three models: a structural model (or conceptual), a navigational model (or media) and a presentational model and (ii) integrate an information retrieval system which uses a domain concepts base. This one includes a base of meta-information and a thesaurus of the domain. It allows an interactive search for relevant documents (or extraction from documents).

In this paper we focus on: i) the generation in XML of the structural model of dataweb from a modelling Entity-Relationship (E-R) expressing the application domain of the site to build and ii) the process of interactive search of information which exploits a base of concepts consisted of a thesaurus of the domain and the base of meta-information.

In the process of information retrieval, we exploit two techniques which are very familiar to the users of the Web: i) initialization of the process of search by simple queries by means of key words; and, ii) exploitation and improvement of obtained results.

The information representation that is based on the document's logical structural integration with its semantic content allows the exploitation of the first technique. In classic information retrieval systems, keywords enable finding a list of documents containing these words. In our approach, the pertinent information (documents, simple parts of documents simply or automatically composed) is found from structural elements of the document (the tags), from the text described by these elements, and also meta-information describing its semantic content.

The second technique used is from the results of an interrogation query. These results are presented on a hypertext card (interactive card of concepts) allowing the user to refine the results by navigation or interrogation in the space of solutions proposed.

The structure of this paper is as follows. To show interest and XML's contribution, we present in the section 2 some works allowing to improve the search process for interrogation of HTML Web sites by integrating meta-information. The section 3 presents some main features of XML to allow the reader to understand the concepts explained in the following sections and the structural model of dataweb.

In the section 4, we present the dataweb architecture founded on a concepts base. We give some elements on the interrogation process which is based on the logical and semantic structures of documents.

Examples are extracted from the SIMEV application (Information System on the Senegal stream Valley's enhancements).

This application is part of the GIRARDEL[1] program that integrates into its actions the conception of a regional observatory of development. An implementation of this approach is currently under way in Java on a Java-enabled platform.

2   Meta-information to Improve the Interrogation Process of HTML Web Sites

Two types of solutions were proposed to integrate meta-information: the annotating of the HTML pages of web sites and the reorganization of the web site data. In all these systems, a web site is seen as a set of HTML pages.

2.1   Annotating of the HTML Pages of Web Sites

This first approach to meta-information integration consists of annotating the pages of an existing site with tags of meta-content. One creates a formal description of the content with HTML documents. Dobson and Burill [8] are the first to propose a semantic marking of the document content; they relied on the Entity-Relationship model that allows one to capture the conceptual level of the site. They introduced a set of three tags that allows one to define the entities or concepts in documents while labeling the sections of the text as attributes of these concepts, and while defining relationships of an entity to another external entity. Ontobroker [10] is an HTML extension that allows one to declare ontological annotations in the web pages. It uses some attributes-values couples in hyperlink tags. The authors of documents use these tags to delimit an element. WebKB [15] integrates some knowledge in the web pages while using knowledge representation languages to represent and to index information in the documents with HTML tags. It exploits the conceptual graph formalism and the ontology of concepts and kinds of relations.

2.2   Reorganization of the Web Sites Data

In this second approach, the researchers proposed semi-structured data models, databases and languages to model, to store and to interrogate the web sites. The interrogation languages like W3SQL, WebSQL [16] are interesting in the setting of precise information retrieval based on strings and structures (tags, link names...) retrieved in the documents. WebSQL introduces the navigable link notion to answer a query.

In other systems the query is done on an uniform structure constructed from the conversion/integration of heterogeneous data sources [7], notably the web, while using a semi-structured data model. The OEM model (Object Exchange Model) [1] has been conceived for semi-structured data management. In OEM the data constitute a graph with an object unique root. The YAT system [7] is based on a data model and a declarative language for integration and conversion of heterogeneous data sources. This model is more powerful than OEM: on the one hand the data are represented in YAT by arbitrary graphs, and, on the other hand, part of it has the capacity to possibly represent some data with their type

3 An XML-Based Dataweb Model

A dataweb is a collection of data: (i) structured such as the ones stored in relational or object databases; and, (ii) semi-structured as the data of Web. A dataweb system is a system that manages a dataweb.

3.1 Introduction to XML

The aim of this section is to give some basic notion about the main features of XML in order to keep the reader able to understand the concepts that we introduce in the next sections. For more details, see Michard [17].

XML has recently emerged as a new standard for representation and exchange of data on the Web [4]. It is a subset of the standard SGML [17], [20]. This meta markup language defines its own system of tags representing the structure of a document explicitly. HTML presents information and XML describes information. Let's consider the piece of XML document below:

```
...
        <ENHANCEMENT>
        <SOIL CODE = "DGX">
                <NAME>sandy</NAME>
                <HYDROLOGY>Important streaming</HYDROLOGY>
        </SOIL>
        <SEASON>Hivernage</SEASON>
        <TECHNIQUE>Rain</TECHNIQUE>
        </ENHANCEMENT>
    ...
```

A document is composed mainly of a tree of elements that forms the content. XML documents are composed by *markup* and *content*. In the above example, the coupling of tags <HYDROLOGY> et </HYDROLOGY> is the markup for the string, Important streaming.

Attributes are another important kind of markup; they are names-values couples that appear after the name of an element in the opening tag. For example, <SOIL CODE="DGX"> is element SOIL with an attribute CODE that has DGX value. A well-formed XML document doesn't impose any restrictions on the tags or attribute names. But a document can be accompanied by a DTD (Document Type Definition) which is essentially a grammar for restricting the tags and structure of a document. An XML document satisfying a DTD is considered a valid document.

A Document Object Model (DOM) defines the logical structure of the document (HTML and XML) and the way a document is accessed and manipulated by application programs. With the DOM one can build documents, navigate their structure, and add, modify, or delete elements and content. It is designed to be used with any programming language.

The structure of a document can be transformed and its contents displayed by using the XSL (eXtensible Style Language) language or a programming language (java, javascript, vbscript,...).

XSL is a declarative language which model refers the data by using patterns. It is limited where one wants retrieve data with specific criteria as one can realize that with the query language XQL (or OQL) for relational databases (or objects). This extension is proposed, on the one hand, by three languages coming from the database community (XML-QL [11] , Lorel [1], [12] and YATL [7]) and, on the other hand by XQL from the Web community.

3.2   The Structural Model

The dataweb design is a complex work that requires an important analyzing and conceptualizing effort. The method design of dataweb we have developed is based on three models:

- A structural (or conceptual) model which allows to identify and to describe the different domain objects and their relationships. E-R relationships can be *one-one*, *one-many* and *many-many,* representing associations between entities.
- A navigational model which describes the media objects and their browsing structures. Many navigational models can be built from a structural model.
- A presentational model that describes the user interface; each media object is associated to a presentation, which specifies how the corresponding information is presented to user.

In this paper, we are interesting to structural model. It allows to catch on the one hand the necessities of the user and the characteristics of the domain of the application, and on the other hand the data of the site .

3.2.1   Methodology
The conception approach of structural model we propose is realized on three stages:

1. The Entity-Relationship (E-R) model is used as conceptual model to capture user needs and characteristics of domain application.
2. The previous E-R model is transcript in a relational schema. The corresponding database is then created.
3. The XML documents base of the dataweb is obtained from the E-R schema above and data of the base.

To illustrate this E-R conversion to XML, we consider the E-R schema below, which is inspired from a pedagogic application at the faculty of sciences of Pau University [6].

We give below an extract of the database corresponding to the E-R schema above:

Etudiants

C_etudiant	Nom	Prenom	C_diplome
E100	Astet	Alain	minfo
E105	Alba	Celine	mim
E106	Conemi	Jean	minfo
...	...	...	

Diplomes

C_diplome	Intitule	Effectif
minfo	Mait_info	60
mim	Mait_maths	23
...	...	...

Modules

C_module	Intitule	departement
mi1	Compilation	informatique
mi2	Bases de Données	informatique
mi3	Génie Logiciel	informatique
mim3	Calcul scientifique	mathematique
mim5	Statistique	mathematique
...	...	...

Comporte

C_diplome	C_module
minfo	mi1
minfo	mi2
mim	mim3
minfo	mi3
minfo	mim3

3.2.2   Mapping between Relational Database and XML Documents

The problem of transformation of data coming from relational databases to XML documents is only approached since a short time and has given some tools like DB2XML [19], PLSXML [18] and XML-DBMS (4). These three systems are java packages and are adopted the same approach: they use a very simple mapping technique based on the DOM. The foreign key's problem is not approached; that implies a redundancy of information in the obtained document. The structure of this document is a tree structure and not oriented graph as thus we propose.

The DOM doesn't directly support the graph structure. But XML use special attribute types to translate the notion of graph. An element can have an attribute with ID type that value allows an unique identifier which can be referred by attributes with IDREF or IDREFS types from another elements.

The XML document is built from relational schema and data of the base in accordance with the general algorithm below:

- Creation of an element root (<ROOT_ELEMENT>)
- Each relation is transformed to parent element,
- For each child element: the different attributes contained in the relation constitute different attributes of these elements,
- The primary key becomes an attribute with ID type. This attribute is specified in the tag of the element,
- The relation's foreign keys are transformed to attributes with IDREF type. Their values appear in the main tag of the element. For example the element *etudiant* will be described as below *<etudiant id= "c_etudiant" diplome="c_diplome">*,
- The keys which appear in the relations which coming from (n-n) associations are transformed to values with IDREFS type; example: *<module id="c_module" diplome="c_diplome1 c_diplome2 ">*.

This approach presents the main advantage to avoid the information redundancy.
The XML document derived from the E-R schema and the database above is :

```
<ROOT_ELEMENT>
<etudiants>
    <etudiant id= "e100 " diplome="minfo">
        <nom> Astet </nom>
        <prenom>Alain</prenom>
    </etudiant>
    <etudiant id= "e105 " diplome="mim">
        <nom> Alba </nom>
        <prenom>Celine</prenom>
    </etudiant>
        ....
</etudiants>
<diplomes>
    <diplome id= "minfo " etudiant="e100 e106
e101" module ="mi1 mi2 mi3 mim3" >
        <intitule> Maitrise informatique </intitule>
        <effectif>60</effectif>
    </diplome>
<diplome id= "mim" etudiant="e105" module
="mim3 mim5" >
        <intitule> Maitrise mathematique </intitule>
        <effectif>23</effectif>
    </diplome>
</diplomes>
<modules>
    < module id= "mi1 " diplome="minfo">
        <intitule> Compilation </intitule>
        <departement>informatique</departement>
    </ module >
    < module id= "mim3 " diplome="mim minfo">
        <intitule> Calcul Scientifique </intitule>
<departement>mathematique</departement>
    </ module >
< module id= "mi2 " diplome="minfo">
    <intitule> Bases de données </intitule>
    <departement>informatique</departement>
    </ module >
        ....
</modules>
</ROOT_ELEMENT>
```

The id attribute has ID attribute, and the attributes etudiant, diplôme and modules have IDREFS attributes.

4   Information Search Process

Information contained in an XML document is accessible through the DOM by exploiting tags and the contents of elements. These elements are therefore potential answers to a query or interrogation. To the logical structure of data, we associate a semantic structure (expressing semantic contents) which acts to: characterize the descriptive information or external attributes; define semantic elementary or compound units; index the content of the document by keywords come from the thesaurus.

4.1   A Dataweb Architecture Founded on a Concepts Base

To improve the research process, a concepts base of the domain is used. The thesaurus and the meta-base are the main constituents of a concepts base [6], [14] (Fig. 1) .

The thesaurus: A thesaurus is a set of concepts connected by hierarchical relations, of equivalence of association [2].

The hierarchical relations define the notions of generalization and specialization. They find a document which treats very specific concepts if one asks a question at a more general level. Inversely, to a very specific question, this type of relationship finds, if need be, more general documents. For example, concept " Enhancement technique" is a sub-concept of "enhancement".

The notion of equivalence (or of synonymy) translates semantic equality between two concepts. For example, the concepts "arable zones" and "cultivable zones" are equivalent.

The associative relations (or neighborhood) are used to cover strictly connected concepts (for example "enhancement" and "kind of soils"). They allow to clarify or to extend the sense of a question. The thesaurus is described in the XML format.

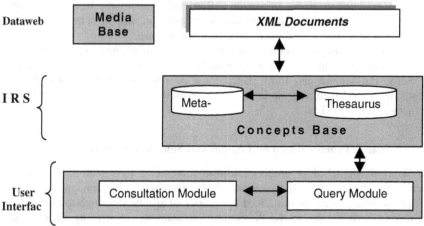

Fig. 1. Dataweb architecture for the information retrieval

The meta-information base : Meta-information constitute a description of the XML documents. These last ones are represented by a meta-information set which is constituted of:

– descriptive information (title, date, author, etc.) or external attributes;
– a set of keywords (tag <Keyword>) making reference to the concepts of the thesaurus. It is the result of the indexation process of documents elements of XML base. We call this part of meta-information, *semantic meta-information*. This aspect is exploited during the interrogation process of the dataweb. We can associate to every element of the document a semantic unit with the attribute SU. These semantic units regroup elements which have a meaning in the context where they are presented together.

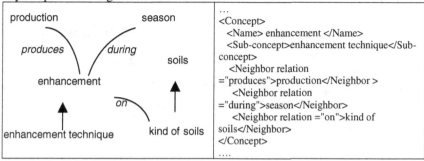

Fig. 2. Example of concept definition and its XML representation in the SIMEV thesaurus

To allow the implementation of our dataweb model, we propose a platform based on Java and XML. This development studio comprises provides a set of tools which allows to implement a dataweb in three main stages:

(i)   Dataweb modeling: the dataweb designer defines the three models (structural, navigational and presentational),

(ii)   Construction of the Information Retrieval System: the base of concepts is realized with the tools of thesaurus construction, and the semantic structuring and indexing tools for the meta-information base,

(iii)   Dataweb implementation: the designer implements the models defined in the modeling stage with the generating XML documents and media base tools.

In reality, the second and third stages must be done together because the XML documents provided during the third stage are used in the second stage to construct the meta-information base.

4.2   A Search Process Based on Logical and Semantic Structure

The dataweb model thus proposed allows interactive information retrieval. This research is done according to two approaches: interrogation, and navigation.

The research process is initialized by queries of simple interrogation by means of keywords which belong to the thesaurus. The semantic structuring of a document retrieves a semantically structured document, i.e., the same document but structured into semantic units instead of elements (logical units). It is on this document that the indexing is done; every semantic unit is indexed by a set of keywords.

For every interrogation, the information retrieval system compares the query with information stored in the dataweb. An interrogation query is treated in two stages: (i) normalization of the query; and, (ii) execution of the normalized query that provides a list of pertinent documents (or extracts of documents). A generalized query is a normalized query which is extended to the semantic relations of the thesaurus. For example, we complete the query by using the synonyms of the present keywords in the query. These keywords are directly accessible in the thesaurus.

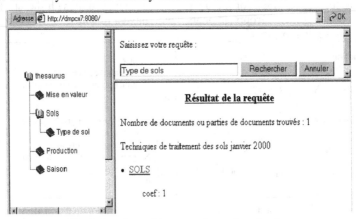

Fig. 3. Interface of interrogation and navigation

The query normalization transforms, by means of the thesaurus and of the DOM, to obtain a query semantically homogeneous with the concepts base and syntactically correct with the DOM.

The results of this query are presented in a hypertext card. This interrogation interface (Fig. 3) permits access at any time to the thesaurus which can be consulted.

The generalization of the query allows one to retrieve semantically interesting documents. The reformulation of query is not automatic but directed by users with invocation of knowledge contained in the thesaurus. The documents are re-stored according to their degree of semantic resemblance with the query.

In our example, if the query is "kinds of soils", all documents indexed by the concept "kind of soils" are provided by the method of questioning. But the generalization of this query, with the help of the thesaurus, takes out, again, the semantic links of this concept with "soils" (generic concept) and "enhancement" (neighboring concept). The documents indexed by concepts "soils" and "enhancement" are then provided in the answer.

5 Conclusion

In this paper, we showed the specificity of HTML applications and their limits in the search for relevant information although the integration of meta-information to improve the quality of information retrieval.

We proposed a dataweb model based on XML and a search process which uses a concepts base.

This approach allows an interactive search of relevant information and opens new perspectives in access, treatment and exploitation of information contained in web sites.

Currently, we are working on another statistical method to improve the degree of semantic resemblance. The specificity of environmental information systems [9], [13] makes environmental observatories potential applications for our dataweb model.

Notes

1. Interdisciplinary Group of research for support regional planning and local development: set by the regional Council of Saint-Louis (Senegal) with Senegalese and foreigner researchers and experts.

References

1. Abiteboul, S.; Quass, D.;.McHugh, J.; Widom, J. & Wiener, J. : The Lorel query language for semi-structured data. Journal of Digital Libraries, 1(1): 68–88.
2. Boughanem, M. : Les systèmes de recherche d'informations: d'un modèle classique à un modèle connexionniste, PhD Thesis of Toulouse III University, 1992

3. Bray, T.; Paoli, J. & Sperbeg-MacQueen, C.:   eXtensible   Markup   Language   (XML) 1.0,W3C Recommendation available at:http://www.w3.org/TR/1998/REC-xml-19980210.
4. R. Bourret, C. Bornhövd, A.P. Buchmann :   A Generic Load/Extract Utility for Data Transfer between XML Documents and Relational Databases, Technical Report DVS99-1, Darmstadt University of Technology. December 1999.
5. Brou, M.K.   Base de concepts: contribution à la représentation et à l'exploitation hypertextuelle de concepts : le système CoDB-Web, PhD Thesis of Universiry of Pau, France, 1997.
6. S. Chenouf and L. Reig : Génération automatique d'applications sur le web. Projet de DESS IMOI, University of Pau, march 2000
7. Cluet, S. & Siméon, J. (1997). Data integration based on data conversion and restructuring. In http// www-rocq.inria.fr/verso/Jerome.Simeon/YAT
8. Dobson, S.A. & Burill, V.A.. (1995). Lightweight databases. In. Proceedings. of the 3th Intern. WWW Conference, Darmstadt, April 1995, pp. 282–288.
9. Dzeakou, P. (1998). Méthodes et architectures des systèmes d'information sur l'environnement, In Proceedings of Cari'98, Dakar, 1998.
10.Fensel, D.; Decker, S.; Erdmann, M. & Studer, R. (1998). Ontobroker: Or how to ernable intelligent access to WWW. In Proceedings. of the 11th Banff Knowledge Acquisition Workshop (KAW98), Banff, Canada, April 1998.
11.Fernandez, M.; Florescu, D.; Levy, A. & Suciu, D. (1997). A query language and Processor for a Web-Site Management System. In SIGMOD Record, 26(3).
12.Gardarin, G. (1999). Internet/Intranet et bases de données. Edition Eyrolles, 1999.
13.Gayte, O.; Libourel, T.; Cheylan, J.P. & Landon, S. (1997). Conception des systèmes d'information sur l'environnement, Hermès, 1997
14.Hocine, A. & Brou, M.K. (1999). Le modèle de représentation et de gestion hypertexte des concepts d'un domaine dans le système CoDB-Web In Proceedings of Terminologie et intelligence artificielle (TI1'99), Nantes, Mai 1999.
15.Martin, P. & Eklund, P. (1999). Embedding knowledge in web documents. In Eighth International WWW Conference, Toronto, May 11–14, 1999.
16.Mendelzon, A.O.; Mihaila, G.A. & Milo, T. (1997). Querying the world wide web. In Journal of Digital Libraries, 1997.
17.Michard. A. (1998). XML, Langage et applications, Editions Eyrolles.
18.Muench, (1999). S. Muench : PLSXML Utilities and demos. Oracle Technical Whitepaper , March 1999.
19.Turau, (1999). V. Turau, Making legacy data accessible for XML applications, Technical Report, FH Wiesbaden, University of Applied Sciences, 1999.
20.Yoshikawa, M.; Ichikawa, O. & Uemura, S. (1996). Amalgamating SGML documents and databases. In Proceeding of EDBT'96, Avignon, March 1996.

A Rule-Based Approach to Represent Spatio-Temporal Relations in Video Data*

Mehmet E. Dönderler, Özgür Ulusoy, and Uğur Güdükbay

Department of Computer Engineering,
Bilkent University, Bilkent, 06533 Ankara, Turkey
{mdonder, oulusoy, gudukbay}@cs.bilkent.edu.tr

Abstract. In this paper, we propose a new approach for high level segmentation of a video clip into shots using spatio-temporal relationships between objects in video frames. The technique we use is simple, yet novel and powerful in terms of effectiveness and user query satisfaction. Video clips are segmented into shots whenever the current set of relations between objects changes and the video frames where these changes have occurred are chosen as key frames. The topological and directional relations used for shots are those of the key frames that have been selected to represent shots and this information is kept, along with key frame intervals, in a knowledge-base as Prolog facts. We also have a comprehensive set of inference rules in order to reduce the number of facts stored in our knowledge-base because a considerable number of facts, which otherwise would have to be stored explicitly, can be derived by these rules with some extra effort.

Keywords: video modeling, rule-based systems, video databases, spatio-temporal relations.

1 Introduction

There is an increasing demand toward multimedia technology in recent years. Especially, first image and later video databases have attracted great deal of attention. Some examples of video search systems developed thus far are OVID [9], Virage [5], VideoQ [2] and VIQS [6].

One common property of the image and video databases is the existence of spatial relationships between salient objects. Besides, video data also has a time dimension, and consequently, objects change their locations and their relative positions with respect to each other in time. Because of this, we talk about spatio-temporal relationships rather than spatial or temporal relationships alone for video data. A spatio-temporal relationship between two objects can be defined on an interval of video frames during which the relation holds.

* This work is supported by the Scientific and Research Council of Turkey (TÜBİTAK) under Project Code 199E025.

T. Yakhno (Ed.): ADVIS 2000, LNCS 1909, pp. 409–418, 2000.

This paper is concerned with the representation of topological and directional relations within video data. We propose a new approach for high level video segmentation based on the spatio-temporal relationships between objects in video. Video clips are segmented into shots whenever the current set of topological and directional relations between video objects changes, thereby helping us to determine parts of the video where the spatial relationships do not change at all. Extraction of the spatio-temporal relations and detection of the key frames for shots are part of our work as well.

We believe that this high level video segmentation technique results in an intuitive and simple representation of video data with respect to spatio-temporal relationships between objects and provides more effective and precise answers to such user queries that involve objects' relative spatial positions in time dimension. Selecting the key frames of a video clip by the methods of scene detection, as has been done in all systems we have looked into so far, is not very well suited for spatio-temporal queries that require searching the knowledge-base for the parts of a video where a set of spatial relationships holds and does not change at all. For example, if a user wishes to query the system to retrieve the parts of a video clip where two persons shake their hands, the video fragments returned by our system will have two persons shaking hands at each frame when displayed. However, other systems employing traditional scene (shot) detection techniques would return a superset where there would most probably be other frames as well in which there is no handshaking at all. The reason is that current methods of scene detection mainly focus on the camera shots rather than the change of spatial relationships in video.

We use a rule-based approach in modeling spatio-temporal relationships. The information on these relationships is kept in our knowledge-base as Prolog facts and only the basic relations are stored whilst the rest may be derived in the process of a query using the inference rules we provide by using Prolog. In our current implementation, we keep a single key frame interval for each fact, which reduces the number of facts stored in the knowledge-base considerably.

The organization of this paper is as follows: In Sect. 2, we describe and give the definitions for spatio-temporal relations. Our rule-based approach to represent topological and directional relations between video salient objects, along with our inference rule definitions, is introduced in Sect. 3. Section 4 gives some example queries based on an imaginary soccer game whereby we demonstrate our rule-based approach. We briefly mention about our performance experiments in Sect. 5. Finally, we present our conclusions in Sect. 6.

2   Spatio-Temporal Relationships

The ability to manage spatio-temporal relationships is one of the most important features of the multimedia database systems. In multimedia databases, spatial relationships are used to support content-based retrieval of multimedia data, which is one of the most important differences in terms of querying between multimedia and traditional databases. Spatial relations can be grouped into mainly three categories: topological relations, which describe neighborhood and incidence, directional relations, which describe order in space, and distance relations that describe range between objects. There are eight distinct topological relations: *disjoint, touch, inside, contains, overlap, covers, covered-by* and *equal*. The fundamental directional relations are *north, south, east, west, north-east, north-west, south-east* and *south-west*, and the distance relations consist of *far* and *near*. We also include the relations *left, right, below* and *above* in the group of directional relations; nonetheless, the first two are equivalent to the relations *west* and *east*, and the other two can be defined in terms of the directional relations as follows:

Above The relation *above*(A,B) is the disjunction of the directional relations *north*(A,B), *north-west*(A,B) and *north-east*(A,B).

Below The relation *below*(A,B) is the disjunction of the directional relations *south*(A,B), *south-west*(A,B) and *south-east*(A,B).

Currently, we only deal with the topological and directional relations and leave out the distance relations to be incorporated into our system in future. We give our formal definitions for the fundamental directional relations in Sect. 2.1. The topological relations are introduced in Sect. 2.2. Further information about the topological and directional relations can be found in [4,7,8,10]. Finally, in Sect. 2.3, we explain our approach to incorporate the time component into our knowledge-base to facilitate spatio-temporal and temporal querying of video data.

2.1   Directional Relations

To determine which directional relation holds between two salient objects, we consider the center points of the objects' minimum bounding rectangles (MBRs). Obviously, if the center points of the objects' MBRs are the same, then there is no directional relation between the two objects. Otherwise, we choose the most intuitive directional relation with respect to the closeness of the line segment between the center points of the objects' MBRs to the eight directional line segments. To do this, we place the origin of the directional system at the center of the MBR of the object for which to define the relation as illustrated in Fig. 1(a).

Even if two objects overlap with each other, we can still define a directional relation between them. In other words, objects do not have to be disjoint to define a directional relation between as opposite to the work of Li et al [8]. Our approach to find the directional relations between two salient objects can be formally expressed as in Definitions 1 and 2.

Definition 1. *The directional relation β(A,B) is defined to be in the opposite direction to the directional line segment which originates from the center of object A's MBR and is the closest to the center of object B's MBR.*

Definition 2. *The inverse of a directional relation β(A, B), β^{-1}(B,A), is the directional relation defined in the opposite direction for the objects A and B.*

Examples of the fundamental directional relations are illustrated in Fig. 1.

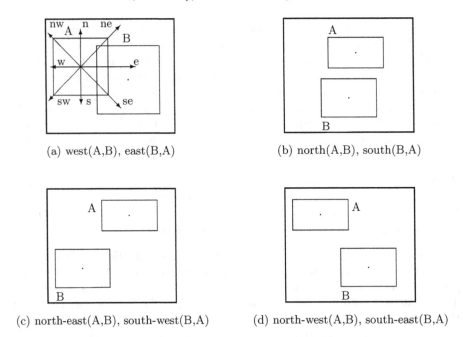

(a) west(A,B), east(B,A)

(b) north(A,B), south(B,A)

(c) north-east(A,B), south-west(B,A)

(d) north-west(A,B), south-east(B,A)

Fig. 1. Directional Relations

2.2 Topological Relations

The topological relations *inside* and *contains* are inverses of each other, and so are *cover* and *covered-by*. In addition, the relations *equal, touch, disjoint* and *overlap* hold in both directions. In other words, if β(A,B) holds where β is one of these four relations, then β(B,A) holds too.

The topological relations are distinct from each other; however, the relations *inside, cover* and *equal* imply the same topological relation *overlap* to hold between the two objects:

- *inside*(A,B) $\implies$ *overlap*(A,B) $\wedge$ *overlap*(B,A)
- *cover*(A,B) $\implies$ *overlap*(A,B) $\wedge$ *overlap*(B,A)
- *equal*(A,B) $\implies$ *overlap*(A,B) $\wedge$ *overlap*(B,A)

We base our definitions for the topological relations on Allen's temporal interval algebra [1] and Fig. 2 gives some examples of the topological relations.

2.3 Temporal Relations

We use time intervals to model the time component of video data. All directional and topological relations for a video have a time component, a time interval specified by the starting and ending frame numbers, associated with them during which the relations hold. With this time component attached, relations are not

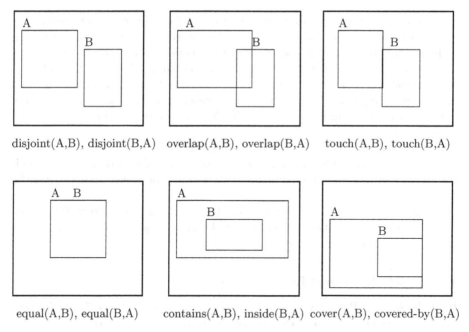

disjoint(A,B), disjoint(B,A)   overlap(A,B), overlap(B,A)   touch(A,B), touch(B,A)

equal(A,B), equal(B,A)   contains(A,B), inside(B,A)   cover(A,B), covered-by(B,A)

Fig. 2. Topological Relations

anymore simple spatial relations, but rather spatio-temporal relations that we use to base our model upon.

The topic of relations between temporal intervals has been addressed and discussed in [1]. There are seven temporal relations: *before, meets, during, overlaps, starts, finishes* and *equal*. Inverses of these temporal relations are also defined and the inverse of the temporal relation *equal* is itself.

3   A Rule-Based Approach for Spatio-Temporal Relations

Rules have been extensively used in knowledge representation and reasoning. The reason why we have employed a rule-based approach to model the spatio-temporal relations between salient objects is that it is very space efficient.

We use Prolog for rule processing and querying of spatio-temporal relations. By rules, we can derive some spatial knowledge which is not explicitly stored within the database; thus, we only store some basic facts in our knowledge-base and let Prolog generate the rest of the knowledge itself. Besides, our rule-based approach provides an easy-to-process and easy-to-understand knowledge-base structure for our video search system.

Our approach greatly reduces the number of relations to be stored in the knowledge-base which also depends on some other factors as well, such as the number of salient objects, the frequency of change in spatial relations and the relative spatial locations of the objects with respect to one another. Nevertheless, we do not currently claim that the set of relations we store in our knowledge-base is a minimum set of facts that must be stored, but we have been working on improving our algorithm considering the dependencies among our rules.

In our system, we define three types of inference rules, *strict directional*, *strict topological* and *heterogeneous directional and topological*, with respect to the relations' types in the rule body. For example, *directional rules* have only directional relations in their body whilst *heterogeneous rules* incorporate rules from both types.

We describe our *strict directional rules* and *strict topological rules* in Sects. 3.1 and 3.2, respectively. Our *heterogeneous topological and directional rules* are given in Sect. 3.3. More elaborate discussion of our rule-based approach can be found in [3].

In defining the rules, we have adopted the following terminology: if the relation r_1 implies the relation r_2, we show it by $r_1 \Longrightarrow r_2$. Moreover, if $r_1 \Longrightarrow r_2$ and $r_2 \Longrightarrow r_1$, we denote it by $r_1 \Longleftrightarrow r_2$.

3.1   Strict Directional Rules

Rule 1 (Inverse Property) The directional relations *west, north, north-west, north-east, right* and *above* are inverses of *east, south, south-east, south-west, left* and *below*, respectively.

$west$(A,B) $\Longleftrightarrow$ $east$(B,A)
$north$(A,B) $\Longleftrightarrow$ $south$(B,A)
$north$-$west$(A,B) $\Longleftrightarrow$ $south$-$east$(B,A)
$north$-$east$(A,B) $\Longleftrightarrow$ $south$-$west$(B,A)
$right$(A,B) $\Longleftrightarrow$ $left$(B,A)
$above$(A,B) $\Longleftrightarrow$ $below$(B,A)

Rule 2 (Transitivity) If $\beta \in$ S, where S is the set of directional relations, then
β(A,B) $\wedge$ β(B,C) $\Longrightarrow$ β(A,C).

3.2   Strict Topological Rules

The *strict topological rules* can be formally described as follows:

Rule 1 (Inverse Property) The topological relations *inside* and *cover* are inverses of *contains* and *covered-by*, respectively.

$inside$(A,B) $\Longleftrightarrow$ $contains$(B,A)
$cover$(A,B) $\Longleftrightarrow$ $covered$-by(B,A)

Rule 2 (Reflexivity) The topological relations *equal* and *overlap* are reflexive.

$equal$(A,A), $overlap$(A,A)

Rule 3 (Symmetry) The topological relations *equal, overlap, disjoint* and *touch* are symmetric.

$equal(A,B) \iff equal(B,A)$
$overlap(A,B) \iff overlap(B,A)$
$disjoint(A,B) \iff disjoint(B,A)$
$touch(A,B) \iff touch(B,A)$

Rule 4 (Transitivity) The topological relations *inside* and *equal* are transitive.

$inside(A,B) \wedge inside(B,C) \implies inside(A,C)$
$equal(A,B) \wedge equal(B,C) \implies equal(A,C)$

Rule 5 The topological relations *inside*, *equal* and *cover* imply the relation *overlap*.

$inside(A,B) \implies overlap(A,B)$
$equal(A,B) \implies overlap(A,B)$
$cover(A,B) \implies overlap(A,B)$

Rule 6 The relationships between *equal* and {*cover, inside, disjoint, touch, overlap*} are as follows:

a) $equal(A,B) \wedge cover(A,C) \implies cover(B,C)$
b) $equal(A,B) \wedge cover(C,A) \implies cover(C,B)$
c) $equal(A,B) \wedge inside(A,C) \implies inside(B,C)$
d) $equal(A,B) \wedge inside(C,A) \implies inside(C,B)$
e) $equal(A,B) \wedge disjoint(A,C) \implies disjoint(B,C)$
f) $equal(A,B) \wedge overlap(A,C) \implies overlap(B,C)$
g) $equal(A,B) \wedge touch(A,C) \implies touch(B,C)$

Rule 7 The relationships between *disjoint* and {*inside, touch*} are as follows:

a) $inside(A,B) \wedge disjoint(B,C) \implies disjoint(A,C)$
b) $inside(A,B) \wedge touch(B,C) \implies disjoint(A,C)$

Rule 8 The relationships between *overlap* and {*inside, cover*} are as follows (excluding those given by Rule 5):

a) $inside(B,A) \wedge overlap(B,C) \implies overlap(A,C)$
b) $cover(A,B) \wedge overlap(B,C) \implies overlap(A,C)$

Rule 9 The relationships between *inside* and *cover* are as follows:

a) $inside(A,B) \wedge cover(C,B) \implies inside(A,C)$
b) $inside(A,C) \wedge cover(A,B) \implies inside(B,C)$
c) $cover(A,B) \wedge cover(B,C) \wedge not(inside(C,A)) \implies cover(A,C)$

3.3   Heterogeneous Topological and Directional Rules

Rule 1 If $\beta \in S$, where S is the set of directional relations, then
$equal(A,B) \wedge \beta(A,C) \implies \beta(B,C)$.

Rule 2 If $\beta \in S$, where S is the set of directional relations, then
$disjoint(A,B) \wedge disjoint(B,C) \wedge \beta(A,B) \wedge \beta(B,C) \implies disjoint(A,C)$.

4   Query Examples

This section provides three query examples based on an imaginary soccer game fragment between England's two soccer teams, *Liverpool* and *Manchester United*. More query examples, along with the results returned by the system, can be found in [3].

Query 1 "Give the number of shots to the goalkeeper of *Liverpool* by each player of *Manchester United* and for the team *Manchester United* as a whole".

In this query, we are interested in the shots to the goalkeeper of *Liverpool* by each player of *Manchester United* except for the goalkeeper. The total number of shots to the goalkeeper of *Liverpool* by the team *Manchester United* will also be displayed.

We find the facts of *touch* to the ball for each player of *Manchester United* except for the goalkeeper. For each fact found, we also check if there is a fact of *touch* to the ball for the opponent team's goalkeeper, whose time interval comes after. Then, we check if there is no other *touch* to the ball between the intervals of the two facts and also if the ball is inside the field during the whole event. If all above is satisfied, this is considered a shot. Then, we count all such events to find the total number of shots to the goalkeeper by each *Manchester United* team member. The total number of shots to the goalkeeper of *Liverpool* by the team *Manchester United* is also computed.

Query 2 "Give the average ball control (play) time in frames for each player of *Manchester United*".

As we assume that when a player touches the ball, it is in his control, we calculate the ball control time for a player with respect to the time interval during which he is in touch with the ball. The average ball control time for a player is simply the sum of all time intervals where the player is in touch with the ball divided by the number of these time intervals. We could also give the time information in seconds provided that the frame rate of the soccer video is known.

To answer this query, we find for each player of *Manchester United*, except for the goalkeeper, the time intervals during which the player touches the ball and sum up the number of frames in the intervals. Divided by the number of facts found for each player, this gives us for each player of *Manchester United* the average ball control time in frames. Since in a soccer game, a player may touch the ball outside the field as well, we consider only the time intervals when the ball is inside the field.

Query 3 "Give the number of kicks outside the field for *David Beckham* of *Manchester United*".

We first find the time intervals when *David Beckham* of *Manchester United* is in touch with the ball while the ball is inside the field. Then, for each time interval found, we look for a fact, whose time interval comes after, representing the ball being outside the field. If there is no *touch* to the ball between these two intervals, then this is a kick outside the field. We count all such occasions to find the total number of kicks outside the field by *David Beckham*.

5   Performance Experiments

We have tested our rule-based system for its performance using some randomly generated data. In conducting these tests, two criteria have been considered: space and time efficiency.

For the space efficiency part, we have looked into how well our system performs in reducing the number of redundant facts. The results show that our inference rules provide considerable improvements in space as the number of salient objects per frame increases. For an example of 1000-frame randomly generated video data where there are 50 objects at each frame, our savings is 53.15%. The ratio for an example with 25 objects at each frame is 40.42%. We have also observed that the space savings ratio is not dependent on the number of frames, but rather the number of salient objects per frame. In addition, we are also certain that our rule-based system will perform better in space efficiency tests if our fact-extraction algorithm is enhanced with a more successful fact-reduction feature.

For the time efficiency part, we have seen that our system is scalable in terms of the number of objects and the number of frames when either of these numbers is increased while the other is fixed.

Unfortunately, we are not able to present our performance test results in detail in this paper due to lack of space.

6   Conclusions

We presented a novel approach to segment a video clip using spatio-temporal relationships assuming that the topological and directional relations of each frame have already been extracted. We extract the topological and directional relations by manually specifying the objects' MBRs and detect the key frames for shots using this relationship information.

In our approach, whenever the current set of directional and topological relations between salient objects changes, we define a new key frame and use that frame as a representative frame for the interval between this key frame and the next one where the spatial relations are the same.

We use a knowledge-base to store the spatio-temporal relations for querying of video data. The knowledge-base contains a set of facts to describe some basic spatio-temporal relations between salient objects and a set of inference rules to infer the rest of the relations that is not explicitly stored. By using inference rules, we eliminate many redundant facts to be stored in the knowledge-base since they can be derived with some extra effort by rules.

We have also tested our system using some randomly generated data for the space and time efficiency measures. The results show that our rule-based system is scalable in terms of the number of objects and the number of frames, and that the inference rules provide considerable space savings in the knowledge-base.

Currently, we are developing the graphical user interface part of our WEB-based video search system. Users will be able to query the system using animated sketches. A query scene will be formed as a collection of objects with different attributes. Attributes will include motion, spatio-temporal ordering of objects and annotations. Motion will be specified as an arbitrary polygonal trajectory with relative speed information for each query object. Annotations will be used to query the system based on keywords. There will also be a category grouping of video clips in the database so that a user is able to browse the video collection before actually posing a query.

References

1. J.F. Allen. Maintaining knowledge about temporal intervals. *Communications of ACM*, 26(11):832–843, 1983.
2. S. Chang, W. Chen, H.J. Meng, H. Sundaram, and D. Zhong. Videoq: An automated content-based video search system using visual cues. In *ACM Multimedia*, Seattle, Washington, USA, 1997.
3. M.E. Donderler, O. Ulusoy, and U. Gudukbay. Rule-based modeling of spatio-temporal relations in video databases. *Journal Paper in Preparation*.
4. M. Egenhofer and R. Franzosa. Point-set spatial relations. *Int'l Journal of Geographical Information Systems*, 5(2):161–174, 1991.
5. A. Hampapur, A. Gupta, B. Horowitz, C-F. Shu, C Fuller, J. Bach, M. Gorkani, and R. Jain. Virage video engine. In *SPIE*, volume vol. 3022, 1997.
6. E. Hwang and V.S. Subrahmanian. Querying video libraries, June 1995.
7. J.Z. Li and M.T. Özsu. Point-set topological relations processing in image databases. In *First International Forum on Multimedia and Image Processing*, pages 54.1–54.6, Anchorage, Alaska, USA, 1998.
8. J.Z. Li, M.T. Özsu, and D. Szafron. Modeling of video spatial relationships in an object database management system. In *Proceedings of the International Workshop on Multimedia DBMSs*, pages 124–133, Blue Mountain Lake, NY, USA, 1996.
9. E. Oomoto and K. Tanaka. OVID: Design and implementation of a video object database system. *IEEE Transactions on Knowledge and Data Engineering*, 5(4):629–643, 1993.
10. D. Papadias, Y. Theodoridis, T. Sellis, and M. Egenhofer. Topological relations in the world of minimum bounding rectangles: A study with R-trees. In *Proceedings of ACM SIGMOD International Conference on Management of Data*, pages 92–103, San Jose, CA, USA, 1996.

Factual and Temporal Imperfection

P. Chountas and I. Petrounias

Department of Computation, UMIST
PO Box 88, Manchester M60 1QD, UK
e-mail: {chountap, ilias}@sna.co.umist.ac.uk

Abstract. Current information systems model enterprises that are crisp. A crisp enterprise is defined as one that is highly quantifiable; all relationships are fixed, and all attributes are atomic valued. This paper is based on precise enterprises, where data are imperfect and where multiple and possibly conflicting sources of information do exist. Imperfection may be related to the actual values or the time that something happened within an organisation. In such domains different sources of information may be assigned different degrees of reliability. Factual imperfection and/or temporal indeterminacy may cause uncertainty. Temporal information may be recurring, periodical or definite. This paper is presenting a conceptual and algebraic framework for a uniform treatment and representation of all these types of information.

1 Introduction

Many real world phenomena require the t representation of both uncertain facts and uncertain time. Someone may have symptoms typical for a certain disease, while the time of occurrence of those symptoms does not support the presupposed disease. In the above example two different kinds of imperfection are derived, factual uncertainty and temporal uncertainty. Whereas factual uncertainty is about evidence available for some proposition, temporal uncertainty is about its occurrence in time. Episodes or sequences of events can express dependencies between local sources, thus determining way that facts interfere with each other. In that way certainty about certainty can be expressed, considering either the dimension of facts or time.

A novel approach to temporal representation and reasoning is proposed in this paper dealing with both uncertain facts and uncertain temporal information at the conceptual level. A point of particular concern is that temporal uncertainty should not influence factual imperfection. The decomposition does not prevent time and facts to have mutual influence. The rest of the paper is organised as follows. Section 2 discusses relevant work in this area. Section 3 provides details on the semantics of temporal information and section 4 discusses the interval based finite time representation. Section 5 presents a conceptual modelling formalism to capture temporal and factual uncertainty and defines the basic elements for a temporal representation. Section 6 describes the instance level. Section 7 summarises the main points of the proposed approach and points out to further research.

T. Yakhno (Ed.): ADVIS 2000, LNCS 1909, pp. 419-428, 2000.

2 Organisation of Relevant Work

In the literature most approaches, which address uncertainty, start form either the perspective of formal temporal theories or probabilistic logic. Relevant work found in the uncertainty reasoning literature is judged by the following criteria: independent representation of factual uncertainty and temporal uncertainty, representation of definite, indefinite, infinite information. As part of a formal temporal theory global and local inequality constraints on the occurrence time of a fact have been added to a temporal data model in [1]. The resulting model supports indefinite instants. However, the model assumes that there is no factual uncertainty. The model cannot represent infinite temporal uncertain or certain information. In probabilistic logic uncertainty is addressed from two different perspectives. The probability of a proposition F at a time "now" or at time t as P (F_{now}) or P (F_t) is expressed [2], [3]. The other orientation is to represent crisp propositions F that may occur at an uncertain time t_p. In these approaches the probability P(t_p at t) is represented [4] [5]. Indefinite temporal information is not explicitly encoded in the facts. Both models do not represent factual uncertainty. Furthermore, the intervalic representation of the time property does not permit the description of infinite temporal information. Choosing either the P (F_t) or P $(t_p$ at t) representation introduces limitations with respect to both the reasoning process and to the kind of knowledge that can be presented [6]. Simultaneous reasoning with both uncertain facts and uncertain time has also not been dealt with in the context of probabilistic data models and databases [6].

3 An Interval Based Finite Time Representation

An activity may be defined over more than one time intervals because of the existence of conflicting sources, or it may be definite, indefinite, recurring, or periodical. *Definite Temporal Information* is defined when all times associated with facts are precisely known in the desired level of granularity. Definite temporal information may be related with factual uncertainty. A possible set of entity instances may be involved in a fact (factual uncertainty). *Indefinite temporal information* is defined when the time associated with a fact has not been fully specified. In valid time indeterminacy [5] it is known that an event did in fact occur but is not known exactly when. If a tuple k in relation R is timestamped with the interval $[t_1...t_2]$, then this is interpreted as tuple k holds at some point t in interval $[t_1...t_2]$ with an indicative belief $[p_1...p_2]$. *Infinite temporal uncertain information* is defined when an infinite number of times are associated with a fact not explicitly defined in the desired level of granularity and factual uncertainty coexists. To be able to present this kind of information effectively, a finite representation is needed. A finite representation can be achieved by limiting the occurrence of a fact in the desired level of granularity.

The central concepts in a temporal representation are a timeline and a time point where the former is comprised of a set of the latter. A timeline is an ordered continuous infinite sequence of time points, whereas a time point is a particular instantaneous point in time. The concepts of duration and time interval are defined with the help of time points. A time interval is defined as a temporal constraint over a linear hierarchy of time units denoted H_r. H_r is a finite collection of distinct time units, with linear order among those units [5]. For instance, $H_1 = day \subseteq month \subseteq year$, $H_2 = minute \subseteq hour \subseteq day \subseteq month \subseteq year$, are both linear hierarchies of time units defined over the Gregorian calendar [5]. A calendar r consists of a linear hierarchy H_r of time units and a validity predicate that specifies a non-empty set of time points. In that way an application may assume the existence of an arbitrary but fixed calendar. The term duration is defined as an absolute distance between two time points. However, the term duration may also imply the existence of two bounds an upper bound and a lower bound. Time is linear. A probability distribution implies ordering of the time alternatives. Ideally it would be preferable to be able to represent any interval predicate, whether finite or infinite. Thus, an interval is described as a set of two time points defined in a linear hierarchy (e.g. $H_2 = minute \subseteq hour \subseteq day \subseteq month \subseteq year$). The lower time point t_1 is described by the equation $t_1 = (c + kx)$. The upper point t_u is described by the equation $t_u = (t_1 + d)$. c is the time point related to an instantaneous event that triggered a fact, k is the repetition factor, or the lexical "every". $k \in N$, x is a random variable, $x \in N$ (including zero) corresponding to the first occurrence of a fact instance restricted by a constraint. The product kx is defined according to a linear hierarchy. d represents the duration of a fact instance, how long it is valid for after it has been introduced. d is also defined in a linear time hierarchy and may be in the range between a lower and upper bound $G_1 \le d \le G_2$ where $G_1 \le G_2 \wedge d \le K$. G_1, G_2 are general or restricted constraints on the time points t_1, t_u. Constraints are built from arbitrary linear equalities or inequalities (e.g. $t_1 = c + 7x$ and $0 \le x \le 5$). Limiting the random variable x results in specifying the lower and upper bound of a time window. The above interval representation permits the expression of the following.

When k=0 definite or indefinite temporal information can be represented. When $0 \le x \le n$, $x \in N$ and k>0, then both infinite and certain or uncertain temporal information can be represented.

4 The Temporal Multisource Belief Model (TMBM)

Factual uncertainty is related to possible semantic propositions. This leads to possible associations between entity instances and not between entity types. Factual uncertainty is the result of ambiguity at the instance level. In our approach there is no ambiguity at the level of concepts, since this considered to be meaningless. An irreducible fact type is defined in the real world over a time interval or time period (valid time). This is called a timestamped fact. Temporal uncertainty in the information context specifies that a semantic proposition in the real world can be defined over a time interval with no explicit duration. A probability distribution and temporal constraints over the time element can be used to express information like "I do not know exactly when" [6]. A time fact is called cognisable if uncertainty about time and fact are independent. In our representation a time fact is cognisable either the timestamped fact or the independent time and fact representations are sufficient to represent all knowledge, provided that correct 2-dimensional representations can be derived from the timestamped fact (3-dimensional space) and not the other way around. In this case the range of possible time points which correspond to the earliest possible instantiation F_s of the fact F_x is constrained, as is the range of time corresponding to the latest instantiation F_e. Similarly the range of the possible instantiations of a fact, corresponding to the earliest time point instantiation t_s of F_x is constrained, as are those fact instantiations corresponding to the latest time t_e. These constraints are summarised in equation (1).

$$F_s \rightarrow [t_s, t_1] \qquad F_e \rightarrow [t_2, t_e] \tag{1}$$
$$t_s \rightarrow [F_s, F_1] \quad t_e \rightarrow [F_2, F_e]$$

In equation (1) t_s, t_1, t_2, t_e are linear points and are applied as constraints to the timestamped fact instantiations. A temporal phenomenon is defined over the union of the time intervals specified by the temporal constraints in equation (1). Therefore a temporal phenomenon is defined over a temporal element. This is described in equation (2):

$$F_x = \{F_s,...,F_e\} \rightarrow \{[t_s, t_1],..., [t_2, t_e]\} \tag{2}$$

A fact type in TMBM is composed of the arguments shown in Figure 1, where n is the arity of the fact type. The way that one can refer to specific entities is through reference labels. If the modelled world is certain then the label value is a single value. The meaning of a stochastic label type is that a label value can take a possible (π) set of values and each member of the set is an alternative value with an indicative belief (probability interval, (p_1, p_2)). Based on the possibility / probability consistency principle [9] a connection between the measure of randomness (p) or observation and compatibility (Π) can be achieved. In this way a fact is presenting information that is observed and testified by one or more information sources therefore a set of alternatives is defined with $p \in [\alpha,\beta]$ where α,β are constrained by $0 \le \alpha,\beta \le 1$, $\alpha < \beta$ and $\Pi = 1$. Let F_1 and F_2 be two fact instances of a fact. F_1 is more accurate than F_2 if $[F_1\alpha_1, F_1\beta_1] \subseteq [F_2\alpha_2, F_2\beta_2]$. F_1 associates a probability interval, which is sharper than that associated with F_2. A singleton probability is expressed with the aid of a dummy interval both ends are identical $[\alpha,\alpha]$ [5]. A fact may also represent information that is compatible with its domain, $p \in [0,0]$ and $0 < \Pi < 1$ based on some specified or unspecified criterion. However, the information source cannot testify these values thus information that is more elementary and less context dependent can be represented The time interval Δt that a fact instance is defined over has a constrained duration since both ends of the interval are unambiguously defined in the time line. The temporal element that a fact is defined over is an alternative from a multiple set of time intervals, with an indicative belief interval. Any instance of the (Disease-Prescription) in Figure 2 must exist during the period (or at the same period) that the corresponding fact type exists. The following relationship between T and T_1 must exist: (T$_1$ during T) or (T$_1$ same as T). In the case of temporal imperfection, the time interval over which a fact instance is valid is accompanied by an indicative belief (probability interval) that the relationship between T and T_1 still holds. The validity lifespan that a concept is defined (e.g. Disease) over is the union of the time intervals that 'Disease' instances are believed to be valid. If an entity type is involved in non-timestamped facts, the interval [now - t_1, now] is awarded to non-timestamped fact types, where t_1 is the smallest granularity of all timestamped facts that the entity type participates. In cases of either value or temporal imperfection the belief dimension is affected by the reliability of the source. The reliability of the source is expressing the 'actors' concern about the identity and trustiness of the source, in forming the conclusive belief for the timestamped fact. The interval [t_e, now] is the valid period that the reliability measure is defined for a particular source

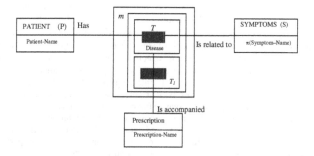

Fig. 1. Fact types

The 'now' upper bound, will be updated to t_{e1}, where t_{e1} is the time point another instantaneous event e_1 is triggered and subsequently modifies the reliability measure of the source.

$$F = \{\{\{\{<E_1, L_1, R_1>\dots <E_n, L_n, R_n>\}, \{\Delta t \leq T\}, m\}$$

Where:
F is a fact type consisting of k fact instances
T is the time interval that an irreducible fact type is defined in the real world.
E_i is the entity type playing a role in the fact type
L_i is the label type (referencing E_i)
R_i is the role of the fact type
Δt is the time interval that an irreducible fact instance is defined in the real world
m is the reliability of the source that circulates a particular fact type. The reliability of the source is a domain independent variable.

Fig. 2. Uncertain Ternary Timestamped Fact

6   The Instance Level –A Recursive NF² Algebra

The 1NF relational model is simple and mathematically tractable but not rich enough to model complex objects. In order to represent complex objects hierarchical structures are used instead of flat tables [10]. A relation schema R is recursively defined as: If $\{A_1, \dots, A_n\} \subset U$ and $A_1 \dots A_n$ are atomic valued or zero order attributes then $R = \{A_1, \dots, A_n\}$ is a relation schema. If $\{A_1, \dots, A_n\} \subset U$ and $A_1 \dots A_n$ are atomic valued attributes and $R_1, \dots, R_n$ are relation schemas or high order attributes then $R = (A_1, \dots, A_n, R_1 \dots R_n)$ is a relation schema .Consider the extracted fact type Disease from (Figure 2). Patient-Name is an atomic value attribute. π(Symptom –Name) declares that a single label value can be a possible (π) set of values. Symptom name is a relation schema or a high order attribute. The time interval $\Delta t \subset T$ that a fact instance is defined over may be explicitly known or may be a set of probable time intervals (π). T is represented by a high order attribute. Figure 3 presents a sample population for the fact type Disease. In it two kinds of value imperfection found in the real world can be modelled: the possible ($\Pi = 1$) and probable $p \in [\alpha, \beta]$ where α, β constrained by $0 < \alpha, \beta \leq 1$, $\alpha < \beta$, or the possible ($0 < \Pi < 1$) and unexpected, improbable ($p = [0,0]$). In Figure 5 Disease is presented as a hierarchical structure. A node can be either an atomic value attribute or a relation.

Selection ($_\sigma$): For all nodes $\in$ node S where $S_a \neq S_b$, if node S_a is a child of an ancestor of a node S_b, then S_a, S_b are called selection comparable nodes ($S_a{}^\sigma \rightarrow S_b$). (ValidTime($R_1$)$^\sigma \rightarrow$Patient–Name) and (ValidTime(R_1)$^\sigma \rightarrow \pi$(Symptom–Name)) are selection comparable notes in Figure 5. Since there is a path between π(Symptom–Name) and ValidTime(R_1) then (R_1) is also comparable to Name/Probability (p). However Valid Δt/(p), Patient–Name are not selection comparable nodes. Selection conditions are comparisons between attributes and constants and may include also membership operators. Temporal Database applications involve large amounts of time sequences [11]. Research has been focused in examining shapes and values repeated over two or time intervals defined in a linear time hierarchy. However it is assumed that the frequency of reappearance between two successive recurrences of a fact is constant, this implies that k is constant. If k is not constant, the interpretation is that a fact does not have a constant duration (d). In such cases a fact is not periodical, but is definitely recurring. Information providers may disagree about the time point (c) that a fact becomes noticeable. Value imperfection suggests the existence of different facts since it is questioning, the "what is " or "structure" of a fact. Scale and shift transformations as part of the select operator, in detecting recurring or periodical facts defined over more than one sequences of possible time intervals with variable belief: [t_l, t_u] / [p1, p2]. Assuming two time intervals of the form: [t_l, t_u], [t_{l1}, t_{u1}], separated by a distance of l where $t_l = c + kx$, $t_u = tl + d$, $t_{l1} = c_1 + k_1 x$, $t_{u1} = t_{l1} + d_1$, (1) and $m \leq x \leq n$. Scaling is defined as multiplying the repetition factor k_x by a factor of α, and $\alpha \in N$. α is defined as the difference of ($k_z - k_x$) and x<z. α is either an odd (2n+1) or even (2n) number. The assumption is that (c_1, c), (d_1, d) are identical and d<k, d_1<k_1 and k<k_1. If the scaling factor α, is not constant for all the time intervals that a fact is defined over, then the specific fact is at least recurring. If the scaling factor is constant and the duration is constant, then the fact is called periodical. Scaling may result in moving in higher linear time hierarchies in the case of recurring patterns. Shifting is defined as adding an offset θ to the upper and lower bounds of a sequence of time intervals. Shifting guarantees that there is no altering in the linear time hierarchy. If the offset θ is constant then a fact is called periodical, otherwise is recurring. Assuming two time intervals of the form: $It_1 = [t_l, t_u]$, $It_2 = [t_{l1}, t_{u1}]$, separated by a distance of l where $t_l = c + kx$, $t_u = tl + d$, $t_{l1} = c_1 + k_1 x$, $t_{u1} = t_{l1} + d_1$, (2) and $m \leq x \leq n$, and (c1≠c, $d_1 \neq d$). In such cases an interval can be derived from another by applying a combination of transformations (shifting the lower and upper bound of It_1 by a factor of α and an offset of θ). The lower bound has to be scaled by and shifted simultaneously as following: $\alpha * x + \theta = (k_1 - k)x + (c_1 - c)$, where $\alpha = (k_1 - k)$, $\theta = (c_1 - c)$. The upper bound of It_1 has to be scaled and shifted by a factor of $\alpha = (k_1 - k)$ and an offset of $\theta = ((c_1 + d) - (c + d_1))$. The combined transformation is of the following form: $\alpha * x + \theta = (k_1 - k) * x + ((c+d) - (c_1 + d_1))$. The transformation operators do not violate the constraints defined by a linear time hierarchy. Conceptually the transformations are showing the frequency of change in the behaviour of an actor through the time line. Thus they can be assumed as reliability measurements.

Timestamped Fact Type Disease					
Multisource Fact Type Disease				Valid Time	
Patient-Name	π (Symptom-Name)			Δt / (p)	Source
Amber Moore	Name / Probability(p) Fever / [0.2, 0.4] Sore throat / [0.3,0.4] Headache / [0.1,0.2]	Source Dr Smith Dr Tan Dr Son	Possibility 1	[t_l, t_u] /[0.6 ,0.8] $t_l = c + 7x$ and $0 \leq x \leq 5$ [t_l', t_u']/ [0.2,0.4] $t_l' = c' + 7x$	Dr (Son, Tan) Dr Smith
	Depression / [0,0]	Dr Son	0.5	and $0 \leq x \leq 5$	

Fig. 3. Multisource Timestamped Fact Type Disease

Fig. 4. Relation SP for the Information provider

Projection (π′): A projection operation is a way of accessing attribute values or relation schemas from the outermost level to the innermost level. Existing projection operators deal only with projection of attribute values based on a selection condition that is defined on the attribute domain. For all nodes ∈ node (S) if two nodes are selection comparable notes then the projection operator is defined. The project operator is defined as an ordered sequence of zero level attributes and relation valued attributes. Projection operators can be either simple or complex. A simple projection involves a one level vertical or horizontal path (e.g. Patient-Name$^{π'}$→π(Symptom-Name)). A complex projection involves the derivation of values through paths in the tree hierarchy (e.g. Patient-Name$^{π'}$→SP (Source Identity/reliability). Duplicates are not eliminated in the case that the values of the timestamps are different or the conclusive beliefs are different.

P Join (^{p}×): The idea behind the extended Join is to combine relations with common high order attributes not only at the top level but also at the subschema level. Let R be the relational relation schema and S be the schema tree of R, the path $P_r = (M_1...M_k)$ is a join-path of R if M_1 is a child of root (S) and M_k is a non-leaf node of S. Path expressions describe routes along the composition hierarchy and expressions describe links between attribute domains. The P join can be extended with multiple path joins, which exploit the more general situation. The source attribute presented by the Source relation (SP) in Figure 4, is evident in two relational subschemes R_1, R_2 of Figure 5. A join path between R_1, R_2 and SP can be defined. In defining the path join the following relationship must exist: Valid Time (SP$_j$)∩Valid Time (Rt$_i$) ≠ ∅

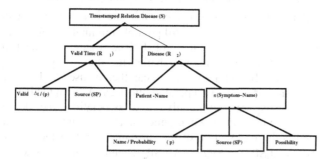

Fig. 5. Nested schema tree for value and temporal imperfection

Behavioural P Join (m^{p}×): A P behavioural Join is defined as an extension of the P-Join and is also an extended natural join. With the behavioural P Join attribute values defined using a probability can encapsulate the belief of the source and express the conclusive belief in a single probability interval, thus forming higher level attributes with complex data types. The conclusive belief for a possible value is defined as the outcome of reliability measure and probability interval C_p= [mα, mβ] and 0≤mα<mβ≤1. When two probability intervals [α,β] &[γ,δ] where α<β & γ<δ have to be joined, the outcome is a new interval C_p which is defined using one of the following strategies. When elements of different possibility distributions are joined then the min ($\Pi_1...\Pi_n$) possibility is the common one, thus expressing the less possible. Sequences of events or episodes [12] are describing the behaviour and action of local sources. An episode is defined to be a collection of facts that occur in a partial order within time intervals. Episodes are triggered by partially ordered sets of events. They can be described as directed acyclic graphs. Dependencies between local sources are expressed as a serial episode. A serial episode occurs in a sequence only if there are events that occur in this order in the sequence relatively close together. In this case the interference between sources is maximal. Serial episodes can be simple or complex. In a complex serial episode individual events may include more than one event. In contrast a parallel event imposes no constraint on the relative order of events therefore the interference in reality is minimal. Figure 7 presents a simple serial episode (a), a parallel episode (b) and a complex serial episode (c). Episode (c) can be seen as the serial combination of two parallel episodes. Let events e_1, e_2 have probabilistic intervals $[\alpha,\beta]$, $[\gamma,\delta]$. If events e_1, e_2 are part of episode (a) then a probabilistic conjunction strategy $(e_1 \wedge e_2)$ is applied to estimate the compound belief. The compound belief is derived using positive correlation (pc) or negative correlation (nc). Positive correlation implies maximal interference between sources, whereas negative correlation implies minimal interference. $(e_1 \wedge e_2)$: $[\alpha,\beta] \oplus _{pc}[\gamma,\delta]=[\min(\alpha,\gamma),$ $\min(\beta, \delta)]$ maximal interference between e_1, e_2 $[\alpha,\beta] \oplus _{nc}[\gamma,\delta]=[\max(0,\alpha+\gamma -1),$ $\max(0,\beta+\delta-1)]$ minimal interference between e_1, e_2. If events e_1, e_2 are part of episode (b) then a probabilistic conjunction strategy $(e_1 \wedge e_2)$ or disjunction $(e_1 \vee e_2)$ can be applied to estimate the compound belief. The compound belief is derived using independence (in) or positive and negative correlation to derive a hypothetical belief. $(e_1 \wedge e_2)$: $[\alpha,\beta] \oplus _{in}[\gamma,\delta] = [\alpha_{\bullet}\gamma, \beta_{\bullet}\delta]$ e_1, e_2 are independent positive and negative correlation have been defined already for episode (a)

$(e_1 \vee e_2)$: $[\alpha,\beta] \oplus _{in}[\gamma,\delta]=[\alpha+\gamma-(\alpha_{\bullet}\gamma),\beta+\delta-(\beta_{\bullet}\delta)]$ e_1, e_2 are independent

$[\alpha,\beta] \oplus _{pc}[\gamma,\delta]=[\max(\alpha,\gamma), \max(\beta, \delta)]$ possible maximal interfering between e_1, e_2

$[\alpha,\beta] \oplus _{nc}[\gamma,\delta]=[\max(1,\alpha+\gamma), \max(1,\beta+\delta)]$ possible minimal interfering between e_1, e_2

It is obvious that for episode (c) the probabilistic strategies defined for episode (b) may be used to derive the compound belief for the two parallel episodes. The probabilistic strategy defined for episode (a) can then be used to derive the final belief of the serial combination of the two parallel episodes. As conjunctive and disjunctive probabilistic strategies are commutative and associative, the definition can be extended to apply too more than two arguments. The time interval that the conclusive belief is defined is the intersection of the time intervals that the sources $(SP_1...SP_n)$ are defined. $\Delta t_{Cp} = \Delta t_1 SP_1 \cap t_2 SP_2 \cap ..... \cap t_n SP_n$. Given two intervals with lower bounds $t_1 = c + kx$ and $t_{1'} = c' + k_1 x$ and upper bounds $t_u = t_1 + d$ and $t_{u'} = t_{1'} + d'$, the time interval for the result is defined as $t_{1''} = c'' + k'x$ and $t_{u''} = t_{1''} + d''$ where $c'' = \max (c, c')$, $d'' = \min (d, d')$ and $k' = \min (k, k_1)$. Therefore, the join of two periodical uncertain facts will be a recurring uncertain fact. Figure 6 presents relation $(Sm^p \times SP)$ after applying the Maybe P Join. Relation S is from Figure 7. Relation SP is from Figure 6.

		$((\Delta t / C_{pl}), (\Delta t / C_{pl},$ Source Identity))
Patient-Name	π (Symptom-Name)	
	$(((Name / C p) (\Delta t / C_p),$ Source Identity)) Possibility)	

Fig. 6. S m^p×SP Behavioural P Join Example

8 Conclusions

A conceptual context for capturing and representing the semantics of factual and temporal uncertainty under three levels of abstraction, the specification level and the instance level is presented. An exteneded relational algebra that integrates both factual and temporal uncertainty is also proposed. An important point for consisderation is that of integrity constraints. What does it mean for a database state to satisfy an integrity constraint, when in fact information reflects a set of possible states, together with an associated probability distribution.   There is a challenge beyond local environments, in heterogeneous environments where intentional inconsistencies between heterogeneous sources have to be resolved, before considering factual and temporal uncertainty as part of a collective environment.

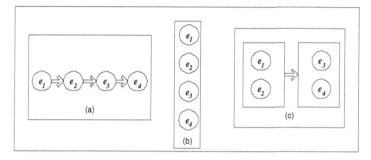

Fig. 7. Different Types of Episodes

References

1. M. Koubarakis, Representation and Querying in Temporal Databases: the Power of Temporal Constraints, Proc. of 9th Conference on Data Engineering, pp 327-334, 1993
2. D. Dey, S. Sarkar, A Probabilistic Relational Model and Algebra, ACM Transactions on Database Systems, Vol. 21, No. 3, 1996
3. D. Barbará, H. Garcia-Molina, D. Porter, The Management of Probabilistic Data, IEEE Transactions on Knowledge and Data Engineering, Vol. 4, No, 5, 1992
4. C. Dyreson, R. Snodgrass, Support Valid-Time Indeterminacy, ACM Transactions on Database Systems, Vol. 23, No. 1, pp. 1-57, 1998
5. A. Dekhtyar, R. Ross, V. S. Subrahmanian, TATA Probabilistic Temporal Databases, I: Algebra, Technical Report CS-TR-3987, University of Maryland, USA, pp 1-81 1999
6. P. Chountas, I. Petrounias, Precise Enterprises and Imperfect Data, Proceedings 4th IEEE International Baltic Workshop on DB&IS, May 2000, Vilnius, Lithuania
7. I Petrounias, A Conceptual Development Framework for Temporal Information Systems, Proc. of ER '97, 16th International Conference on Conceptual Modelling, Los Angeles, California, USA, November 1997
8. F. Kabanza, J-M. Stevenne, P. Wolper, Handling Infinite Temporal Data, Proceedings of ACM Symposium on Principles of Database Systems pp 392-403, (PODS), 1990
9. M. Delgado, S. Moral, On the concept of Possibility-Probability Consistency in Fuzzy Sets For Intelligent Systems, in D. Dubois, H. Prade and R. Yager (eds), Morgan Kaufman Publishers, pp 247-250, 1993

10. H. Liu, K. Ramamohanarao, Algebraic Equivalences Among Nested Relational Expressions, Proc. of ACM Conference on Information and Knowledge Management (CIKM), pp 234-243, 1994
11. Lin, T. Risch, Querying Continuous Time Sequences, Proceedings of 24th Conference on Very Large Databases (VLDB), pp. 170-181, New York, 1998
12. Mannila, H. Toivonen, A. Verkano, Discovering Frequent Episodes in Sequences, Proc. of 1st Conference on Knowledge Discovery in Databases (KDD'95), pp 210-215, Canada, 1995

An Agent-Based Information Retrieval System

Suat Ugurlu and Nadia Erdogan

Istanbul Technical University
Computer Engineering Dept.
Ayazaga, 80626, Istanbul, TURKEY
Sugurlu@egebank.com.tr, Erdogan@cs.itu.edu.tr

Abstract. This paper presents the design and implementation details of an information retrieval system. A multi-agent architecture has been adopted to allow for extended flexibilities over similar traditional systems. Since the system is consisted of various types of agents a short agent information will be given first. It will also explained that how these agents communicate with each other in order to accomplish information gathering through an agent messaging router which is a part of a software package called JATlite (Java Agent Template). JATLite provides the basic infrastructure for creation of agents and their communication.

1 Introduction

Each year, Internet-oriented applications become more popular. To fully use the Web's advantages, new design and programming paradigms need to be developed. Traditional distributed applications assign a set of processes to a given execution environment that, acting as local resource managers, cooperate in a network unaware fashion. In contrast, the agent-based programming paradigm defines applications as network-aware cooperating clusters of agents.

Several information retrieval systems are widely used on the web. Many of them use a central database. When a browser sends a request, a cgi or asp program on the server side executes a search on the database to process the request. This approach has several inadequacies which can easily be overcome by a design that conforms to the agent-based programming paradigm. Utilization of a distributed database architecture, extended user facilities, reduced network consumption are some of the advantages gained by an agent-based design. In this paper we explain in detail the implementation issues of a multi-agent information retrieval system, which is developed using the JATLite package as the basic agent template to create agents and an agent message router for agent communication.

2 Software Agents

It is difficult to find a simple and universally accepted definition for an agent that includes all its attributes. We can define an agent as a software module, eventually equipped with AI mechanisms, capable of solving autonomously or in cooperation with other agents — a certain problem or carrying out a particular task .Attempts to characterize agents have resulted in a list of attributes.

T. Yakhno (Ed.): ADVIS 2000, LNCS 1909, pp. 429-436, 2000.

- Autonomous — acts independently
- Adaptive/Learning — adapt to evolving environment
- Mobility — across machine boundaries
- Intelligence — can reason
- Persistent — over time
- Goal oriented
- Communicative / Collabrative — interacts with other agents
- Flexible

Agents are able to perform numerous functions or activities without external intervention over extended periods of time. They also tend to be small in size. They do not constitute a complete application, instead they form a part of the system by working with other agents.

2.1 Appication Domains of Agents

Their interesting features make agents an attractive topic in computer science. Objects and distributed object arhitectures, adaptive learning systems, artificial intelligence, expert systems, genetic algorithms, distributed processing, distributed algorithms, security and social environments are just a few areas in computer science that agent can be integrated to easily. There are basically three different domains where mobile agents have potential deployment. One is data-intensive application where the data is remotely located, is owned by the remote service provider, and the user has specialized needs. Here, the user sends an agent to the server storing the data. The second domain is where agents are launched by an appliance — for example, shipping an agent from a cellular phone to a remote server. The third is for extensible servers, where a user can ship an install an agent representing him more permanently on a remote server.

2.2   Problems Agents Solve

An agent arhitecture solves many classical   problems. Mobile agents solve client/server network bandwith problem. When performing a client/server query it may be necessary to create many transactions and each create network traffic and decrease available network bandwith. By creating an agent and sending the agent to the server side, network consumption is reduced. The agent will do all the queries at the server side and will come back with the results.

An agent architecture also solves unreliable network connections. In most applications today, the connection must be alive during the process. If the connection goes down, the client has to start the transaction again. Agent technology allows a client to dispatch a transaction when it is online. The client then goes offline while agents process the job for the client, and when the client gets online again, an agent sends the results to the client.

3   JATLite

JATLite (Java Agent Template, Lite) is a package of programs written in the Java language that allow users to quickly create new software agents that communicate robustly over the Internet. JATLite provides a basic infrastructure in which agents register with an Agent Message Router facilitator using a name and password, connect/disconnect from the Internet, send and receive messages, transfer files, and invoke other programs or actions on the various computers where they are running. Fig. 1 depicts the JATLite approach.

JATLite facilitates especially construction of agents that send and receive messages using the emerging standard communications language, Knowledge Querying and Manipulation Language (KQML). The communications are built on open Internet standards, TCP/IP, SMTP, and FTP. However, developers may easily build agent systems using other agent languages using JATLite.

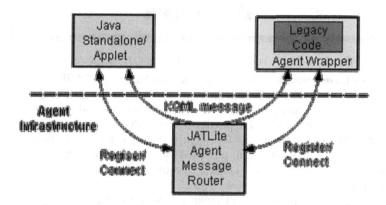

Fig. 1. JATLite Approach to wrapping legacy software

3.1   JATLite Architecture

The JATLite architecture is organized as a hierarchy of increasingly specialized layers shown in Fig.2, so that developers can select the appropriate layer from which to start building their systems. Thus, a developer who wants to utilize TCP/IP communications but does not want to use KQML can use only the Abstract and Base layers as described below

The **Abstract Layer** provides the collection of abstract classes necessary for JATLite implementation. Although JATlite assumes all connections to be made using TCP/IP, one can implement different protocols such as UDP by extending the Abstract Layer.

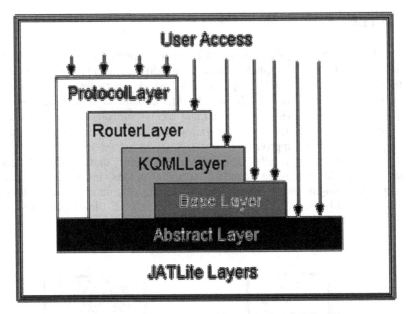

Fig. 2. JATLite is built as a series of increasingly specialized layers

The **Base Layer** provides basic communication based on TCP/IP and the abstract layer. There is no restriction on the message language or protocol. The Base Layer can be extended, for example, to allow inputs from sockets and output to files. The Base Layer can also be extended to provide agents with multiple message ports, etc.

The **KQML Layer** provides for storage and parsing of KQML messages. Extensions to the KQML standard, proposed by the Center for Design Research,are implemented to provide a standard protocol for registering, connecting, disconnecting, etc.

The **Router Layer** provides name registration and message routing and queuing for agents. All agents send and receive messages via the Router, which forwards them to their named destinations. When an agent intentionally disconnects, or accidentally crashes, the Router stores incoming messages until the agent is reconnected. The Router is particularly important for applet agents, which can only initiate socket connections with the host that spawned them, due to WWW and Java security restrictions.

The **Protocol Layer** on top of Router Layer will support diverse standard internet services such as SMTP,FTP,POP3,HTTP, etc both for stand alone applications and applets. Current beta version supports SMTP and FTP but other protocols can be easily extended from Protocol Layer. If an agents is expecting to transfer non-sentential, lengthy data or needs to send KQML message through email, Protocol Layer will be a good starting point.

4   The Information Retrieval System

The Information Retrieval System is an agent based framework that gathers information from a computer network on behalf of a user. To support an interface in which heterogenous components can interoperate and appear homogenous, we need a variety of agents. Different agents are needed for each of the different components. The system architecture is elaborated on four types of agents, each with a specialized function: user, broker, resource and backup agents. There exist only one broker and backup agents, while numerous resource and user agents take part in the system. Fig. 3 depicts the system architecture.

Each agent has a unique name across the system. As user agents represent users who wish to gather information, the user name is used as the unique identification of the corresponding user agent. An agent which wants to enter the system registers itself with its unique name and a password. A password needs to be supplied for security reasons, in order to prevent other agents to enter the system with the same name. After being registered, an agent may connect to the system anytime by providing its name and password and be on-line, and disconnect at a later time. It is also possible for an agent to unregister but this feature is not enabled on the user side. Below we expain what each agent is responsible of and how they interact with each other in order to fulfill user requests.

4.1   User Agent

A user agent acts as an intermediary between the user and the information retrieval system, providing access to information resources, running as an applet on the user's web browser. It supports a graphical user interface and displays the results. It's primary task is to determine user requests (queries) and to pass them to the broker agent. It is also responsible of displaying the results of a request. A user agent interacts only with the user and has no knowledge about the source of the information it gets. There may be several user agents connected in the system at a time, each representing a different user.

4.2   Broker Agent

The broker agent is the central decision maker and is unique in the system. It receives a user's request (through a user agent) and then satisfies it by managing its processing. It implements directory services for locating appropriate agents with appropriate capabilities and relevant information, to which it directs user requests to be processed. Under its control, a request might be sent to one or more databases, which are managed by resource agents

The broker agent's world (its environment and other agents) continuously evolves, with new agents signing in and out. It interacts with it environment by sending periodically certain control messages to resource agents, updating information on their status and thus gaining knowledge to help it to make decisions on the most appropriate action in the current context. It directs user requests to those resource agents most capable of processing it, thus minimizing network traffic and search time.

The broker agent keeps track of agents registered to the system, those connected on line at a certain point in time, and details related to results of a user's request, such as, when the result is expected (immediately/at a future time) and how the result is to be sent (displayed on the screen/ via e-mail). It runs as a Java application and all other agents have to know its unique name in order to communicate with it.

4.3 Resource Agent

Resource agents provide access to information stored in specific information database resources. They search their database for user requests and send the results back directly to user agents or to via supplied SMTP addresses. Each resource agent may reply to a specific type of information request, called its category. When a new resource agent registers to the system, it supplies its category information. New resource agents that are similar to existing ones in their implementation of data storage and search techniques are easily added to the system by creating a copy of the existing resource agent and making the necessary modifications in its configuration file which holds management information.

There may be several resource agents in the system, each one accessing data in the same or in a different category. Resource agents do not communicate with each other. Each database agent runs as a java application. The implementation of each resource agent may be completely different. Issues about how and where it reaches the data, which search tecnique it uses and how the data is stored are related to a specific resource agent and are not known by other agents in the system. The agents should only conform to a specific request and reply message format.

4.4 Backup Agent

We maintain the persistence and reliability of the system through a backup agent which is assumed to replace the broker agent in case of a failure. This agent checks continuously if the broker agent is still working properly and is online. In case of failure, it creates a new broker agent with the same properties and knowledge base. The backup agent communicates only with the broker agent. It runs as a java application and does not have to be on the same computer the broker agent executes.

5   User Flexibilities

The system provides the user with certain flexibilities not found in classical information retrieval systems. He can make a choice on the arrival time of the results — he may wish to see the results immediately after making the request or he can register the request at a futur time. The results will either be displayed on his screen, or will be sent in a message to a priorly provided e-mail address, according to his preference. In case the user disconnects before the results arrive, the system keeps them for him while he is off-line and sends them to him as soon as he reconnects.

Information is classified into categories and a user may request a search in a certain category. The user may have the categories among which he may make a selection be listed. As new resource agents register to the system, this list will expand. If the user can not decide on the category, he may just enter the keywords. In this case, the broker agent directs this request to certain resource agents that have registered to serve general requests, without a specific category information.

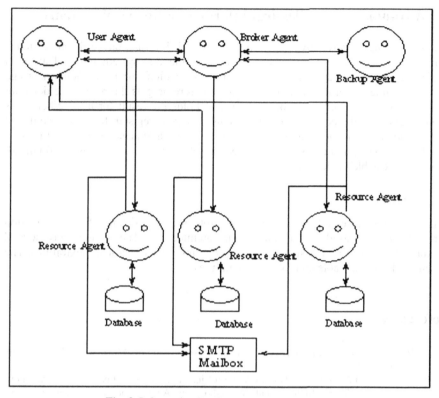

Fig. 3. Information Retrieval System Architecture

6   System Characteristics

The basic characteristics of the agent based information retrieval system are as follows.

Autonomous and Intelligent : An autonomous agent is within and part of an environment and acts on it , over time , in pursuit of its own agenda and so as to affect what it senses in the future. We easily see that the broker agent satisfies these requirements. We can also say that it is intelligent since it has the ability to decide, can think on behalf of the user. Its actions can change over time as agents in the system change their behaviour.

Learning: We shortly discussed the questions broker agent can respond. This could be possible with learning. When the broker agent first gets on-line, it knows nothing but with control messages it sends to other agents, it begins gathering information and in a short time it becomes fully informed about the status of the entire system. Its information base is up to date since these control messages are sent periodically.

Communicative/collaborative: The system is based on the Jatlite agent message router and consists of multiple agents that try to respond to user requests by communicating and working together.

Flexible: Besides the flexibilities provided to the user, the system can easily be expanded with the integration of new resource agents.

7   Advantages over Classical Information Retrieval Systems

The system has many advantages over classical information retrieval systems. Traditional system rely on a central database while this system has a distributed database architecture. Utilization of category information minimizes network usage and search time. Even if a user line is broken before he gets the results, he does not need to make the search again because system holds the result for him. When he is online again he will receive the results. The user can also register the request so that at a later time, he can retrieve the results. When he issues a request, he does not need to wait for the results and can start a new search before his prior requests are finalized. This is not possible in classical systems.

8   Results

This paper has presented the design and implementation details of an information retrieval system based on a multi-agent architecture. The system has several advantages over similar classical systems, providing the user various flexibilities. The system is also expandable and tolerant to failures.

References

1.   M.R. Genesereth and S.P.Ketchpel."Software Agents", Comm. ACM, Vol. 37, No. 7, July 1994, pp.48-53.
2.   S. Franklin and A. Graesser "Is it an Agent, or just a program? A taxonomy for Autonous Agents" Proc. Third International Workshop on Agent Theories, Architecures,and Languages, Springer-Verlag,1996.
3.   D. Milojicic, "Mobile Agent Applications", IEEE Concurrency, Vol. 7, No. 3, July-Sept. 1999, pp. 80-90.
4.   4.JATLite (Java Agent Template, Lite) http :// cdr.stanford.edu /ProcessLink / Papers / Jatl.html
5.   Knowledge Querying and Manipulation Language- KQML standard http://www.cs.umbc.edu/kqml/
6.   Extensions to the KQML standard — http://cdr.stanford.edu/ProcessLink/kqml-proposed.html

An Intelligent Tutoring System for Student Guidance in Web-Based Courses

Bülent Özdemir and Ferda N. Alpaslan

Computer Engineering Department, Middle East Technical University
06531, Ankara, Turkey
{ozdemir,alpaslan}@ceng.metu.edu.tr

Abstract. In this study, an intelligent agent to guide students throughout the course material in the internet is defined. The agent will help students to study and learn the concepts in the course by giving navigational support according to their knowledge level. It uses simplified prerequisite graph model as domain model and simplified overlay model for modeling student. The system adapts the links in the contents page to help students for easy navigation in the course content. In link-level adaptation hiding and annotation technologies which effectively support gradual learning of the learning space are used.

1 Introduction

The advent of the Internet provides a new and interesting environment for new learning environments. Among these new teaching architectures, simulation-based learning by doing, incidental learning, learning by reflection, case-based teaching, learning by exploring, and goal-directed learning are the most attractive ones [1]. It is a fact that almost all of these teaching architectures need software designed for the specific teaching architecture. The obvious advantage of the Internet is that it is time and location independent. These aspects have made the Internet the most appealing media for education for the last few years. At the beginning, the courses on the Internet were not much different than an electronic multimedia textbook. Internet has a disadvantage that it looks as a bunch of links where you can be mixed up in a minute if you are not guided well. From this view, a paper or book can do better than a web-based course. So, there is a need for more effective learning than traditional hypertext [2].

For this purpose, several different tools with different features have been implemented at various universities. University of Illinois at Urbana-Champaign's SCALE, SUNY Plattsburgh's Top Classes System, Washington State University's Virtual Campus, University of Illinois' The Mallard Learning Environment, Heron Laboratories' SAFARI are some of these systems. Most of these systems are generally group of programs used for preparation and presentation of the course material. They have also some facilities such as automatic grading, conferencing with students, making announcements on bulletin board systems, etc.

In an intelligent tutoring system, reasoning about the user actions and responding immediately and accordingly are essential. To achieve this, a user model should be generated for each user to make assumptions about the user's state of knowledge and

T. Yakhno (Ed.): ADVIS 2000, LNCS 1909, pp. 437-448, 2000.
© Springer-Verlag Berlin Heidelberg 2000

learning needs. SAFARI developed at Heron Laboratories is an intelligent tutoring system with a multi-agent architecture, defined   for the pedagogical component. These systems use multiple learning strategies and cognitive agents that can model the human behavior in learning situations [3].

Experiences show that creating an intelligent tutoring system still remains difficult for various reasons. There are many different components in such a system and stabilization of these components is difficult. The structure and type of knowledge to be used are of different granularity and complex to handle. This makes design and implementation of the reasoning system a difficult task. Another problem is that the components' design and implementation requires multiple expertise from different fields such as computer and cognitive science. Integration of these expertise is a challenging task [4].

Although difficulties are many, the advantages of an intelligent tutoring system having more or less intelligent features are undeniable. In this study, an intelligent agent for student guidance on web-based courses developed is explained. The aim was to create an agent that would present concepts in a course in a coherent way such that the learner is guided throughout the courseware. While doing this, the student's knowledge level of the subject being taught is followed. Mainly, individualized tutoring is used.

Course material which is in the form of .html files is represented by a course curriculum. A curriculum is composed of concepts and each concept is a task to be completed by the student and connected to some other concepts with prerequisite relation. The resulting structure is a *conceptual network*. The student is presented with a new concept if (s)he completes the concepts that are the prerequisites of the concept to be learned. This is the first point where the system reasons about the user actions.

In a web-based course, students' learning skills usually are tested with tests and case studies. In the system proposed, the students are given a "pretest" which tests the concepts (s)he must know well before taking a real test or continue with the following concept in curriculum. If there are incorrect answers, the student is advised to navigate backwards and complete the tasks related to the pretest. This is the second point where the system reasons about the user actions.

The intelligent agent designed and implemented in this study has been integrated with the BADE authoring tool developed at METU Informatics Institute in the context of BADE Project. (AFP-06-04DPT.98K1225). The objective of the project is to develop an authoring tool for web-based course generation and administration, and offer on/off campus web-based courses.

2 Conceptual Model

The model that will be explained in this section uses the idea similar to the components of *prerequisite graph model* in [5]. Prerequisite graph model is composed of four components:

- ❑   conveniently partitioned hyperspace
- ❑   a prerequisite graph,
- ❑   a comprehension measure and
- ❑   an interaction formula.

Conveniently partitioned hyperspace and a prerequisite graph are the two components that are primary concerns of the model explained in this section. The main features of the prerequisite graph model are to help students navigate in course content hyperspace; to follow preprogrammed learning strategies and support student's goal-oriented efforts.

2.1   Conceptual Network and Curriculum

A course is defined by its content. A course content is composed of several concepts to be learned, examples to support the concepts, and the tests to measure the student's knowledge level. The course content is modeled by hierarchical sub-units. The division is shown in the Table 1. In this division entities *course* and *chapter* are logical subunits. They are composed of physical sub-units named *concept*. Concept is a physical entity because each concept is associated with an .html file in the hyperspace. A concept can be one of the five different types: it can be a theorem, a definition, a lemma, or a test.

Table 1. Hierarchical Division of a Course Content

Unit	Composed of
Course	Chapter
Chapter	Concept
Concept	Theorem | lemma | definition | test

The relationship between concepts is represented in a network expressing the prerequisite relation. The resulting semantic network is called *conceptual network*. In this network, a node is either a non-terminal node or a terminal node. A terminal node has no successors. A non-terminal node is either an AND node or an OR node. Each node in the graph is connected to one or more nodes with a direction. A node is said to be solved if all the AND nodes it is connected are solved [6]. A sample graph is given in Fig.1.

Defining the course content with this model makes the student learn concepts step-by-step fashion upgrading their knowledge level gradually. The logical coherence is established, and a student feels comfortable, because (s)he knows everything necessary to learn a concept right before studying a new concept. This model also makes the navigation chaos disappear by giving the right guidance according to each student's knowledge level individually.

After the division is done, and the conceptual network is generated the sequence of concepts should be presented by a curriculum. The curriculum is a simple list of concepts with the level information. If a concept is a subtopic in a curriculum, its level is higher than its container topic. In Fig.2. there is a part of the curriculum for the web-based Artificial Intelligence course.

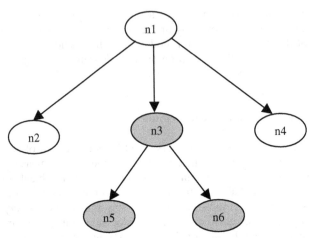

Fig. 1. A Sample Conceptual Network. In the figure, the shaded nodes are AND nodes, the others are OR nodes. The node n3 is said to be solved if the nodes n5 and n6 are solved. The node n1 is said to be solved only if n3 is solved, because n3 is the only AND node, the other nodes n2 and n4 which n1 has a path to are OR nodes and need not to be solved. In other words, n5 and n6 are the prerequisites of node n3, and n3 is the only prerequisite of node n1.

ARTIFICIAL INTELLIGENCE
    CHAPTER 1
        Introduction-definition of AI
        Why Study AI?
        The Turing Test
        Mental and Physical Machines
        A Quick Survey
        .
        ..

Fig. 2. A Sample Curriculum

2.2   Pretest

Knowledge level of a student on a particular concept is measured by small tests, called *pretests*. A pretest is composed of two types of questions: multiple choice or true/false. The questions in a pretest reflect the basic concepts that the student should necessarily know right before (s)he starts studying a new concept. According to the incorrect answers given to the pretests, the student will be advised to restudy the concept(s) (s)he lacks enough knowledge.. This idea is a simplification of the idea presented as explicit evidence in overlay modeling in the [7]. There assumed to be interactive examples or questions in the course material also. During the attempt to solve a problem, test, or example student may be supplied with a list of help topics. Help topics will be of course the concepts relating to the problem at hand. Help topics are nothing but the nodes in the conceptual network.

2.3   Adaptive Hypermedia and User Modeling

A system is said to be an adaptive hypermedia (AH) system if it adapts the information and links being presented to the particular user by using the user model derived from the user's knowledge level, goals or preferences. An AH system can also assist user in navigation through the hyperspace.

There exist six kinds of hypermedia systems, which are used at present as application areas resulting from the research projects on adaptive hypermedia. These are educational hypermedia, on-line information systems, on-line help systems, information retrieval hypermedia systems, institutional information systems, and systems for managing personalized views [8]. Among these systems, educational hypermedia system is the most popular one as a hypermedia research area and the system developed in the thesis study fits in this class. An educational hypermedia system has relatively small hyperspace, which is composed of course material. The main goal of the user is to learn this material. In such a system the knowledge level of the user is the most important user feature. The course material should be presented according to the knowledge level of the student. The system should also supply navigational help for the user to not get lost in the hyperspace.

The knowledge level of the user is not the only feature of the user. User's goal, background, hyperspace experience and preferences are the other features. In this study, only user's knowledge level is taken into consideration.

User knowledge of the subject is represented by an overlay model. In this model, for every concept presented in the domain, there is a value for the user knowledge level of that concept. This value can be just a binary value (known/not-known), a qualitative measure (good/average/poor), or a quantitative measure, such as the probability that the user knows the concept [8]. In this study, binary value is used as the value for the user knowledge level of a specific concept. The information of the user knowledge level is obtained from the user's answers given to the pretests.

Thus, the system takes into consideration the user's knowledge level when providing adaptation. There are two things in an AH system that can be adapted: the content of the course material's building blocks that is pages (content-level adaptation) or the links used to reach that pages (link-level adaptation). Link-level adaptation has been used in this study.

In general, there are two information objects in a hypermedia system: the .html pages and the links that connects them. The links can be placed in a separate page, which is called an index page, content page, or global map. Link-level adaptation adapts the presentation or permission of the links in the system. The link-level adaptation is also named as "adaptive navigation support" in [8]. As the name implies, it provides help for users to find their ways in the hyperspace. There are five technologies for adapting link presentation: direct guidance, sorting, hiding, annotation and map adaptation [8].

In this study, the links are in a content page and the techniques used for adapting link presentation are based on hiding and annotation. The idea of hiding is concealing the links, which are not relevant to the user's goal. This idea is adapted in the system by not allowing the user open a page if the user does not have sufficient background, which means that the prerequisites of the page have not, learned yet. Hiding can effectively support gradual learning throughout the learning space. The adaptive annotation technology mentioned in [8], [9] and [10] suggests the augmentation of the links with some form of comments, so that user can see the current state of the link.

This idea is adapted in this study by putting a *plus* sign next to a link whose associated html page is visit

3 Agent Architecture

The major tasks associated with the agent are as follows:

- ❑ Task 1: curriculum generation and maintenance
- ❑ Task 2: pretest generation and maintenance
- ❑ Task 3: presentation of curriculum to the student
- ❑ Task 4: presentation of pretest to the student
- ❑ Task 5: reasoning about student actions on the curriculum
- ❑ Task 6: reasoning about student actions on the pretest
- ❑ Task 7: giving the results of reasoning to the student
- ❑ Task 8: database operations
- ❑ Task 9: user interface tasks

There are two types of users in the system: *teacher* and *student*. It is obvious that the curriculum and pretests are created and maintained by the teacher are presented to the student. The reasoning mechanism reasons about the student actions, such as navigation in the hyperspace, success in the pretests, etc. The results of the reasoning are presented to the student in a user-friendly way. The architecture of the agent is represented by Fig.3.

According to the figure the system works as follows:
1. Teacher creates the curriculum with a user interface in the curriculum generation module.
2. Teacher creates the pretests with a user interface in the curriculum generation module.
3. Teacher can update the existing curriculum and pretests.
4. All the data regarding to curriculum and the pretests are stored in the database.
5. The curriculum and the pretests are presented to the student with a user interface in the personal assistant module.
6. Student can study the course material and pretests by using the interface in the personal assistant module.
7. Student actions are evaluated by the reasoning mechanism and the database is updated accordingly.
8. The reasoning mechanism gives unsolved concept list to the personal assistant module so that personal assistant module can do advises and gives navigational support to the student according to the student action.
9. The reasoning mechanism gives .html file pathname to the personal assistant module so that personal assistant module can open the .html page for the student to study.

Conceptual network is created by the teacher after dividing the course content into hierarchical sub-units and determining the concepts in the course. This work should be done carefully, because the effectiveness of the system depends on the design of the conceptual network.

The second step of curriculum generation is relating the conceptual network with the corresponding .html files in the database. After the curriculum is prepared and .html file pathnames is stored in the database, teacher can input the course curriculum data into the system. The user interface shown in Fig. 4. allows the teacher to create a new curriculum, open or save an existing one, and exit.

Node operations menu (Fig.5.) is used to add a new node to the curriculum or update node properties. Test menu (Fig.6.) is used to create a new test by inserting the questions and answers in it (Fig.7.) or to update the existing tests.

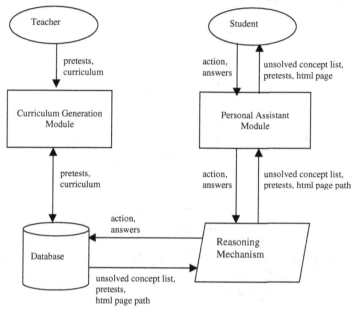

Fig. 3. Agent Architecture. Ovals indicate the users in the system while rectangles indicate the modules in the system. Arrows show the direction of the data flow and the labels on the arrows specifies the data. The cylinder indicates the database where the data resides. The trapezoid represents reasoning mechanism working on the user actions and the database

A student follows the course material in the Internet using the intelligent user interface in the personal assistance module shown in Fig.8. The student solves the tests/pretest via the same module.

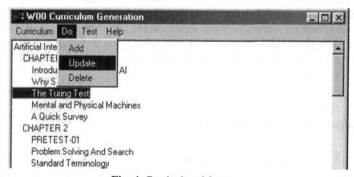

Fig. 4. Curriculum Menu

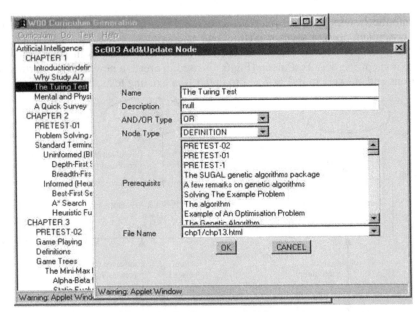

Fig. 5. Node Operations Menu

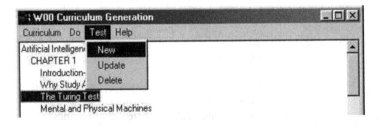

Fig. 6. Test Menu

The system was implemented as a standalone application with Microsoft J++ 6.0 with the features of Microsoft Developer Studio 6.0. Some of the WFC (Windows Foundation Classes for Java) developed by Microsoft were used in the system. The program was then converted to Java Applets. The database management system is Mini SQL. MsqlJava package (A Java Class Library for mSQL) is used for database operations such as executing SQL query statements, connecting to the database and closing the database.

4 Conclusion and Future Work

The system described in this study integrates some of the techniques and methods used in adaptive hypermedia, intelligent tutoring systems, and personal assistants. It uses simplified prerequisite graph model as domain model that is well suited for user modeling and adaptive navigation support. The simplified overlay model is used for student modeling. Overlay model takes into consideration the user's knowledge level, which can be taken from the answers of the pretests when providing adaptation. The system adapts the links in the contents page to help students for easy navigation in the course content. In link-level adaptation hiding and annotation technologies which effectively support gradual learning of the learning space are used.

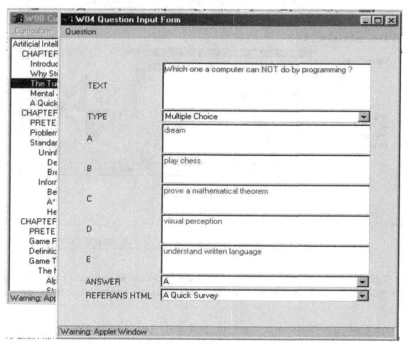

Fig. 7. Question Input Form

Teaching via the Internet requires special skills of educators, just as receiving instruction through the Internet requires different types of preparation and use by learners. An effective distance learning is a good combination of qualified educators, appropriate technologies, effective materials designed for the Internet, and learners who appreciate and need this type of instruction [11]. The effectiveness of the tutoring system discussed here is closely related with the preparation of the domain model. Educator should partition the system in such a logical and coherent way that all prerequisite relations among the concepts should be stated correctly and completely. An incomplete or incorrect domain model would give rise to disadvantages rather than advantages for the students. Pretests should be prepared in such a way that they meet the needs and expectations of the students. Effectiveness, sufficiency, and completeness of the course material is another important component determining the efficiency of the system.

Although the system satisfies its main objective, which is forming a base for an adaptive hypermedia system, there are some improvements when added to the system will make it more complete. As stated earlier, building an intelligent tutoring system is not an easy task for various reasons. The previous experiences and this thesis study show that there are many different components in such a system. The components' design and implementation requires multiple expertise from different fields of computer and cognitive science.

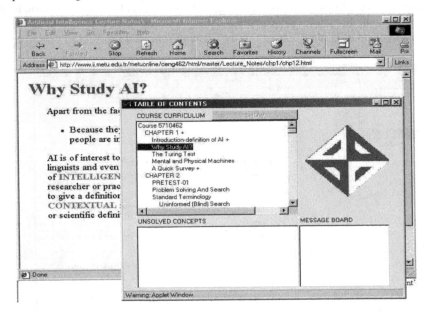

Fig. 8. Intelligent User Interface for Students. The nodes in the curriculum have a plus (+) sign next to the node's name if the user solved (studied) the concept represented with that node before.

Two types of questions are used in pretests: multiple-choice and true/false questions. More sophisticated questions that can be solved in a number of episodes can be prepared and the solutions to these questions can be analyzed to get more information about the student knowledge level. This is the real adaptation of the overlay model and used in ELM-ART II [9] as episodic learner model.

The value used for the user knowledge level is binary in this study. The user either knows the subject or does not know the subject. With the addition of more sophisticated questions, this value can be a quantitative which , then, can be used in content-level adaptation. Moreover, user preferences and background can be taken into consideration to hide the irrelevant information in the hyperspace. The user can inform the system directly about the preferences which are the links or .html pages preferred over others. The user background can contain the user's profession, experience of work in related areas etc. The user background can, then, be used for adaptive navigation support.

In long term, some of the information retrieval techniques can be adapted to the system for letting students to ask questions to the system. The question asked by the student can be similar to the question asked in the classroom. The system will parse the question and provide information asked by sorting: most relevant information first and least relevant information last.

Actually, the applications in this research area show that the list of improvements can be quite much in long term. However, we should emphasize once more that the feasibility of implementing these improvements depends on the integration of the multiple expertise from different fields of computer and cognitive science.

An educational adaptive hypermedia system with a personal assistant is best evaluated and tested by the students. The system implemented in this study was evaluated by a group of students enrolled in a regular Artificial Intelligence course for successfulness and effectiveness of navigational support and pretests. The students also evaluated the time consideration as well as the design and the user-friendliness of the user interfaces. The numbers in the table indicates the number of students.

Table 2. Evaluation Results

Evaluation Criteria		Above Average	Average	Below Average	Time needed (Average)
Time	Curriculum Load		18	4	2 min.
	Page Load	21		1	2 sec.
	Question Form Load	20		2	2 sec.
User Interface	Design	12	7	3	
	User Friendliness	14	3	5	
	Navigational Support	18	2	2	
	Effectiveness of Pretests	18	2	2	

The results obtained shows that:
- The time needed to display the curriculum is considerably long.
- The time needed to load a page is quite fast when a link in the curriculum is clicked.
- The design of the user interface satisfactory
- Question form should be redesigned.
- Navigational support is effective and necessary.
Pretests are effective and necessary.

References

1. Schank, Roger C.: Active Learning Through Multimedia. IEEE Multimedia. 1 (1994) 69-78.
2. Hammond, N., McKendree, J., Reader, W., Trapp, A., Scott, P. J.: The Psycle Project: Educational Multimedia for Conceptual Understanding. Multimedia-95, San Fransisco, (1995).
3. Frasson, C., Mengelle, T., Aïmeur Esma, Gouardères, G.: An Actor-based Architecture for Intelligent Tutoring Systems, ITS'96 Conference, Lecture Notes in Computer Science, Vol. 1086, Springer Verlag, Berlin Heidelberg New York (1996) 57-65.
4. Gecsei, J. and Frasson, C.: SAFARI: an Environment for Creating Tutoring Systems in Industrial Training. EdMedia, World Conference on Educational Multimedia and Hypermedia, Vancouver, (1994) 25-30.
5. Nykänen, Ossi.: Work in Progress: User Modeling in WWW with Prerequisite Graph Model. In: Proceedings of the Workshop on Adaptive Systems and User Modeling on the World Wide Web, Sixth International Conference on User Modeling, Chia Laguna, Sardinia, 2-5. (1997).

6. Chakrabarti, P. P.: Algorithms for Searching Explicit AND/OR Graphs and their Applications to Problem Reduction Search. Artificial Intelligence. **65** (1994) 329-345.
7. Carr, Brian P., Goldstein Ira, P.: Overlays: A Theory of Modelling for Computer Aided Instruction. AI memo. **406** (1997) 1-23.
8. Brusilovsky, P.: Methods And Techniques Of Adaptive Hypermedia. User Modeling and User-Adapted Interaction, Special issue on: Adaptive Hypertext and Hypermedia. **6** (1996) 87-129.
9. Weber, G. , Specht, M.: User Modeling and Adaptive Navigation Support in WWW-based Tutoring Systems. In: Proceedings of User Modeling'97, (1997) 289-300.
10. Brusilovsky, P., Schwarz, E., Weber, G.: A Tool for Developing Adaptive Electronic Textbooks on WWW. In: Proceedings of the WebNet '96 World Conf. of the Web Society, Part Six, HCI: Education and Training, AACE (1996) 64-69.
11. 11Porter, L.R. : Creating the Virtual Classroom. John Wiley and Sons, Inc. New York (1997).

An Approach to Embedding of Agents into Knowledge Representation and Processing System[*]

Yuriy Zagorulko[1], Ivan Popov[1], and Yuriy Kostov[2]

[1]Russian Research Institute of Artificial Intelligence
Lavrentiev av., 6, Novosibirsk, 630090, Russia
{zagor, popov}@iis.nsk.su
http://www.rriai.org.ru/
[2]Institute of Informatics Systems SB RAS
Lavrentiev av., 6, Novosibirsk, 630090, Russia
kostov@iis.nsk.su
http://www.iis.nsk.su/

Abstract. The paper presents an approach to the development of a software environment for creation of intelligent systems. This environment is based on an integrated knowledge representation model that combines both classic knowledge representation means and constraint programming methods with agent-based techniques. In contrast to other systems following constraint programming methods, the environment presented here allows one to manipulate objects with imprecise (subdefinite) values of attributes. The process of refining such subdefinite attributes is performed by means of a special data-driven mechanism. Another feature of the environment is the use of an agent-based technique instead of the traditional production rule technique, in order to increase efficiency of the processes of logical inference and data processing. Due to a natural combination of the data-driven and event-driven mechanisms, the environment can be used for the development of efficient intelli-gent systems for various applications.

1 Introduction

There are a few well-known paradigms of knowledge representation and processing. Each of them has demonstrated its advantages for solving certain classes of problems in *fixed* subject and/or problem domains (SPD). None of these paradigms taken separately, however, can cope with all the difficulties that arise while solving problems in heterogeneous SPD.

Our approach is based on the simple idea that the problems caused by heterogeneity of SPD can be solved by means of integration (or synthesis) of the paradigms and methods that can effectively solve the problems of each type. We will try to show the advantages of such a "synthetic" approach. Our ultimate, strategic goal is to create a universal modeling system, i.e., a system that will enable us to build

---

[*] This work is partially supported by Russian Foundation for Basic Research (grant N 99-01-00495).

models of any SPD in a natural and easy way and to use them to solve the problems efficiently.

In this paper, we consider an approach used in the development of a programming environment SemP-A, intended for design of a wide range of intelligent systems. Within the framework of this environment, we will try to integrate some of the well-known paradigms in a *natural* and *cohesive* way. This attempt is another step towards the creation of tools that will permit solving problems in *any* SPD.

The system SemP-A presented here is a further development of the programming environment described in [1], which is based on an integrated model of knowledge representation that unifies the classic means such as frames, semantic networks, and production rules, as well as methods of constraint programming [2]. Inclusion of methods of agent technology [3,4], which is a natural evolution of the object-oriented approach, opens up new opportunities for improving descriptive power and efficiency of the knowledge representation and processing systems produced with the help of the SemP-A environment.

Unlike other systems using the techniques of constraint programming, this system allows us to manipulate objects with imprecise or partially defined (in other words, *subdefinite*) values of attributes (slots). Automatic refinement of the values is based on constraints that can be linked with the objects and relationships presented in a semantic network. The method of subdefinite computational models proposed by Narin'yani [5,6] is used to implement the mechanism of constraint satisfaction.

Another important feature of the system is the use of the agent-based technique rather than of the more traditional rule-based one. The use of the agents provides important advantages both for convenience of describing inference and data processing and for improving its efficiency. In particular, the agent-based technique makes the inference and data processing more efficient due to the use of the event-driven mechanism in which every agent reacts only to relevant events.

General Model of Knowledge Representation and Processing in SemP-A

The general model of knowledge representation and processing in the SemP-A environment consists of the following components:

- semantic network,
- functional network,
- agent network.

The semantic network is represented by an oriented graph that consists of objects connected by relations. Objects can be any entities of the subject domain; they are represented by frames. Instances of relations in the semantic network differ from objects only in their meaning: relations are objects that stand for links between other objects. Therefore, relations are treated as a special class of objects. Relations over relations are also easy to implement, since relations in the semantic network are represented by objects. Thus, in what follows the term "object" will refer either to a usual object or to an instance of a relation, unless otherwise noted.

Objects with identical properties are combined into classes. Classes may inherit properties of other classes (in this case, the former are called subclasses, and the latter are superclasses), with a possibility of multiple inheritance. Some properties of a superclass can be redefined in subclasses.

The operations allowed in a semantic network are the following: creation of objects or relations (***new***), removal of objects/relations (***delete***) or modification of the values of the objects' slots (***edit***). The values of the slots of an object can be either exact or subdefinite.

The functional network consists of computational models connected with all objects and relations of the semantic network. Each computational model represents a set of constraints defined on the values of the slots of some object. For a relation, the computational model is defined on the slots of the objects linked by the relation. The goal of the functional network is to refine the subdefinite values of the objects' slots or to signal contradiction if the set of constraints is inconsistent. The functional network operates on the basis of a data-driven procedure of constraint propagation. It is activated after any operation with objects of the semantic network, and operates until it becomes stable. Execution of operations over the semantic network (creating, deleting or editing objects) may result in modification of the structure of the functional network (addition/deletion of computational models or change of links between them).

The agent network is represented by a set of agents tracking changes in the semantic network. The agents respond to creation, deletion or any change of the objects and relations of the semantic network. Activation of the agents is based on an efficient event-driven procedure. The agents are ordered into a hierarchy based on the criterion of "general-special," may have priorities and may be divided into groups to simplify the control of their activation.

The semantic and functional networks in the SemP-A environment are structured and operate just as in the SemP-TAO environment [1], so further attention will be focused mainly on the organization and operation of the agent network.

Agent Model and Event-Driven Control

Structure of Agents

The agents in the SemP-A environment are regarded as a special type of active objects, although agents are generally a richer notion. The general structure of an agent can be presented by the following diagram (Fig.1).

This diagram shows that each agent consists of the following components:
- a unique name A that allows one to identify the agent;
- a set of parameters that represents its internal data;
- a set of arguments that determine the classes of objects which the agent will respond to;
- a set of conditions defined on the arguments and parameters of the agent; the agent is activated when all of these conditions are satisfied;
- actions to be executed by the agent.

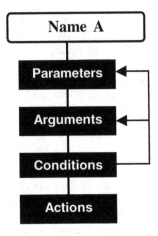

Fig. 1. General agent scheme

Like objects, agents with identical properties are combined into classes. All agents of one class have an identical structure of parameters, arguments and actions performed by the agents. They differ only in the values of their parameters and possibly arguments. Classes of agents may inherit properties of other agent classes. It allows one to organize agents into a hierarchy based on the criterion "general-special."

Structure of agents can be presented in more detail by means of the following scheme (Fig.2).

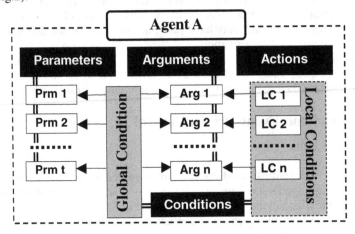

Fig. 2. Detailed structure of agents

There is a fundamental difference between parameters and arguments of agents. Parameters are characteristics or properties of the agent viewed as an object, while arguments define the objects in the semantic network that the agent is tracking (observing), i.e. its "field of vision". Each argument has a unique name and a class of objects corresponding to this argument.

Local conditions on the arguments are in fact conditions that filter "visible" objects. The definition of local conditions makes it possible to reduce significantly the number of combinations of objects for which the global condition will be calculated. Note that the global condition may contain the object's parameters as well.

An agent should be activated for all combinations of its arguments for which the global condition is true. Activation of an agent results in execution of the actions defined in the agent's body. These actions can lead to creation, modification or deletion of objects and relations of the semantic network, activation of other agents, execution of input-output operations, etc.

Activation of an agent for one combination of objects is regarded as an indivisible action. At the same time, activation of one agent for different combinations of objects occurs asynchronously, in general non-deterministically, and may be alternated with activation of other agents for their combinations of objects. This algorithm allows one, on the one hand, to simulate parallel execution of agents and, on the other hand, to avoid the conflicts that arise when several agents manipulate shared objects.

To control activation of objects and resolution of conflicts, we use a system of priorities and tools for controlling activation of agents used.

Priorities and Hierarchy of Agents

The arrangement of agents into a hierarchy defines a natural specialization of agents and a priority of their execution.

Each agent can inherit properties of another agent, i.e., its parameters, arguments, conditions of agent activation, and actions. Moreover, the derived agent can contain its own parameters and arguments, as well as refine (specialize) the classes of arguments of the base agent and expand its global and local conditions. Thus, the base agent can describe a more general fragment of the semantic network and define more general actions for this fragment, while the derived agent can describe a more special fragment and define more specialized actions, respectively. Therefore, derived agents have higher activation priorities than base agents.

This approach allows the system to react to particular situations and to perform general actions only if no reaction to specific situations has been defined. Besides, new derived agents describing special situations can be easily added to the system.

The definition of an agent's priority of activation according to its position in the general hierarchy is a quite natural way to specify processes of knowledge inference. However, it is not always sufficient, for instance, to resolve conflicts between agents from different branches of the hierarchy. In this case it is allowed to set more conventional, numeric priorities for the agents. Thus, conflicts between agents from the same branch of the hierarchy are resolved in favor of the derived agents, whereas for agents from different branches the conflict is resolved depending on the numeric priorities of the agents.

Resolution of the conflicts is also possible through activation and deactivation of the agents. This approach allows one to split a task into subtasks and to solve each subtask with a separate group of agents that are activated when needed. Furthermore, reduction of the set of active agents increases the efficiency of the overall inference and data processing.

Event-Driven Control

The general algorithm of operation of the agents is based on an event-driven control mechanism. This technique ensures coherent operation of all agents in both asynchronous and concurrent mode.

All agents are activated by events. An event is defined as appearance, disappearance or modification of objects in the semantic network. Each agent is tracking the events that correspond to its arguments, i.e., the events that are connected with objects whose classes are associated with the arguments of the agent.

The reactions of the agent to the various events can be different. For instance, when an object is created it is added to the set of objects associated with the corresponding argument of the agent. At the same time, this generates the set of combinations of this object with objects corresponding to other arguments of the agent, which can trigger activation of the agent. Conversely, when an object is deleted, it is removed from the set of objects associated with the corresponding argument of the agent. All the "trigger" combinations involving this object are removed as well. When an object is modified, actions equivalent to deletion of an object with the old values and creation of an object with new values are performed.

Thus, each creation, deletion or modification of an object in the semantic network results in activation and execution of agents. In turn, execution of agents can lead to creation, deletion or modification of objects, which in turn results in activation of other or the same agents, and so on. This process, triggered by the initial state of the semantic network, will continue until complete stabilization of the agent network.

Example

The Semp-A environment can be used to describe complex subject domains and to model various processes in them. A sample problem is modeling operation of a business. The complexity of such a simulation is in the following.

Firstly, a modern industrial enterprise consists of a large number of entities that are autonomous, though work coherently. They are, for example, workshops, departments, people and automatic devices. Each of these entities neither has to know nor really knows anything about the operation of entities both at a higher level and at the same level of the hierarchy. Secondly, an industrial enterprise is a dynamic system, whose product structure can vary in time, which results in quantitative and qualitative (structural) changes in the model of the enterprise. Thirdly, the enterprise is functioning in the real world, which is characterized by a multitude of uncertain and subdefinite parameters.

To illustrate the capabilities of Semp-A, the following model example is used. Though extremely simple, it clearly demonstrates the main ideas of the approach described. Suppose there is a manufacturer (e.g., a furniture factory) that makes certain products (chairs and tables) and consumes certain resources (wood and dye). The appropriate subject domain is represented by a semantic network with the following hierarchy of object classes (Fig.3).

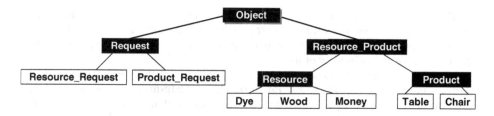

Fig. 3. Classes of objects

The black boxes denote virtual classes, and the white ones designate real classes of the semantic network. Let us present a fragment of description of this hierarchy in the SemP-A language.

Virtual classes look as follows:

```
class ResourceProduct
   price_per_unit: integer;
end;

class Resource ( ResourceProduct )
   amount: integer;
end;

class Product ( ResourceProduct )
   style: Style;
   color: Color;
   dye_per_unit: integer;
   wood_per_unit: integer;
end;

class Request
   amount: Integer;
   from_agent: string;
end;
```

Next, we introduce three types of resources, *Money*, *Dye* and *Wood*, which are derived from the *Resource* class. One can see here how constraints are defined on the values of object slots:

```
class Money ( Resource )
constraints
   price_per_unit = 1;
end;

class Dye ( Resource )
   color: Color;
constraints
   color <> none;
end;
class Wood ( Resource )
end;
```

The classes *Table* and *Chair* are derived from the class *Product* and represented similarly. The classes *Resource_Request* and *Product_Request* are derived from the class *Request*; their purpose is to simulate requests for products and resources. We omit their description in view of its simplicity.

Now let us turn to the description of agents. Assume that there are two kinds of workers in the factory: *Joiners* and *Painters*. They transform resources into products and are paid certain wages. There are also two kinds of managers: *Supplier*, who provides workers with the necessary resources, and *Inspector*, who accepts final products and pays wages to the workers. Suppose there is a *Customer* that cooperates with *Inspector*. *Customer* places orders and pays for the products. The set of agents can be decomposed into classes as follows, for example (Fig.4).

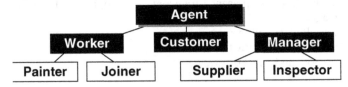

Fig. 4. Classes of agents

Here the black boxes denote virtual types of agents. We will provide a description for only one agent, *Painter*, in order not to overload the description of the example.

```
agent Painter
parameters
   Name: string;
arguments
   product_request: Product_Request;
   product: Product;
   dye: Dye;
conditions
   product_request.color <> none;
   product_request.amount > 0;
   product.color = none;
   dye.color = product_request.color;
   dye.amount > product.dye_per_unit;
actions
   edit product_request: Request( amount: amount - 1 );
   edit dye: Dye( amount: amount - product.dye_per_unit
);
   edit product: Product( color: product_request.color
);
end;
```

The description of this agent consists of four parts, as we can see in the example. Let us consider each part in detail.

The set of parameters of this agent consists of one parameter, *Name*, that contains the name of the specific *Painter*.

The arguments of the agent, declared in the section "arguments," describe the objects that are necessary for the *Painter*'s work: request for the product, the product itself and dye.

The following conditions presented in the section "conditions" should be satisfied:
- the color of the product should be specified in the request,
- the number of units of the product must be positive,
- the product must be not painted yet,
- the color of dye should coincide with the color of the product specified in the request,
- the amount of dye should be enough to paint one unit of the product.

When these conditions become true, the following actions are executed:
- amount of the product in the request is decreased,
- amount of the dye is decreased by the amount to be used for painting the product,
- the color of the product is changed to the required one.

The agent *Painter* will work autonomously and will react to various situations. For example, a request for the product arrives and there are unpainted products and enough of dye; or the request already exists and unpainted product or missing dye appears. Regardless of the situation, the agent will be activated and perform its work when the required conditions are satisfied.

The remaining classes of agents are described similarly. The use of these descriptions makes it possible to create one or several copies of agents of the each class. After the creation, each agent joins the common activities and performs its functions independently of the other agents. This process will be controlled by the following events: arrival of requests, creation or change of products and so on.

A number of various problems can be solved using the model of the furniture factory described above. For example, it is possible to study the evolution of the manufacturer under different conditions, to choose the most efficient variants, etc.

Summary

Let us consider the main advantages of incorporating agents in a knowledge representation and processing system.

Using a higher level knowledge representation. It becomes possible to describe in a natural way objects of the subject domain which can be active. Agents can be regarded as ordinary objects of the semantic network; namely, we can create or delete instances of agents, change the values of their parameters, organize them into a hierarchy, derive specialized agents through inheritance of their properties. All of these opportunities provide one with additional flexibility in building models and structuring evaluation and inference.

The use of hierarchy of agents makes it possible to solve the problem of sequencing activation of the agents. It frees the knowledge engineer from the task of assigning priorities to the agents. According to this scheme, the higher level agent will execute only if a lower level agent cannot be executed for this event. Thus, the declarative information on the place of an agent in the hierarchy defines a natural priority of execution.

Encapsulation of algorithms. The imperative component of the model is decomposed into independent parts, which are stored inside the agents (*Actions*). Thus, each agent "knows" when and under what circumstances it should perform those actions. This simplifies development of the program through breaking it up into independent parts. Cooperative work of the agents is ensured by the event-driven control mechanism.

Parallelism. Encapsulation of algorithms allows one to organize in a natural way their concurrent operation on shared data.

All these advantages simplify creation of models for complex subject and problem domains.

Conclusions

Within the framework of the SemP-A environment, a natural combination of various knowledge representation and processing paradigms with the agent-based technology is achieved. This system may be used to create efficient intelligent applications for various subject domains. It also greatly increases productivity of a knowledge engineer (user) by providing him with additional functional and descriptive capabilities.

Thanks to the natural integration of both data-driven (constraint-based technique) and event-driven (agent-based technique) mechanisms, the environment can be used to create efficient intelligent systems requiring a combination of logical inference and computations over imprecise values.

Further development of the agent-based approach incorporated in the SemP-A environment will target strengthening of dynamics, extension of activity and increase of "intelligence" of agents.

References

[1] Yu.A. Zagorulko, I.G. Popov. Knowledge representation language based on the integration of production rules, frames and a subdefinite model. Joint Bulletin of the Novosibirsk Computing Center and Institute of Informatics Systems. Series: Computer Science, 8 (1998), NCC Publiher, -Novosibirsk, 1998. -P.81-100.

[2] Mayoh B. Constraint Programming and Artificial Intelligence, Constraint Programming. Springer-Verlag, 1993. NATO ASI Series F: Computer and Systems Sciences, Vol. 131, pp. 17-50.

[3] M.Wooldridge. Issues in Agent-Based Software Engineering. In: Cooperative Information Agents: First International Workshop, CIA-97 (LNAI Volume 1202), Springer-Verlag, Berlin, 1997, pp. 1-18.

[4] C.Meghini. A reactive logical agent. Ibid., pp. 148-158.

[5] Narin'yani A.S. Sub-definiteness and Basic Means of Knowledge Representation, Computers and Artificial Intelligence, 1983, Vol. 2, N 5, pp. 443-452.

[6] Narin'yani A.S. Subdefinite Models: A big jump in Knowledge Processing Technology // Proceedeings East-West Conference on AI: from theory to practice, EWAIC'93, September 7-9, Moscow, 1993, P. 227-231.

Author Index